Avian Urban Ecology

Avian Urban Ecology

Behavioural and Physiological Adaptations

EDITED BY

Diego Gil and Henrik Brumm

OXFORD
UNIVERSITY PRESS

Great Clarendon Street, Oxford, OX2 6DP,
United Kingdom

Oxford University Press is a department of the University of Oxford.
It furthers the University's objective of excellence in research, scholarship,
and education by publishing worldwide. Oxford is a registered trade mark of
Oxford University Press in the UK and in certain other countries

First Edition published in 2014
Impression: 1

Published in the United States of America by Oxford University Press
198 Madison Avenue, New York, NY 10016, United States of America

British Library Cataloguing in Publication Data
Data available

Library of Congress Control Number: 2013943367

ISBN 978–0–19–966157–2 (hbk.)
ISBN 978–0–19–966158–9 (pbk.)

Printed in Great Britain by
Clays Ltd, St Ives plc

DG dedicates this book to Hélène, Pablo, and Oscar; a meagre but sincere return for all the time stolen from them.

Foreword

The wonderful thing about being an integrative organismal biologist is the opportunity to apply so many disciplines to the organism you eventually come to love. You can study ecology, evolution, behavior, physiology, genetics, and development, all with a focus on how your organism interacts with its environment and how its many attributes relate to fecundity and viability. You can work in the lab, in the field, and meet people with similar interests from all over the world.

But as we all know, love isn't free. It comes with concern for the thing you love and a desire to keep it safe. If you are an ornithologist of a certain age, you know from experience that many birds are not as numerous as they once were. So in my case, while I remain as curious about the natural world as I have ever been, I find that I also want to learn things that might prove useful to the birds I study and encourage others to do the same.

Diego Gil and Henrik Brumm are serious scientists who happen also to care about birds. In this volume entitled *Avian Urban Ecology*, they have gathered together a highly talented group of bird biologists to produce a volume about birds in the city. Urban habitats are expanding as non-urban habitats contract, and the realization has taken hold that urban environments are not only destroying ancestral ones; they are critical to study in their own right.

What do we want to know about urban ecology? This stimulating volume provides important answers by offering summaries of current knowledge and strong guideposts for what to address next. Drawing from the disciplines of behavioral, physiological, evolutionary, and population ecology, the volume raises and answers more questions than we might have expected, indicating what a growth area for research urban ecology has become. The volume also makes clear that there is much we don't yet understand.

Ecology is the study of interactions between organisms and their environment and how these interactions determine the distribution and abundance of organisms. Ecologists interested in community composition often consider trophic structure—who's eating whom—and will concern themselves with whether urban bird communities are structured top down or bottom up. We would all like to know what the future will bring. Will the number of bird species that ultimately adjusts to urban living be more a function of the distribution and abundance of predators or the availability of appropriate food? Or would it be more informative to focus on parasites and pathogens, employing concepts from disease ecology? When applying concepts from non-urban ecology, will urban birds be best understood as instances of range expansion, colonization, or invasive species?

To the degree that cities resemble islands, we might employ island biogeography to predict future numbers of urban bird species based on the balance between rates of immigration and extinction. Large islands tend to have more species than small islands, and islands close to their mainland source tend to have more species than more distant islands. Will bird diversity be greater in large cites than small cities, or will big cities be more like far islands because their centers are farther from their sources? Urban ecology promises new opportunities to test our understanding of the forces influencing biodiversity.

Study of how attributes of organisms are associated with their relative ability to colonize and persist in the city is also part of urban ecology. Many in this field of inquiry are asking whether successful urban bird species are pre-adapted to urban life owing to their evolutionary history, such that the same attributes that permit success in their native habitats also provide a good fit with the urban environment. Or, are successful species those that can more readily alter their behavior and physiology in the face of environmental change? Birds in the city are confronted with numerous demands including light at night, noise, people, novel parasites, novel predators, and more. Observed attributes that may represent changes among successful colonists include variations in song that favor transmission, response to novel stimuli consistent with reduction in fear, and flexibility in diet to match resource availability. To date we know that city birds tend to breed earlier and sometimes for longer too. Life history theory would predict that greater reproductive effort should come at the cost of survival unless resource availability is great enough to mask the trade-off, but we have much to learn about evolution in action in the city.

How do phenotypes develop in urban environments? This question, which until recently has probably been the Tinbergen question to receive the least attention, is now front and center. For those seeking a conceptual framework for quantifying the importance of flexibility in phenotypic development to population establishment and persistence, the network of ideas captured in phenotypic plasticity, reaction norms, genetic accommodation, and genetic assimilation is proving to be enormously useful. We have moved quite rapidly from an intellectual environment in which if it wasn't genetic, it wasn't interesting, to one in which it has become critical to model and understand how environmental variation interacts with mechanisms to give rise to phenotypic variation. Said in another way, we are living in an intellectual climate where changes in gene expression are gaining precedence over changes in gene sequence. If we are to understand adaptation to rapidly changing environments, there is a growing consensus that we should focus on development.

To conclude, *Avian Urban Ecology* contains many ideas that avian biologists will be able to apply to their own systems. In my case I hold long-term interests in the role of hormones in mediating trade-offs in life histories and in migratory behavior in the dark-eyed junco. Juncos have only recently begun to breed in urban environments in southern California, and numerous differences in phenotype have been documented between urban juncos and those from their nearby montane ancestral habitat. These changes include alterations in song, stress, fear, parental behavior, aggression, body size, plumage coloration, and steroid hormones. In the years to come, stimulated by this volume and working with other researchers who study juncos, I hope to learn how all these changes have been accomplished so rapidly, whether hormones have played a pivotal role, and what the changes predict about the junco's future in urban environments. I thank the editors and authors of *Avian Urban Ecology* for pointing the way.

Ellen D. Ketterson
Bloomington, Indiana

Contents

Contributors

Valentin Amrhein Zoological Institute, University of Basel, 4051 Basel, Switzerland, and Research Station Petite Camargue Alsacienne, 68300 Saint-Louis, France

Alexander V. Badyaev Department of Ecology and Evolutionary Biology, University of Arizona, USA

Daniel T. Blumstein Department of Ecology and Evolutionary Biology, 621 Young Drive South, Los Angeles, CA 90095-1606, USA

Martyna Boruta Department of Integrative Biology, University of South Florida, Tampa, FL 33620, USA

Henrik Brumm Max Planck Institute for Ornithology, Communication and Social Behaviour Group, Eberhard-Gwinner-Str., 82319 Seewiesen, Germany

Pilar Carbó-Ramírez Arctic Centre, University of Lapland, PO Box 122, 96101-FI Rovaniemi, Finland and Centro de Investigaciones Biológicas, Universidad Autónoma del Estado de Hidalgo, Km. 4.5 carretera Pachuca-Tulancingo s/n, Col Carboneras, Mineral de la Reforma, Hidalgo, C.P. 42184, Mexico

Scott Davies School of Life Sciences, Arizona State University, Tempe, AZ 85287-4501, USA

Kathleen Semple Delaney National Park Service, Santa Monica Mountains National Recreation Area, 401 W. Hillcrest Dr., Thousand Oaks, CA 91360, USA

Pierre Deviche School of Life Sciences, Arizona State University, Tempe, AZ 85287-4501, USA

Renée A. Duckworth Department of Ecology & Evolutionary Biology, University of Arizona, Tucson, AZ 85721, USA

Richard A. Fuller School of Biological Sciences, The University of Queensland, St Lucia, Queensland 4072, Australia

Diego Gil Departamento de Ecología Evolutiva, Museo Nacional de Ciencias Naturales (CSIC), José Gutiérrez Abascal 2, 28006 Madrid, Spain

Wouter Halfwerk Institute of Biology, Leiden University, PO Box 9505, 2300 RA Leiden, The Netherlands

Marja-Liisa Kaisanlahti-Jokimäki Arctic Centre, University of Lapland, PO Box 122, 96101-FI Rovaniemi, Finland

Jukka Jokimäki Arctic Centre, University of Lapland, PO Box 122, 96101-FI Rovaniemi, Finland

Lynn B. Martin Department of Integrative Biology, University of South Florida, Tampa, FL 3620, USA

Anders Pape Møller Laboratoire d'Ecologie, Systématique et Evolution, CNRS UMR 8079, Université Paris-Sud, Bâtiment 362, F-91405 Orsay Cedex, France

Raoul A. Mulder Animal Behaviour Group, Department of Zoology, The University of Melbourne, Australia

Erwin Nemeth Max Planck Institute for Ornithology, Communication and Social Behaviour Group, Eberhard-Gwinner-Strasse, 82319 Seewiesen, Germany and BirdLife Austria, Museumsplatz 1/10/8, 1070 Vienna, Austria

Kirsten M. Parris School of Botany, The University of Melbourne, Australia

Jesko Partecke Max Planck Institute for Ornithology, 78315 Radolfzell and University Konstanz, 78464 Konstanz, Germany

Dominique A. Potvin Animal Behaviour Group, Department of Zoology, The University of Melbourne, Australia

Danielle F. Shanahan School of Biological Sciences, The University of Queensland, St Lucia, Queensland 4072, Australia

Ellen Kettersson Department of Biology, Indiana University, 1001 E. 3rd St., Bloomington, IN 47405, USA

Hans Slabbekoorn Institute of Biology, Leiden University, P.O. Box 9505 2300 RA Leiden, The Netherlands

Kamiel Spoelstra Department of Animal Ecology, Netherlands Institute of Ecology (NIOO-KNAW), P.O. Box 50, 6700 AB Wageningen, The Netherlands, and Nature Conservation and Plant Ecology Department, Wageningen University, P.O. Box 47, 6700AA Wageningen, The Netherlands

Michael W. Strohbach Department of Environmental Conservation, University of Massachusetts, Amherst, MA 01003-9285, USA, and Thünen-Institute of Biodiversity, Bundesallee 50, 38116 Braunschweig, Germany

Paige S. Warren Department of Environmental Conservation, University of Massachusetts, Amherst, MA 01003-9285, USA

Marcel E. Visser Department of Animal Ecology, Netherlands Institute of Ecology (NIOO-KNAW), P.O. Box 50, 6700 AB Wageningen, The Netherlands

Sue Anne Zollinger Max Planck Institute for Ornithology, Communication and Social Behaviour Group, Eberhard-Gwinner-Strasse, 82319 Seewiesen, Germany

Introduction

Diego Gil and Henrik Brumm

'Urban birds? Who cares?' was the comment of a friend when he heard that we were editing this book. And it is true that, given the choice, few people would go birdwatching to a city park instead of a beautiful natural habitat, such as a Neotropical forest, the Arctic tundra in summer, or an African savannah. As ornithologists and nature lovers, it is difficult to see the appeal of the simplified, noisy, and polluted urban habitats over the more species-rich natural habitats. However, there are some good reasons to focus our binoculars on the birds just outside our doorsteps. Birds are an important model in the study of ecology and evolutionary biology, and urban areas allow a unique insight into how animals interact with their environment. Moreover, urbanization is one of the most pressing global issues, leading to a huge reduction in the extent of natural habitat available. In this context, the study of birds helps to understand the impact that urbanisation has on biodiversity. Next to climate change, the United Nations considers urbanization the biggest environmental challenge of our time (United Nations Secretariat, 2012). The growth of urban areas is expected to attain unprecedented levels in the coming decades and this will have many detrimental effects on ecosystems, including for example the extinction of many species. In 2009, the percentage of the human population living in cities reached 50%, and demographic projections predict that this proportion will continue to grow at a large rate. Whether wild animals will be able to survive in an increasingly urbanized world will depend on their capacity to adapt to these new environmental conditions (Candolin & Wong, 2012). Rather than preserving pristine places, conservationists are increasingly addressing the problem of whether and how other organisms can cohabit with us.

Furthermore, cities offer researchers the possibility of 'natural' experiments that can be used to test evolutionary hypothesis and explore the interface between behaviour, physiology and adaptation. Although cities were obviously not designed with experimental aims in mind, they nonetheless have the advantage of allowing researchers to use multiple replicates, studying contrasting avifaunas and conducting powerful comparative analyses. In terms of conservation, it is important to understand how urbanization affects biodiversity, and in particular which factors limit the viability of populations in urban areas.

Last but not least, birds and their enchanting songs increase the quality of life for many people living in cities. Thus, knowing what makes an urban habitat attractive to a wide variety of bird species would help city planners and policy makers increase human wellbeing in an increasingly urbanized world (Richter & Weiland, 2011).

There is a growing body of literature on how birds respond to urban life (e.g. Lepczyk & Warren, 2012; Marzluff et al., 2001). These studies often underline the contribution of behavioural and physiological plasticity to their responses to habitat change (Candolin & Wong, 2012). Whether this plasticity primes later genetic selection, or whether selection itself drives the change is often a matter of debate. This volume specifically adopts an evolutionary framework to explore how pre-existing differences in life history, behaviour, and physiology may determine the course of adaptations. Several books on the effects of urbanization on wildlife have been published in recent years (e.g. Gaston, 2010; Marzluff et al., 2008; Niemelä, 2012). The justification for the present volume comes from the fact that we would like to concentrate not on the differences between urban and natural habitats per se, but rather on the mechanisms that enable birds to survive in and adapt to urban habitats. Therefore, after investigating patterns of urban bird ecology this book will mainly explore the ways in which birds actually survive in our urban jungles.

We have divided the book in four different sections; the first three contain reviews presenting (1) the urban habitat and how its characteristics affect birds, (2) the behavioural and physiological differences that may allow biological adaptation to the urban habitat, and (3) a discussion of evolutionary mechanisms at the heart of these differences. The fourth section presents recent case studies that explore particular topics in detail.

The first section introduces the urban environment from an ornithological perspective. In Chapter 1, Danielle Shanahan, Michael Strohbach, Paige Warren, and Richard Fuller enumerate and analyse the challenges that the urban environment poses for bird populations. They explore the different causes that lead urban bird assemblages to species homogenization, underlining the role of human drivers in many of these changes. In Chapter 2, Kamiel Spoelstra and Marcel Visser examine the effects of light pollution on avian behaviour and reproduction. Artificial lighting is one of the major transformations that humans impose in their habitat, and only recently have researchers started to understand its consequences in terms of bird behaviour and reproduction. In Chapter 3, Valentin Amrhein focuses on the potential effects that artificial feeding may have for some bird species. The love of humans for birds, which in many countries takes the form of feeding wild birds, can exert an important pressure in selecting which species are successful in the urban habitat and influence body condition, breeding season and even expansion in distribution ranges.

The second section deals with modifications in behaviour and physiology that may be adaptations to the urban environment. In Chapter 4, Daniel Blumstein demonstrates how antipredator behaviour can provide us with important insights into how birds may respond to urbanization. By combining perception studies with behavioural analyses he proposes a useful framework for predicting behavioural change. In Chapter 5, Anders Møller conducts a meta-analysis of fear responses and ecological characteristics that distinguish successful urban bird immigrants from unsuccessful ones, and concludes from these results patterns of microevolutionary adaptation to urban habitats. In Chapter 6, Diego Gil and Henrik Brumm explore the difficulties of acoustic communication in the urban habitat, the different solutions that birds may use to circumvent them, and the possible consequences that this may have for sexual selection and speciation. In Chapter 7, Wouter Halfwerk and Hans Slabbekoorn review how noise pollution affects avian fitness, stressing the disruption of acoustic communication as major factor of this perturbation, and describe the complex interactions between noise masking, behavioural adjustment and song perception with data from their particular study species. In Chapter 8, Pierre Deviche and Scott Davies analyse the physiological modifications that may connect urban abiotic changes with adaptive reproductive patterns in urban colonisers, stressing how different drivers affect the birds' physiology and functional response in different ways. Finally, in Chapter 9, Lynn Martin and Martyna Boruta study the epidemiological implications of avian life in the city, exploring the ways in which parasite and disease transmission may be affected by the change in life habits experienced in urban populations.

The third part of this volume focuses on the evolutionary processes that may mediate the adaptation of birds to urban life. In Chapter 10, Jesko Partecke discusses the predictions of different mechanisms, from phenotypic plasticity to microevolutionary processes, and offers a synthesis of these in a historic framework that also considers the timing and trajectory of urbanization. In Chapter 11, Kathleen Delaney reviews evidence concerning the genetics of urban and isolated avian populations. These data are unfortunately scarce so far, but necessary to predict possible directions of change, as well as identifying selective forces in urban populations. Chapter 12, by Alexander Badyaev, explores the developmental and evolutionary origins of convergent beak modifications in the house finch

(*Carpodacus mexicanus*) that have been linked to the exceptional success of this species in urban environments. His research combines detailed data from developmental gene expression and ecological selection regimes to underline the role of phenotypic plasticity at the heart of microevolutionary change.

Finally, in section four we present a series of case studies that provide pertinent, real-world examples of some of the topics reviewed in the first three sections. In Chapter 13, Dominique Potvin, Raoul Mulder, and Kirsten Parris examine the acoustic, morphological and genetic differences of urban populations of the silvereye (*Zosterops lateralis*) across Australian populations, suggesting that modifications in song characteristics underline the ability of this bird to become a successful urban adapter. In Chapter 14, Renée Duckworth explores how populations of two species of bluebirds (*Sialia sialia* and *S. sialis*) respond to human-induced changes in the environment, reporting interesting insights into species differences in their resilience to human modifications. Chapter 15 presents a case study by Erwin Nemeth and Sue Anne Zollinger in which the possible impact of a projected motorway on avian communication in an endangered population of stone-curlews (*Burhinus oedicnemus*) is predicted by means of signal transmission modelling. This represents a good example of the integration of physiological and behavioural data with a direct application to conservation biology. In a similar vein, Chapter 16 by Jukka Jokimäki, Marja-Liisa Kaisanlahti-Jokimäki, and Pilar Carbó-Ramírez analyses possible factors affecting breeding bird communities in wooded urban green areas in northern Finland, highlighting those characteristics that maximise urban bird conservation and that could be used to inform urban management decisions.

Editing this volume has been a great experience, and we have been particularly surprised to discover how much is known and how large the field that we are covering has become. However, we also found that there are crucial gaps in our knowledge. We sincerely hope that this volume of a bird's eye view of urban ecology helps to inspire future biologists and policy makers as to the need to understand the mechanisms by which birds colonise or are excluded from urban environments.

Acknowledgements

This book was initiated by Helen Eaton of Oxford University Press, after she attended a symposium organised by us on the same topic at the European Conference on Behavioural Biology in Ferrara (Italy). We would like to thank her for inviting us to edit this volume, as well as Lucy Nash and Ian Sherman, our editors at OUP, for their help and patience. In addition, we are most grateful to all the reviewers for their comments on the chapters. Finally, many of our immediate colleagues and respective research groups helped with discussions.

References

Candolin, U. & Wong, B. B. M. (eds) (2012). *Behavioural Responses to a Changing World*. Oxford University Press, Oxford.

Gaston, K. J. (ed.) (2010). *Urban Ecology*. Cambridge University Press, Cambridge.

Lepczyk, C. A. and Warren, P. S. (eds.) 2012. *Urban Bird Ecology and Conservation*. University of California Press, Berkeley, CA.

Marzluff, J. M., Bowman, R., and Donnelly, R. (eds.) (2001). *Avian Ecology and Conservation in an Urbanizing World*. Kluwer Academic, Boston.

Marzluff, J. M., Shulenberger, E., Endlicher, W., Alberti, M., Bradley, G., Ryan, C., Zumbrunnen, C., and Simon, U. (eds) (2008). *Urban Ecology: An International Perspective on the Interaction between Humans and Nature*. Springer, New York.

Niemelä, J. (ed.) (2012). *Urban Ecology: Patterns, Processes, and Applications*. Oxford University Press, Oxford.

Richter, M. and Weiland, U. (eds) (2011). *Applied Urban Ecology*. Wiley-Blackwell, Oxford.

United Nations Secretariat (2012). *World Urbanization Prospects: The 2011 Revision*. Department of Economic and Social Affairs, United Nations, New York.

PART 1

The Urban Environment

CHAPTER 1

The challenges of urban living

Danielle F. Shanahan, Michael W. Strohbach, Paige S. Warren, and Richard A. Fuller

Urbanization embodies one of the most dramatic and irreversible human transformations of natural ecosystems (McKinney, 2002). It typically results in the wholesale change of habitats through the burial of previously vegetated surface under impervious materials or the conversion of original vegetation or farmland to parklands or backyards (Er et al., 2005). Superimposed upon this physical conversion of habitat is an intense occupation of the landscape by people. While all forms of anthropogenic habitat loss or degradation present challenges for birds, urbanization is exceptional because of the sheer intensity of the habitat transformation and the dense occupation of urban landscapes by people (McKinney, 2006). Impacts such as disturbance, noise, and light pollution can directly affect the ability of birds to survive and reproduce, and there are indirect impacts on bird populations through management of urban habitats such as agricultural plots, backyards, streetscapes, or parklands (Barber et al., 2010; Luck & Smallbone, 2010; Slabbekoorn & Peet, 2003). Many urban-dwelling people take specific actions aimed at encouraging or discouraging certain animal or plant species, such as horticulture, pest control, or bird feeding (Davies et al., 2009; Evans et al., 2011). Although many bird species decline in abundance once an area is urbanized, other species increase in abundance taking advantage of the new opportunities and the altered patterns of competition and predation that attend a dramatic shift in assemblage composition (Catterall et al., 2002).

The challenges that urban environments pose have a profound influence on which bird species can persist, colonize or thrive in our towns and cities (Chace & Walsh, 2006). Ultimately, those that can exploit or adapt to the challenges will increase, and those that cannot will decline or disappear. Such filtering processes give rise to novel urban assemblages that often have no natural analogue, and because urbanization around the world results in similar changes to the environment, urban assemblages are often more similar to each other than to their neighbouring natural habitats; this phenomenon is known as biotic homogenization (McKinney, 2006). In this chapter we will survey some of the challenges that birds face in urban landscapes, and consider some of the ways in which urbanization filters bird communities. We will also explore how the interactions between birds and people shape avian assemblages in towns and cities. The responses of birds to urbanization provide an insight into how species can cope with rapid environmental change through mechanisms ranging from individual behavioural plasticity through to rapid evolutionary adaptation.

1.1 The challenges for birds living in urban environments

1.1.1 Habitat loss

The process of urbanization involves clearing much or all of the original vegetation to make way for buildings, roads and other infrastructure (Er et al., 2005; Marzluff & Ewing, 2001). These changes are generally irreversible and often intensify over time (McKinney, 2006). Consequently, urban landscapes typically comprise a highly altered, scattered mosaic

Avian Urban Ecology. Edited by Diego Gil and Henrik Brumm

of habitats, with some small areas of farmland, remnant native vegetation or other natural features mixed with constructed areas of varying form and function, and the configuration of these remnant green spaces strongly affects bird distributions (Fuller et al., 2010). Urban landscapes can also often be characterized by an increasing proportion of non-native plant species (McKinney, 2006). Given these significant changes the most immediate challenge for many native bird species in urbanizing environments is habitat loss, where fewer resources such as food or nesting habitat are available for survival; as a result populations decline and extinctions occur (Er et al., 2005). The negative effect of initial habitat loss is compounded by the fact that urbanization often occurs in areas with naturally high species richness (Marzluff et al., 1998).

Where habitat does remain, degradation of that habitat is an on-going challenge (Castelletta et al., 2000; Er et al., 2005; Friesen et al., 1995). Pressures that degrade habitat can include invasion of non-native plant species and even human activities such as weeding, which can alter local habitat characteristics such as the structural complexity of vegetation (Benvenuti, 2004; Evans et al., 2009; Luck & Smallbone, 2010). The spatial configuration of residential development can also profoundly influence bird occurrence within towns and cities. A recent modelling study in Brisbane, Australia, compared the impacts of sprawling and compact urban design on the patterns of bird occupancy. While urban growth of any type reduced bird distributions, compact development where new housing was built within existing residential areas substantially slowed these reductions at the city scale. Urban-sensitive species particularly benefited from compact development because large green spaces were left intact, whereas the distributions of non-native species were predicted to expand as a result of sprawling development (Sushinsky et al., 2013).

An additional challenge for urban birds is that any new vegetation planted by people is unlikely to resemble pre-settlement habitat. This could be due to both social and environmental factors (Cook et al., 2012). For example, private backyards in Sheffield, UK, support more than twice as many plant taxa as any other habitat type in the country, and two-thirds of those garden plants were non-native (Thompson et al., 2003). This highlights the important ecological role that people can have not just in removing habitat from a landscape, but in the kinds of habitat they create.

Several studies have attempted to disentangle the effects of direct habitat loss from factors such as habitat degradation or fragmentation, and in general highlight that the amount of habitat alone is a very poor predictor of bird species richness and abundance in urban environments (Friesen et al., 1995). For example, studies on Neotropical migrant and resident bird species have found that even when forest patch size and width is controlled or accounted for, species richness and bird abundance still declines with increasing residential habitation in the surrounding landscape (Friesen et al., 1995), but that life-history traits such as diet and nesting preferences can be mediators of this response (Kennedy et al., 2010). The reasons for these declines in Neotropical species are attributed to factors such as increased predation in urban areas or changes in the availability of resources such as food or nesting habitat (Friesen et al., 1995; Kennedy et al., 2010). Such studies clearly highlight that though habitat loss has a profound influence on species, there are a range of other factors introduced by the intensive land-use of urbanization that influence bird communities; we discuss many of these below.

1.1.2 Habitat fragmentation

Habitat fragmentation is closely linked to (and often a product of) habitat loss; it is the process where the connectivity between patches of remnant habitat is reduced (Fahrig, 2003; Saunders et al., 1991). Fragmentation can be particularly severe in urban environments as habitat patches are isolated by intensive land-use such as roads, buildings or other structures (Marzluff & Ewing, 2001; see Figure 1.1 for an illustration). This fragmentation can negatively impact bird populations by restricting movement between remnant vegetation patches. Since most birds are highly mobile, it is often behavioural rather than physical inhibition of movement (Harris & Reed, 2002).

A consequence of fragmentation can be smaller, more isolated populations of species, and these populations are likely to have reduced long-term

Figure 1.1 Urban landscapes in Brisbane, Australia illustrating the gradient of habitat loss and fragmentation that can occur across a city. Urbanization can range from low density housing within largely intact remnant vegetation patches (a) through to much more densely occupied landscapes in which there are few significant areas of remnant vegetation, with what is left being separated by built structures (c) and (d). Note that many of the trees present in higher density urban areas occur within backyards and other forms of private green spaces.

viability as they may more readily succumb to factors such as disease or catastrophic events (Fahrig, 2003; Levins, 1970; Saunders et al., 1991). It is likely that life-history traits such as mobility, habitat specialization, and characteristics of the landscape in which a species evolved are likely to mediate the effects of fragmentation on birds, with some species able to cross even intensely used landscapes (Fahrig, 2003; Shanahan et al., 2011b; Tremblay & St Clair, 2011). However, the effects of habitat fragmentation can be highly variable from city to city (Chace & Walsh, 2006), as it also depends on the configuration of remaining habitat in the landscape. For example, the size of remaining urban habitat patches is generally thought to have a greater influence on species richness than isolation (Evans et al., 2009; Oliver et al., 2011), in part because larger vegetation patches will on average support a greater diversity of habitat types than small patches (Chamberlain et al., 2007). However, connectivity between remnants is also important for long-term viability of many bird species, and connectivity is likely to be important for the colonization of revegetated areas (Shanahan et al., 2011a). This is an important consideration for urban conservation efforts as colonization processes are critical to the successful recovery of bird populations.

There are many local-level impacts of fragmentation that add to the complexity of understanding the challenges of urbanization on birds. For example, the dissection of habitat by roads can result in significant levels of road-related fatalities for birds and this is likely to be more of a problem in high-traffic areas (Mumme et al., 2000). Higher edge-to-area ratios of the remaining small habitat patches can also accelerate habitat degradation through, for example, invasion of weed species or changed light conditions, again factors that are likely to be more significant in urban environments (e.g. see Gascon & Lovejoy, 1998). Edge effects can also more directly influence birds through altering the abundance of competitors, predators or parasites in habitat remnants (Marzluff & Ewing, 2001). For example, Marzluff and Ewing (2001) conclude that urban environments favour corvids, small to medium sized mammals, brood parasites, and some raptors. These species can have a particularly significant impact on the nests of birds at habitat edges, and as a result few native bird species can reproduce well enough to survive along those edges (Robinson & Wilcove, 1994).

1.1.3 Alteration of resource flows

Urban areas can constitute rich foraging grounds for many bird species, except at the very highest intensities of urbanization (Chace & Walsh, 2006). Though natural food availability might be significantly reduced in urban areas due to the loss of native vegetation, resources are commonly provided by people through increased scavenging opportunities, the introduction of a wide range of exotic plant species and direct intentional bird feeding (Chace & Walsh, 2006; Davies et al., 2009; Thompson et al., 2003; see also Chapter 3). Thus, urban environments are likely to favour species with dietary requirements and foraging traits that match the new resource regime, or species that have the ability to adapt. Food availability is widely considered a key driver of differences in bird species communities between landscapes (Chamberlain et al., 2009).

The novel food resources mentioned above present a challenge to which bird species must adapt, but they also present a significant opportunity; this is because many of these resources are available with greater reliability across the year than some seasonal natural resources (Shochat et al., 2006). An intriguing outcome of this is the possibility that winter feeding may change the migratory status of birds, particularly in northern latitudes. One case is the European robin *Erithacus rubecula*, in which the males are resident in urban habitats in Belgium and migratory in nearby woodlands, perhaps as a result of year-round food availability in urban parks and gardens (Adriaensen & Dhondt, 1990). Similarly, studies suggest that due to bird feeding residential areas provide important non-breeding grounds for birds in Rovaniemi, Finland (Jokimäki & Kaisanlahti-Jokimäki, 2012).

Changes in food availability and the buffering of temporal variation in that food supply in urban areas are probably key mechanisms that lead to changes in avian community structure (Shochat et al., 2004, 2006). Studies suggest that these changes in some instances could result in increased competition for food (Shochat et al., 2004) or resource matching where individuals are distributed based on the availability of resources (Rodewald & Shustack, 2008b). However, it is evident that species vary enormously in their level of adaptability to novel food resources, and those species that are more able to innovate in the ways they acquire food can thrive in novel urban environments (Møller, 2009). Indeed, omnivory is one of the common factors uniting many of the species most able to occupy urban environments (Evans et al., 2009).

1.1.4 Limiting resources

For some species, urban environments simply do not have sufficient resources such as food or habitat, and as a result populations can decline or disappear. One of the most fundamental habitat factors influencing avian assemblages is the structure and composition of vegetation communities (e.g. Orlowski et al., 2008). For example, a Tasmanian study found that native bird species tended to prefer native plant species in urban habitats (Daniels & Kirkpatrick, 2006), and other studies have found that increasing plant species richness has a positive influence on bird species richness (Huste & Boulinier, 2007). A common feature of urban plant communities in comparison with surrounding habitats is a much higher diversity, but also a much greater abundance of exotic species (Luck & Smallbone, 2010). Exotic plants generally support fewer insect species than native plants, and foliage gleaning insectivores may suffer reduced abundances in towns and cities as a consequence (Rosenberg et al., 1987).

This said, research in Sheffield, UK, revealed the key driver of invertebrate species richness in domestic backyards was the density of large trees irrespective of their native status (Smith et al., 2006).

Food provision and directed habitat management can increase avian abundance and richness dramatically, suggesting that overall food availability is indeed limiting at least in some cases (Evans et al., 2011). In some countries the provision of resources by local householders is remarkably high and constitutes a significant additional food resource for birds. In the UK, approximately 12.6 million households provide supplementary food for birds and there is a minimum of 4.7 million nest boxes within gardens across the country (Davies et al., 2009). This equates to one bird feeder for every nine potentially feeder-using birds, and at least one nest box for every six breeding pairs of cavity nesting birds. Bird feeding is such a popular activity in the city of Sheffield that in some parts of the city densities of nearly 1,000 bird feeders per km^2 are achieved (Fuller et al., 2012). Surveys in the United States find comparable or even higher proportions of households feeding birds (66% in Michigan and 43% in Phoenix, Arizona; Lepczyk et al. 2012). Bird feeding can positively influence the abundance, breeding success and survival of some species (Evans et al., 2009; Fuller et al., 2008). The density of bird feeders in Sheffield is positively correlated with the abundance of species that are known to take supplementary food, although avian species richness does not increase in places where bird feeding is intense (Fuller et al., 2008). It is worth being cognisant of some potential negative effects associated with bird feeding, for example increased disease transmission at feeding stations or problems arising from the provision of nutritionally poor foods (Jones & Reynolds, 2008).

1.1.5 Pollution

Many pollutants are associated with the industrial and residential activities that characterize urban environments, and as a consequence urbanization is a major cause of pollution both globally and locally (Grimm et al., 2008). Although many urban areas in old industrialized countries have become much cleaner in recent years as a result of deindustrialization, cleaner technologies, and regulations, the levels of pollution that urban birds are exposed to are likely to remain high and will even increase in many parts of the world. For example, anthropogenic noise has been increasing in both average levels and peak intensity within many towns and cities (Berglund & Lindvall, 1995). This is likely to pose an on-going driver that continues to filter bird communities into the future.

The direct and indirect effects of light, noise, and chemical pollution are increasingly being studied (e.g. Fuller et al., 2007; Gorissen et al., 2005). Of these pollutants, light is commonly linked to disruptions in timing of reproductive or other behaviours. For example, a seven year study in Vienna discovered that artificial light influences the singing and breeding timing of forest songbirds, with females living near street lights laying eggs on average 1.5 days earlier than birds in darker areas (Kempenaers et al., 2010). Another study found that in natural areas illuminated by streetlights, waders with a visual foraging strategy increased their effort and mixed foragers even shifted their effort to favour visual strategies (Santos et al., 2010). Although artificial light could thus be seen as a possible management tool, it also highlights that the effects of light pollution can be pervasive, potentially invading natural or semi-natural areas surrounding cities.

Noise pollution can mask acoustic signals made by birds and alters the way animals perceive sounds such as the calls of other birds or the sounds made by potential predators (Barber et al., 2010). A range of facultative behavioural responses to anthropogenic noise have been documented, including alterations to the amplitude, frequency, timing, and duration of signals to minimize acoustic competition (e.g. Brumm, 2004, 2006; Slabbekoorn & Peet, 2003; Wood & Yezerinac, 2006; reviewed in Chapter 6). Shifts in the timing of bird song can be dramatic in response to urban noise. For example, European robins *Erithacus rubecula* in urban Sheffield sang at night in places that were noisy during the day, suggesting they were signalling at a time when those signals could most easily be heard by receivers (Fuller et al., 2007). Daytime noise was a better predictor of nocturnal singing behaviour than night-time light pollution to which nocturnal singing behaviour is often attributed (Fuller et al., 2007).

Urban noise pollution is typically concentrated at the lower end of the audible frequency spectrum (Lohr et al., 2003). A number of studies have found that some species shift adjust their song to higher frequencies (see Chapter 6). However, there is the potential for bird assemblages to be filtered by noise, such that those species vocalizing within the typical frequency range of noise pollution are excluded (Francis et al., 2011). By examining habitat use, a recent study in New Mexico found that noisy areas exhibit reduced abundances of large-bodied species that vocalize at low frequencies (Francis et al., 2011). Conversely, smaller-bodied species that produce higher frequency vocalizations were able to persist around the noisy areas. This research highlights that though some species can adapt to challenges, others may be excluded from an area entirely.

A wide range of chemical pollutants affects urban environments and several have been demonstrated to have negative effects on bird health. For example, heavy metal pollution may reduce the size and survival of nestlings (Nam & Lee, 2006). Run-off that includes wastewater, fertilizer, sewage and storm water all have demonstrated effects on urban waterways (Bibby & Webster-Brown, 2005; Paul & Meyer, 2001). This pollution inevitably impacts any species that rely on aquatic or riverside habitats within urban areas, but also the ecosystems downstream (e.g. Rosselli & Stiles, 2012). Similarly, air pollution can have negative effects on the abundance of invertebrate species (McIntyre, 2000), potentially influencing bird species that rely on them for food such as aerial insectivores. Chemical pollution can also indirectly affect bird behaviour. For example, a 2005 study compared the singing behaviour of great tits *Parus major* at varying distances from a heavy metal pollution source (Gorissen et al., 2005). The study found that males closest to the pollution source had a more limited song repertoire and were less frequent singers; both effects were assumed to be a result of impaired individual health caused by the pollution.

1.1.6 Species interactions

The function of ecosystems relies largely on interactions between species. There are many ways that species interact, for example through competition for resources, predation, or even brood parasitism. Little is known about how urbanization changes the interactions between species and they are very difficult to predict due to complexity of ecosystem function and trophic chains (Faeth et al., 2005; Faeth et al., 2011; Shochat et al., 2010). An example of a human behaviour that has an unintended impact on the interaction between plants and birds in an urban environment is the provision of supplementary food for hummingbirds and other nectarivores. When these species feed on plant flowers they often have a very important and very specific role in pollination. A study in Mexico found that supplementary feeding can reduce pollination rates and consequently impair seed production in plant species dependent on bird pollination (Arizmendi et al., 2008).

Changes to bird-invertebrate interactions in urban environments are also likely to be inevitable in urban environments, but they remain understudied. Researchers have commonly observed differences in invertebrate assemblages associated with urban areas, for example between native and non-native vegetation (Burghardt et al., 2009; Helden et al., 2012), and many invertebrate groups have been found in reduced abundances in urban areas (Jones & Leather, 2012). These kinds of changes are likely to directly influence bird species that forage on these invertebrates, as studies have linked bird fitness to the abundance as well as the quality of their invertebrate food source (Burke & Nol, 1998). Less directly, however, changes to invertebrate communities, or even the connectedness of the landscape for invertebrates, can affect the pollination of plants by insects (Cranmer et al., 2012). In some areas this may have significant implications for birds that rely on specific plant species for food or habitat, though in some instances introduced invertebrate species could fulfil this ecological role (Lomov et al., 2010). These potential changes in species interactions do, however, highlight the profound effect that urbanization can have on even the most fundamental aspects of ecosystem function.

Predators of birds such as domestic cats can reach very high densities in urban environments (Lepczyk et al., 2004; Sims et al., 2008), and the abundance of other predatory species (such as chipmunks) has also been found to increase in some urban locations (Richmond et al., 2011; Shochat et al., 2004). Free ranging domestic cats alone are believed to be the greatest source of mortality for birds in the USA (Loss et al., 2013). This clearly has the potential to cause significant changes to predator–prey relationships, consequently altering bird communities (Chace & Walsh, 2006; Major et al., 1996). There is, however, a paradox emerging from research into predation rates on birds in urban areas; in many studies even though vertebrate predator numbers increase, predation rates decline (Fischer et al., 2012). There are a wide range of hypotheses as to why this might be the case. To name just two, prey itself may be hyperabundant in urban areas, which reduces overall predation rates (Fischer et al., 2012), or the prey species that remain in highly altered urban environments may be better adapted to predators (Fischer et al., 2012; Shochat et al., 2004).

Competitive interactions among birds can be dramatically altered by urbanization. For example, the noisy miner *Manorina melanocephala* is a highly aggressive honeyeater that spends a great deal of time chasing other birds, particularly small-bodied passerines, from its territory. The noisy miner prefers edge habitats, seemingly benefiting from woodland fragmentation and degradation, and is now highly abundant in urban landscapes throughout eastern Australia (Catterall et al., 2002). Evidence is mounting that the presence of the noisy miner significantly reduces the species richness of smaller passerines in both urban and rural habitats through aggressive exclusion (Montague-Drake et al., 2011; Piper & Catterall, 2003), and thus provides an example of where human alteration of the environment indirectly affects bird communities via the changed abundance of an aggressive species.

Considering that urbanization impacts habitat, resource availability and even species interactions, it is not surprising that it has also been found to influence wildlife disease infection patterns (Bradley & Altizer, 2007). One example is the West Nile virus that was accidentally introduced to the USA in 1999 and rapidly spread through the country. It is transmitted by mosquitoes and causes high avian mortality, but transmission is lower in diverse bird communities (Bradley et al., 2008; Hamer et al., 2011). Increased infection rates in cities could therefore be a result of lower bird diversity and altered communities (Bradley et al., 2008). Common urban birds such as American robin *Turdus migratorius* and house sparrow *Passer domesticus* appear to be 'super-spreaders' of the virus (Hamer et al., 2011). Anthropogenic landscape change additionally influences the vector of the disease by changing mosquito habitat. For example, neglected swimming pools of abandoned and foreclosed houses provide excellent conditions for mosquito larvae and are believed to having contributed to an increase of West Nile virus cases in Bakersfield, California, and the invasion of a new host of the disease (Reisen et al., 2008).

Space permits treatment of only a few different kinds of species interactions in this section. In reality, changes to species interactions and the community ecology of urban areas can result from any of the challenges listed above. For example, habitat fragmentation can result in the invasion of a weed species, which could consequently out-compete an important food or nesting plant for a bird. Conversely that weed could provide important food or nesting habitat. As such, when considering how and why particular species persist or otherwise in an urban environment, it is critical to think about how those species interact with other species and with their environment.

1.2 Urbanization as a filter

Because the challenges outlined above will affect each species differently, there has been much interest in discovering any general patterns in how urbanization filters bird communities. There are at least two distinct approaches to tackling this question. First, one can measure how the species richness and abundance of birds changes in response to urbanization. This is usually achieved by studies that substitute space for time by identifying plots located in areas with different degrees of urbanization and comparing the structure of bird assemblages. Second, one can take a comparative approach to identify how urban-adapted species differ from those that are sensitive to urbanization. We now briefly review each of these approaches in turn.

1.2.1 Species richness and abundance

Researchers have commonly observed lower species richness in urban areas relative to that of the surrounding rural landscape (e.g. Sandström et al., 2006). However, many other studies that have investigated the change in bird species richness across the full rural to urban gradient (rather than just urban versus rural settings) have found that species richness often follows a hump-shaped curve, where areas with intermediate levels of urbanization exhibit the highest species richness (Blair, 1996; McKinney, 2002; Tratalos et al., 2007). While the details vary, there is mounting evidence that this is a general pattern (Luck & Smallbone, 2010). One of the most complete investigations of this issue was by Tratalos et al. (2007), who measured how the richness and abundance of breeding birds changes with housing density among 2,132 1 km × 1 km squares across Britain. Species richness increased from low to moderate household densities and then declined at greater household densities (Figure 1.2a). Abundance showed a very different pattern (Figure 1.2b,c), with the total number of individuals increasing consistently with household density, but standardized abundances consistently declining for urban-sensitive species and consistently increasing for urban specialists. In other words, a small number of highly abundant species were occupying urban habitats. Indeed, species that reached the highest abundances within the city of Sheffield were also those that were most abundant across Britain as a whole (Fuller et al., 2009). Marzluff and Rodewald (2008) argue that the form of the richness pattern depends on the relative importance of colonization and extinction dynamics along the gradient and, therefore, may vary among cities and ecological regions. Comparative studies on this issue are still needed, particularly in tropical regions (Ortega-Alvarez & MacGregor-Fors, 2011).

One hypothesis is that the humped shape of the curve species richness pattern is due to increased habitat heterogeneity and a higher availability of resources at intermediate levels of urbanization; the tree-lined street verges, gardens and parks that are found in suburban areas are likely to provide a diversity of habitats for a greater range of species (Blair, 1996, 2004; Gaston et al., 2007). Another complementary idea is that although there may be greater food availability supporting greater numbers of birds particularly in higher density urban landscapes, the competition for these resources is much greater and this can act to exclude many species (Shochat et al., 2006). However, some studies suggest that birds resource-match across urban gradients, where their

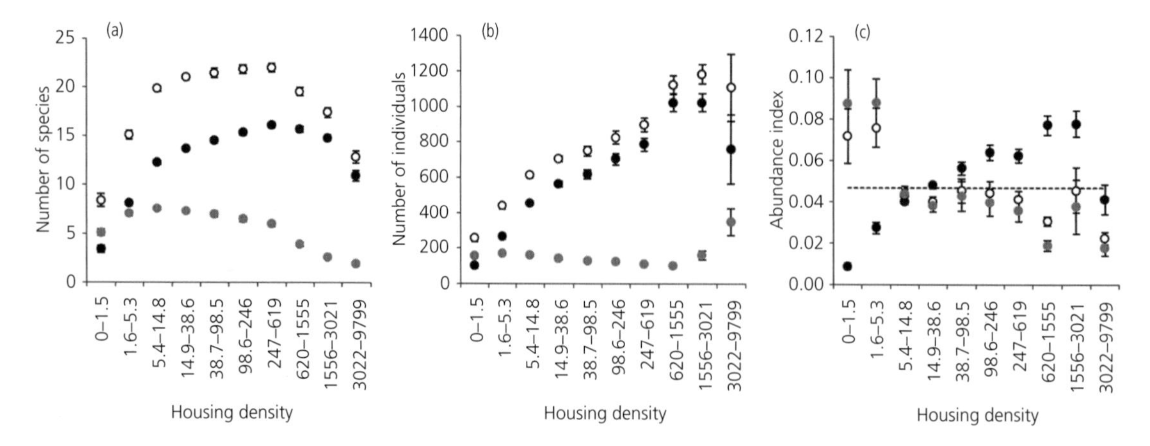

Figure 1.2 Relationships between housing density in 2,132 1 km × 1 km squares across Britain and (a) mean number of species, (b) mean number of individuals, and (c) abundance index. Housing density (households per square kilometre) has been binned into equal interval classes based on a $\log_e$-transformation. Open circles are for all species, black circles for 27 urban indicator species (see Tratalos et al., 2007), and grey circles for the remaining species. Error bars are ±1 SE. Broken line indicates the average abundance index for all species across all squares. Reproduced with permission from Tratalos et al. (2007).

abundance reflects the availability of food resources (Rodewald & Shustack, 2008b). Other behavioural mechanisms may also have an important influence on species richness, such as dispersal limitation, settlement biases, and heterospecific attraction (Patten & Bolger, 2003; Rodewald & Shustack, 2008a; Rodewald, 2012). These forms of species interaction have received less attention from ecologists, and present an important avenue for behavioural and evolutionary ecologists to contribute to an understanding of urban bird communities.

A number of studies do show that the overall abundance and biomass of birds often increases from rural to urban settings, with just a few species contributing the majority of individuals (Blair, 2004; Chace & Walsh, 2006). In effect, urbanization acts to homogenize bird communities as habitat variability declines, food productivity and competition increase, and fewer species can cope (Blair, 1996; Chace & Walsh, 2006; Luck & Smallbone, 2010; Shochat et al., 2006; Tratalos et al., 2007). These more abundant species can include exotic species or well-adapted native species (Catterall, 2009; Chace & Walsh, 2006). Abundance has generally only been found to decline at the highest density of buildings (Tratalos et al., 2007).

1.2.2 Which species disappear or decline?

Some studies group bird species into 'urban adapted' or 'urban sensitive' categories (e.g. see Blair, 1996). However, generalizations about which bird species groups decline and which increase are extremely difficult to make. This is because ecological traits and local factors play such a significant role in how species respond to the changed environment, and also influence whether a species can adapt or cope with change (Faeth et al., 2011; Luck & Smallbone, 2010). For example, cavity nesters have been found to increase with increasing urbanization in some locations such as Colorado (Chace & Walsh, 2006; Miller et al., 2003). However, in Australia the opposite is the case (Luck & Smallbone, 2010). This is because Australian cavity nesters rely on tree hollows formed primarily by invertebrate activity and they are unable to excavate their own nesting hollows or commonly use buildings (Harper et al., 2005). Tree hollows are increasingly rare in Australian urban landscapes because hollow-bearing trees are commonly removed owing to safety concerns (Harper et al., 2005). This highlights the need to consider local factors and specific mechanistic drivers when interpreting the results of urban bird studies. Using a small number of selected examples, Table 1.1 highlights some of the great variation in documented responses to urbanization by birds across the world.

Reflecting the loss of particular habitat types, a number of studies have shown that urbanization is often associated with a loss of ground-nesting species, habitat specialists, shrub-nesters, old forest specialists, and species that require large areas of intact habitat (Chace & Walsh, 2006; Lim & Sodhi, 2004; Luck & Smallbone, 2010; McKinney, 2006; Rosenberg et al., 1987).

1.2.3 Which species do well?

There has been a great deal of interest in determining whether urban-adapted species have particular characteristics in common that shed light on the reasons for their success. Those species most successfully exploiting urban environments include exotic species, widely distributed species, sedentary or resident species, tree nesters, habitat generalists, omnivores, granivores (particularly those attracted to feeders), and cavity-nesting species (Beissinger & Osborne, 1982; Chace & Walsh, 2006; Kluza et al., 2000; Lancaster & Rees, 1979; Luck & Smallbone, 2010; Miller et al., 2003; Pidgeon et al., 2007; Jokimäki and Kaisanlahti-Jokimäki, 2012a). A recent study by Møller (2009) found that urban birds in the Western Palearctic were characterized by large breeding ranges, high propensity for dispersal, high rates of feeding innovation, short flight distances when approached, and high annual fecundity and high adult survival rates. However, most studies have divided species into urban-adapted and urban-sensitive groups without considering important interspecific variation in the urbanization response. When this variation was accounted for in a recent UK study, many of the associations disappeared, with only niche generalism, herbivory, and nesting above ground predicting a positive response to urbanization (Evans et al., 2010). In a range of more specific cases, the increased abundance of some bird species

Table 1.1 Selected examples of drivers of differences between urban and rural environments in abundance (A) and species richness (R) of birds drawn from studies around the world. Increase means that the urban environment has a higher value for richness or abundance than the rural environment. Note the variation within and among the various guilds, and the range of drivers that have been implicated.

Difference	Suggested driver	Study location	Source
Cavity nesters			
Increase (A, R)	Habitat availability	Front Range, Colorado	Miller et al. 2003
Decrease (R)	Nesting sites and foraging opportunities	USA	Pidgeon et al. 2007
Increase (A)	Habitat availability	Vancouver, BC	Lancaster and Rees 1979
Increase (A)	–	Western Massachusetts	Kluza et al. 2000
Decrease (A, R)	Amount and quality of habitat	Örebro, Sweden	Sandström et al. 2006
Low-nesting species (ground/shrub)			
Decrease (A, R)	Habitat structure and composition, disturbance	Front Range, Colorado	Miller et al. 2003
Decrease (R)	Predation	USA	Pidgeon et al. 2007
Decrease (R)	Predation	Wisconsin, USA	Lindsay et al. 2002
Decrease (A)	Predation	Western Massachusetts	Kluza et al. 2000
Shrub/tree nesters			
Decrease (A, R)	Nesting sites	Singapore	Lim and Sodhi 2004
Frugivores			
Increase (A, R)	Possible increase in ornamental fruit plants	Singapore	Lim and Sodhi 2004
Decrease (R)	–	Palmas, Brazil	Reis et al. 2012
Insectivores			
Decrease (R)		Palmas, Brazil	Reis et al. 2012
Decrease (R)	Competition?	Wisconsin, USA	Lindsay et al. 2002
Decrease (A)	–	Vancouver, BC	Lancaster and Rees 1979
Decrease (A)	–	Oxford, Ohio	Beissinger and Osborne 1982
Granivores			
Increase (R)	Bird feeding	Wisconsin, USA	Lindsay et al. 2002
Decrease (R)	–	Palmas, Brazil	Reis et al. 2012
Increase (A)	–	Oxford, Ohio	Beissinger and Osborne 1982
Increase (A)	Habitat and food availability	Vancouver, BC	Lancaster and Rees 1979
Increase (A)	Habitat and microclimate	Ohio, USA	Leston and Rodewald 2006
Increase (A)	Consumer resource matching	Ohio, USA	Rodewald and Shustack 2008b
Omnivores			
Decrease (R)	–	Palmas, Brazil	Reis et al. 2012
Increase (A)	–	Oxford, Ohio	Beissinger and Osborne 1982
Increase (R)	Ability to exploit a range of resources	Finland	Jokimäki and Suhonen 1998
Carnivores			
Decrease (R)	–	Palmas, Brazil	Reis et al. 2012
Nectarivores			
Decrease (R)	–	Palmas, Brazil	Reis et al. 2012

in urban environments can be attributed to the ability to utilize abundant and often unusual food sources and habitats that occur in human-modified environments (Møller, 2009; Robertson et al., 2010).

It has been shown in several bird species that individuals in urban environments have a greater tolerance for disturbance and shorter flight initiation distances than their rural counterparts (e.g. Møller, 2008). Birds in urban green spaces that are heavily visited tend to have shorter flight distances when disturbed than those in less heavily visited areas (Fernandez-Juricic & Telleria, 2000). Some studies suggest this ability to tolerate disturbance reflects habituation by individuals over time (Møller, 2008; Møller, 2009), while others suggest it reflects selection of individuals from species that have a variable response regarding fear of people (Carrete & Tella, 2011).

Brain size has been associated with behavioural innovation in birds (Lefebvre et al., 2004), although the importance of brain size in predicting the response to urbanization is somewhat equivocal. A study comparing 12 cities found some indication that the passerine species that had reinvaded at least one city were more likely to belong to big-brained lineages (Maklakov et al., 2011), although relative brain size was not significantly associated with the response to urbanization in a comprehensive study of British breeding birds (Evans et al., 2011).

The lack of consistency among studies of the traits of urban birds (Francis & Chadwick, 2013) raises the intriguing possibility that there might be several pathways to colonization and establishment of species in urban areas—a collection of distinct urban syndromes, as it were. Analyses conducted to date have not been designed to test this hypothesis, instead focusing on the general features described above. Longitudinal studies of bird communities in cities are few, particularly in very old cities, but a variety of species with quite different life histories ultimately colonize urban areas. A few conditions seem to be necessary but not sufficient for such colonization of cities (e.g. widespread geographic distribution or abundance in adjacent non-urban areas; Evans et al., 2010), after which taxonomic groups may differ in the life history traits that allow them to succeed in the city. Time is also an important factor; urban colonization (and likewise extinction) is an on-going process with some species only becoming urban relatively recently (Evans et al., 2010), and the success of urban species in Europe appears to be a function of their time since colonization of urban areas (Møller et al. 2012). Given this dynamic nature of urban bird communities, a set of general features describing a single urban syndrome is unlikely to be identified (Evans et al., 2010). More revealing may be global analyses of temporal and spatial variation in the patterns of urban-adapted birds.

1.3 Adaptive responses to urbanization

Despite the interest in how urbanization filters bird species, we know that animals are not simply static entities to be either filtered out or pre-adapted to urbanization. Instead, as several chapters of this book point out, the distinct biotic and abiotic features of urban environments can act as selective forces (Chapters 6–10 and 13). These can then alter behaviour, physiology, and in some cases, lead to evolutionary changes in species' traits. It has become well-documented, for example, that the altered acoustic environment of the city leads to altered acoustic signals in birds (Slabbekoorn & Peet, 2003; Warren et al., 2006, and reviewed in Chapter 6). Altered resource levels, stressors such as human disturbance, and altered social environments may underpin changes observed in hormone profiles and other aspects of reproductive physiology (Chapter 7–8). Fewer studies have documented evolutionary changes (Chapters 10, 13). The first documentation of genetic evolution in response to urbanization came from a population of dark-eyed juncos *Junco hyemalis* in San Diego, California. The urban population was shown to have evolved reduced amounts of white markings on their tail feathers, though the selective mechanism for this remains unclear (Yeh, 2004). An increasing number of other studies are finding genetic differentiation between urban and non-urban populations (Takami et al., 2004; C. Johnson personal communication). The high mobility of many bird species may lead to higher rates of gene-flow between urban and non-urban populations, preventing the genetic differentiation seen in other non-avian taxa. However, with sufficiently strong selection or barriers to dispersal, differentiation and trait evolution may still occur, even over small distances (Chapter 11). The adaptive responses of birds to urbanization thus represents an important avenue for future research.

1.4 Importance of human drivers

Within cities, human behaviour determines much of the vegetation distribution and composition, which leads to a heterogeneous distribution of food and foraging resources (Collins et al., 2000). Private gardens make up large parts of many cities and gardening practices and preferences strongly determine the resource availability for birds (Cannon, 1999; Chamberlain et al., 2004; Lerman & Warren, 2011). Leisure activities of urban dwellers demand large areas of green space, beyond residential gardens, for sports fields, golf courses, parks, etc. Such areas are often intensively managed in order to fulfil their purpose, but they can provide resources for wildlife (Petersen et al., 2007).

Humans also import resources into the city that are exploited by birds. For example, fountains provide water, and feeders and trash provide additional food sources (Chamberlain et al., 2004; Lerman & Warren, 2011; Peris, 2003). These resource inputs not only lead to the aforementioned patterns of richness, diversity and abundance along the rural–urban gradient, but also to very heterogeneous patterns within the city, in some cases even reflecting socioeconomic patterns (e.g. Kinzig et al., 2005; Strohbach et al., 2009). The drivers of these heterogeneous patterns are complex suites of social processes, ranging from 'bottom up' individual level consumer choices (Kinzig et al., 2005; Larsen & Harlan, 2006) to 'top down' influences of institutions like municipal policy, large development corporations, or local institutions such as Homeowners Associations (Lerman et al., 2012; Warren et al., 2010). In conclusion, human behaviour creates novel, complex and heterogeneous ecosystems within cities, with challenges for many bird species but also opportunities.

Cities can provide ample resources and unoccupied niches for some species. Species or populations of species that can adapt to urban conditions are sometimes called *synurbic* and the process of adaptation *synurbization*, terms derived from *synanthropization* (adaptation of animal populations to human conditions) and *urbanization* (Luniak, 2004). What allows birds to become synurbic can be very case specific. Some is linked to behavioural, physiological and reproductive plasticity; some is directly linked to human behaviour. Common blackbirds *Turdus merula* for example started to colonize urban parks in southern Germany in the early 19th century and have since then colonized cities in many parts of their range. It is not known what triggered urban colonization at this particular time and location, but it was possibly the combination of a sufficiently sized population (higher than in northern Europe) and the advantage of wintering in towns (higher than in southern Europe; Luniak et al., 1990). After the first colonization, an increase of park area and similar habitat in many cities throughout the 19th and 20th century, but also changes in human attitude (feeding, reduced persecution), probably facilitated its establishment and spread through multiple colonizations of cities (Evans et al., 2010). One example where first settlement, expansion and establishment have been well documented and analysed is the establishment of the northern goshawk *Accipiter gentilis* in Hamburg, Germany. Rutz (2008) analysed observation data from 1946 to 2002 and concluded that colonization of the city in the 1980s was probably triggered by a combination of biotic, abiotic and anthropogenic factors: an increase of hunting pressure in nearby rural areas; an increase of prey abundance in the city; and a succession of severe winters. An example where human behaviour directly led to the successful establishment of a bird species in cities is the peregrine falcon *Falco peregrinus*, a bird that was actively introduced in cities in Europe and the US (Luniak, 2004). The strong increase of the white stork *Ciconia ciconia* population in Spain between 1984 and 1994 can be mainly attributed to the exploitation of a new food source that is unintentionally provided by humans—urban refuse dumps (Peris, 2003).

All these examples show the diversity of possible consequences of human behaviours for urban wildlife and the diversity of urban ecosystems that it creates. Trying to understand and manage urban ecosystems requires not only considering biotic, abiotic factors, and temporal factors, but also individual and social human behaviour and its outcome.

1.5 Closing remarks

Urban environments pose great challenges to many bird species, ranging from habitat loss and fragmentation, to disruption of resource flows, reduction of key resources and pollution. In this way, urbanization acts as a filter, limiting those species that colonize the city, and changing the densities of those that do not manage. Yet, urban environments also present new opportunities for adaptation for those species that can colonize and make use of the sometimes abundant food and water resources provided by human activities. Ultimately, human behaviours are the primary underlying drivers of ecological change and thus set the stage for the behavioural, physiological and community composition changes described in the chapters of this volume. Spatial heterogeneity in human activities ranging from consumer level decisions about bird feeding or gardening as well as in municipal or neighbourhood level policies are thus important determinants of the conditions experienced by birds in urban areas. An understanding of avian biology in urban areas is thus incomplete without reference to these underlying human drivers. We advocate for more behavioural and physiological ecologists to make explicit use of the variation within urban areas to understand underlying human drivers of changes in bird communities.

Acknowledgements

D.F.S. is supported through ARC Discovery Grant DP120102857 and R.A.F. holds an ARC Future Fellowship. Some of this material is based upon work supported by the National Science Foundation under Grant No. DEB-0423704, Central Arizona-Phoenix Long-Term Ecological Research (CAP LTER). Support for M.W.S. came from Massachusetts Agricultural Experiment Station and the Department of Environmental Conservation under Project No. MAS009584, paper number 3468 and the National Science Foundation under Grant No. BCS-0948984. Any opinions, findings and conclusions or recommendation expressed in this material are those of the author(s) and do not necessarily reflect the views of the National Science Foundation (NSF).

References

Adriaensen, F. and Dhondt, A. A. (1990). Population-dynamics and partial migration of the European robin (*Erithacus rubecula*) in different habitats. *Journal of Animal Ecology*, **59**, 1077–1090.

Arizmendi, M. D., Lopez-Saut, E., Monterrubio-Solis, C., Juarez, L., Flores-Moreno, I. and Rodriguez-Flores, C. (2008). Effect of nectar feeders over diversity and abundance of hummingbirds and breeding success of two plant species in a sub-urban park next to Mexico City. *Ornitologia Neotropical*, **19**, 491–500.

Barber, J. R., Crooks, K. R. and Fristrup, K. M. (2010). The costs of chronic noise exposure for terrestrial organisms. *Trends in Ecology and Evolution*, **25**, 180–189.

Beissinger, S. R. and Osborne, D. R. (1982). Effects of urbanization on avian community organisation. *Condor*, **84**, 75–83.

Benvenuti, S. (2004). Weed dynamics in the Mediterranean urban ecosystem: ecology, biodiversity and management. *Weed Research*, **44**, 341–354.

Berglund, B. and Lindvall, T., eds. (1995). *Community noise*, World Health Organization, Stockholm, Sweden.

Bibby, R. L. and Webster-Brown, J. G. (2005). Characterisation of urban catchment suspended particulate matter (Auckland region, New Zealand); a comparison with non-urban SPM. *Science of the Total Environment*, **343**, 177–197.

Blair, R. (2004). The effects of urban sprawl on birds at multiple levels of biological organisation. *Ecology and Society*, **9**, 2.

Blair, R. B. (1996). Land use and avian species diversity along an urban gradient. *Ecological Applications*, 6, 506–519.

Bradley, C. A., and Altizer, S. (2007). Urbanization and the ecology of wildlife diseases. *Trends in Ecology and Evolution*, **22**, 95–102.

Bradley, C. A., Gibbs, S. E. J., and Altizer, S. (2008). Urban land use predicts West Nile virus exposure in songbirds. *Ecological Applications*, **18**, 1083–1092.

Brumm, H. (2004). The impact of environmental noise on song amplitude in a territorial bird. *Journal of Animal Ecology*, **73**, 434–440.

Brumm, H. (2006). Signalling through acoustic windows: nightingales avoid interspecific competition by short-term adjustment of song timing. *Journal of Comparative Physiology a-Neuroethology Sensory Neural and Behavioral Physiology*, **192**, 1279–1285.

Burghardt, K. T., Tallamy, D. W. and Shriver, W. G. (2009). Impact of native plants on bird and butterfly biodiversity in suburban landscapes. *Conservation Biology*, **23**, 219–224.

Burke, D. M. and Nol, E. (1998). Influence of food abundance, nest-site habitat, and forest fragmentation on breeding ovenbirds. *Auk*, **115**, 96–104.

Cannon, A. (1999). The significance of private gardens for bird conservation. *Bird Conservation International*, **9**, 287–297.

Carrete, M. and Tella, J. L. (2011). Inter-individual variability in fear of humans and relative brain size of the species are related to contemporary urban invasion in birds. *Plos One*, **6**, e18859.

Castelletta, M., Sodhi, N. S. and Subaraj, R. (2000). Heavy extinctions of forest avifauna in Singapore: Lessons for biodiversity conservation in Southeast Asia. *Conservation Biology*, **14**, 1870–1880.

Catterall, C. P. (2009). Responses of faunal assemblages to urbanization: global reseach paradigms and an avian case study. In M.J. McDonnell, A.K. Hahs and J.H. Breuste, eds. *Ecology of Cities and Towns—A Comparative Approach*, pp. 129–155. Cambridge University Press, Cambridge.

Catterall, C. P., Piper, S. D. and Goodall, K. (2002). Noisy miner irruptions associated with land use by humans in south-east Queensland: causes, effects and management implications. In A. Franks, J. Playford and A. Shapcott, eds. *Landscape Health of Queensland*, pp. 117–127. Proceedings of the Royal Society of Queensland, Brisbane.

Chace, J. F. and Walsh, J. J. (2006). Urban effects on native avifauna: a review. *Landscape and Urban Planning*, **74**, 46–69.

Chamberlain, D. E., Cannon, A. R. and Toms, M. P. (2004). Associations of garden birds with gradients in garden habitat and local habitat. *Ecography*, **27**, 589–600.

Chamberlain, D. E., Cannon, A. R., Toms, M. P., Leech, D. I., Hatchwell, B. J. and Gaston, K. J. (2009). Avian productivity in urban landscapes: a review and meta-analysis. *Ibis*, **151**, 1–18.

Chamberlain, D. E., Gough, S., Vaughan, H., Vickery, J. A. and Appleton, G. F. (2007). Determinants of bird species richness in public greenspaces. *Bird Study*, **54**, 87–97.

Collins, J. P., Kinzig, A., Grimm, N. B., Fagan, W. F., Hope, D., Wu, J. and Borer, E. (2000). A new urban ecology: modeling human communities as integral parts of ecosystems poses special problems for the development and testing of ecological theory. *American Scientist*, **88**, 416–425.

Cook, E. M., Hall, S. J. and Larson, K. L. (2012). Residential landscapes as social-ecological systems: a synthesis of multi-scalar interactions between people and their home environment. *Urban Ecosystems*, **15**, 19–52.

Cranmer, L., Mccollin, D. and Ollerton, J. (2012). Landscape structure influences pollinator movements and directly affects plant reproductive success. *Oikos*, **121**, 562–568.

Daniels, G. D. and Kirkpatrick, J. B. (2006). Does variation in garden characteristics influence the conservation of birds in suburbia? *Biological Conservation*, **133**, 326–335.

Davies, Z. G., Fuller, R. A., Loram, A., Irvine, K. N., Sims, V. and Gaston, K. J. (2009). A national scale inventory of resource provision for biodiversity within domestic gardens. *Biological Conservation*, **142**, 761–771.

Er, K. B. H., Innes, J. L., Martin, K. and Klinkenberg, B. (2005). Forest loss with urbanization predicts bird extirpations in Vancouver. *Biological Conservation*, **126**, 410–419.

Evans, K. L., Hatchwell, B. J., Parnell, M. and Gaston, K. J. (2010). A conceptual framework for the colonisation of urban areas: the blackbird *Turdus merula* as a case study. *Biological Reviews*, **85**, 643–667.

Evans, K. L., Chamberlain, D. E., Hatchwell, B. J., Gregory, R. D. and Gaston, K. J. (2011). What makes an urban bird? *Global Change Biology*, **17**, 32–44.

Evans, K. L., Newson, S. E. and Gaston, K. J. (2009). Habitat influences on urban avian assemblages. *Ibis*, **151**, 19–39.

Faeth, S. H., Bang, C. and Saari, S. (2011). Urban biodiversity: patterns and mechanisms. In R.S. Ostfeld, and W.H. Schlesinger, eds. *Year in Ecology and Conservation Biology*, **1223**, 69–81.

Faeth, S. H., Warren, P. S., Shochat, E. and Marussich, W. A. (2005). Trophic dynamics in urban communities. *Bioscience*, **55**, 399–407.

Fahrig, L. (2003). Effects of habitat fragmentation on biodiversity. *Annual Review of Ecology Evolution and Systematics*, **34**, 487–515.

Fernandez-Juricic, E. and Telleria, J. L. (2000). Effects of human disturbance on spatial and temporal feeding patterns of Blackbird *Turdus merula* in urban parks in Madrid, Spain. *Bird Study*, **47**, 13–21.

Fischer, J. D., Cleeton, S. H., Lyons, T. P. and Miller, J. R. (2012). Urbanization and the predation paradox: the role of trophic dynamics in structuring vertebrate communities. *Bioscience*, **62**, 809–818.

Francis, C. D., Ortega, C. P. and Cruz, A. (2011). Noise pollution filters bird communities based on vocal frequency. *PLoS ONE*, **6**, 8.

Francis, R.A. and Chadwick, M.A. (2013). *Urban Ecosystems: Understanding the Human Environment*. Routledge, London.

Friesen, L. E., Eagles, P. F. J. and Mackay, R. J. (1995). Effects of residential development on forest dwelling neotropical migrant songbirds. *Conservation Biology*, **9**, 1408–1414.

Fuller, R. A., Warren, P. H. and Gaston, K. J. (2007). Daytime noise predicts nocturnal singing in urban robins. *Biology Letters*, **3**, 368–370.

Fuller, R. A., Warren, P. H., Armsworth, P. R., Barbosa, O. and Gaston, K. J. (2008). Garden bird feeding predicts the structure of urban avian assemblages. *Diversity and Distributions*, **14**, 131–137.

Fuller, R. A., Tratalos, J. & Gaston, K. J. (2009). How many birds are there in a city of half a million people? *Diversity and Distributions*, **15**, 328–337.

Fuller, R. A., Tratalos, J., Warren, P. H., Davies, R. G., Pępkowska, A. & Gaston, K. J. (2010). Environment and Biodiversity. In M. Jenks and C. Jones, eds. *Dimensions of the Sustainable City*, pp. 75–103. Springer, London.

Fuller, R. A., Irvine, K. N., Davies, Z. G., Armsworth, P. R. and Gaston, K. J. (2012). Interactions between people and birds in urban landscapes. *Studies in Avian Biology*, **45**, 249–266.

Gascon, C. and Lovejoy, T. E. (1998). Ecological impacts of forest fragmentation in central Amazonia. *Zoology-Analysis of Complex Systems*, **101**, 273–280.

Gaston, K. J., Fuller, R. A., Loram, A., Macdonald, C., Power, S. and Dempsey, N. (2007). Urban domestic gardens (XI): variation in urban wildlife gardening in the United Kingdom. *Biodiversity and Conservation*, **16**, 3227–3238.

Gorissen, L., Snoeijs, T., Van Duyse, E. and Eens, M. (2005). Heavy metal pollution affects dawn singing behaviour in a small passerine bird. *Oecologia*, **145**, 504–509.

Grimm, N. B., Foster, D., Groffman, P., Grove, J. M., Hopkinson, C. S., Nadelhoffer, K. J., Pataki, D. E. and Peters, D. P. C. (2008). The changing landscape: ecosystem responses to urbanization and pollution across climatic and societal gradients. *Frontiers in Ecology and the Environment*, **6**, 264–272.

Hamer, G. L., Chaves, L. F., Anderson, T. K., Kitron, U. D., Brawn, J. D., Ruiz, M. O. et al. (2011). Fine-scale variation in vector host use and force of infection drive localized patterns of West Nile virus transmission. *PLoS ONE*, **6**, e23767.

Harris, R. J., and Reed, J. M. (2002). Behavioral barriers to non-migratory movements of birds. *Annales Zoologici Fennici*, **39**, 275–290.

Helden, A. J., Stamp, G. C. and Leather, S. R. (2012). Urban biodiversity: comparison of insect assemblages on native and non-native trees. *Urban Ecosystems*, **15**, 611–624.

Harper, M. J., Mccarthy, M. A. and Van Der Ree, R. (2005). The use of nest boxes in urban natural vegetation remnants by vertebrate fauna. *Wildlife Research*, **32**, 509–516.

Huste, A. and Boulinier, T. (2007). Determinants of local extinction and turnover rates in urban bird communities. *Ecological Applications*, **17**, 168–180.

Jokimäki, J. and Suhonen, J. (1998). Distribution and habitat selection of wintering birds in urban environments. *Landscape and Urban Planning*, **39**, 253–263.

Jokimäki, J. and Kaisanlahti-Jokimäki, M. L. (2012). Residential areas support overwintering possibilities of most bird species. *Annales Zoologici Fennici*, **49**, 240–256.

Jones, E. L. and Leather, S. R. (2012). Invertebrates in urban areas: A review. *European Journal of Entomology*, **109**, 463–478.

Jones, D. N. and Reynolds, S. J. (2008). Feeding birds in our towns and cities: a global research opportunity. *Journal of Avian Biology*, **39**, 265–271.

Kempenaers, B., Borgstrom, P., Loes, P., Schlicht, E. and Valcu, M. (2010). Artificial night lighting affects dawn song, extra-pair siring success, and lay date in songbirds. *Current Biology*, **20**, 1735–1739.

Kennedy, C. M., Marra, P. P., Fagan, W. F. and Neel, M. C. (2010). Landscape matrix and species traits mediate responses of Neotropical resident birds to forest fragmentation in Jamaica. *Ecological Monographs*, **80**, 651–669.

Kinzig, A. P., Warren, P., Martin, C., Hope, D. and Katti, M. (2005). The effects of human socioeconomic status and cultural characteristics on urban patterns of biodiversity. *Ecology and Society*, **10**, 23.

Kluza, D. A., Griffin, C. R. and Degraaf, R. M. (2000). Housing developments in rural New England: effects on forest birds. *Animal Conservation*, **3**, 15–26.

Lancaster, R.K. and Rees, W.E. (1979). Bird communities and the structure of urban habitats. *Canadian Journal of Zoology-Revue Canadienne De Zoologie*, **57**, 2358–2368.

Larsen, L. and Harlan, S. L. (2006). Desert dreamscapes: Residential landscape preference and behavior. *Landscape and Urban Planning*, **78**, 85–100.

Lefebvre, L., Reader, S. M. and Sol, D. (2004). Brains, innovations and evolution in birds and primates. *Brain Behavior and Evolution*, **63**, 233–246.

Lepczyk, C. A., Mertig, A. G. and Liu, J. G. (2004). Landowners and cat predation across rural-to-urban landscapes. *Biological Conservation*, **115**, 191–201.

Lepczyk, C. A., Warren, P. S., Machabee, L., Kinzig, A., and Mertig, A. (2012). Who feeds the birds? A comparison between Phoenix, Arizona and Southeastern Michigan. In C. Lepczyk and P. Warren, eds, *Urban Bird Ecology and Conservation*, pp. 267–286. University of California Press, Berkeley, CA.

Lerman, S. B., Turner, K. and Bang, C. (2012). Homeowner associations as a vehicle for promoting native urban biodiversity. *Ecology and Society* **17**(4): 45.

Lerman, S. B. and Warren, P. S. (2011). The conservation value of residential yards: linking birds and people. *Ecological Applications*, **21**, 1327–1339.

Leston, L.F.V. and Rodewald, A.D. (2006). Are urban forests ecological traps for understory birds? An examination using Northern cardinals. *Biological Conservation*, **131**, 566–574.

Levins, R. (1970). *Extinction*, American Mathematical Society, Providence, Rhode Island.

Lim, H. C. and Sodhi, N. S. (2004). Responses of avian guilds to urbanization in a tropical city. *Landscape and Urban Planning*, 66, 199–215.

Lindsay, A. R., Gillum, S. S. and Meyer, M. W. (2002). Influence of lakeshore development on breeding bird communities in a mixed northern forest. *Biological Conservation*, **107**, 1–11.

Lohr, B., Wright, T. F. and Dooling, R. J. (2003). Detection and discrimination of natural calls in masking noise by birds: estimating the active space of a signal. *Animal Behaviour*, **65**, 763–777.

Lomov, B., Keith, D. A. and Hochuli, D. F. (2010). Pollination and plant reproductive success in restored urban landscapes dominated by a pervasive exotic pollinator. *Landscape and Urban Planning*, **96**, 232–239.

Loss, S. R., Will, T., and Marra, P. P. (2013). The impact of free-ranging domestic cats on wildlife of the United States. *Nature Communications*, **4**, 1396.

Luck, G. W. and Smallbone, L. T. (2010). Species diversity and urbanization: patterns, drivers and implications. In K. J. Gaston, ed. *Urban Ecology*, pp. 88–119, Cambridge University Press, Cambridge.

Luniak, M. (2004) Synurbisation—adaptation of animal wildlife to urban development. Proceedings of the 4th International Urban Wildlife Symposium, Tuscon, Arizona, pp. 50–55.

Luniak, M., Mulsow, R. and Walasz, K. (1990). Urbanization of the European Blackbird—expansion and adaptations of urban population. In M. Luniak, ed. *Urban ecological studies in Central and Eastern Europe*, pp. 187–198, Ossolineum, Wroclaw.

Major, R. E., Gowing, G. and Kendal, C. E. (1996). Nest predation in Australian urban environments and the role of the pied currawong, *Strepera graculina*. *Australian Journal of Ecology*, **21**, 399–409.

Maklakov, A. A., Immler, S., Gonzalez-Voyer, A., Ronn, J. and Kolm, N. (2011). Brains and the city: big-brained passerine birds succeed in urban environments. *Biology Letters*, **7**, 730–732.

Marzluff, J. M. and Ewing, K. (2001). Restoration of fragmented landscapes for the conservation of birds: A general framework and specific recommendations for urbanising landscapes. *Restoration Ecology*, **9**, 280–292.

Marzluff, J. M. and Rodewald, A. D. (2008). Conserving biodiversity in urbanising areas: Nontraditional views from a bird's perspective. *Cities and the Environment*, **1**, 6.

Marzluff, J. M., Gehlbach, F. R. and Manuwal, D. A. (1998). Urban environments: influences on avifauna and challenges for the avian conservationist. In J. M. Marzluff and R. Sallabanks, eds. *Avian Conservation: Research and Management*, pp. 283–305. Island Press, Washington.

Mcintyre, N. E. (2000). Ecology of urban arthropods: A review and a call to action. *Annals of the Entomological Society of America*, **93**, 825–835.

Mckinney, M. L. (2002). Urbanization, biodiversity, and conservation. *Bioscience*, **52**, 883–890.

Mckinney, M. L. (2006). Urbanization as a major cause of biotic homogenisation. *Biological Conservation*, **127**, 247–260.

Miller, J. R., Wiens, J. A., Hobbs, N. T. and Theobald, D. M. (2003). Effects of human settlement on bird communities in lowland riparian areas of Colorado (USA). *Ecological Applications*, **13**, 1041–1059.

Møller, A. P. (2008). Flight distance of urban birds, predation, and selection for urban life. *Behavioral Ecology and Sociobiology*, **63**, 63–75.

Møller, A. P. (2009). Successful city dwellers: a comparative study of the ecological characteristics of urban birds in the Western Palearctic. *Oecologia*, **159**, 849–858.

Møller, A. P., Diaz, M., Flensted-Jensen, E., Grim, T., Ibanez-Alamo, J. D., Jokimäki, J., Mand, R., Marko, G., and Tryjanowski, P. (2012). High urban population density of birds reflects their timing of urbanization. *Oecologia*, **170**, 867–875.

Montague-Drake, R. M., Lindenmayer, D. B., Cunningham, R. B. and Stein, J. A. (2011). A reverse keystone species affects the landscape distribution of woodland avifauna: a case study using the Noisy Miner (*Manorina melanocephala*) and other Australian birds. *Landscape Ecology*, **26**, 1383–1394.

Mumme, R. L., Schoech, S. J., Woolfenden, G. W. and Fitzpatrick, J. W. (2000). Life and death in the fast lane: Demographic consequences of road mortality in the Florida Scrub-Jay. *Conservation Biology*, **14**, 501–512.

Nam, D. H. and Lee, D. P. (2006). Reproductive effects of heavy metal accumulation on breeding feral pigeons (*Columba livia*). *Science of the Total Environment*, **366**, 682–687.

Oliver, A. J., Hong-Wa, C., Devonshire, J., Olea, K. R., Rivas, G. F. and Gahl, M. K. (2011). Avifauna richness enhanced in large, isolated urban parks. *Landscape and Urban Planning*, **102**, 215–225.

Orlowski, G., Martini, K. and Martini, M. (2008). Avian responses to undergrowth removal in a suburban wood. *Polish Journal of Ecology*, **56**, 487–495.

Ortega-Álvarez, R. and Macgregor-Fors, I. (2011). Dusting-off the file: A review of knowledge on urban ornithology in Latin America. *Landscape and Urban Planning*, **101**, 1–10.

Patten, M. A. and Bolger, D. T. (2003). Variation in top-down control of avian reproductive success across a fragmentation gradient. *Oikos*, **101**, 479–488.

Paul, M. J. and Meyer, J. L. (2001). Streams in the urban landscape. *Annual Review of Ecology and Systematics*, **32**, 333–365.

Peris, S. J. (2003). Feeding in urban refuse dumps: ingestion of plastic objects by the White Stork (*Ciconia ciconia*). *Ardeola*, **50**, 81–84.

Petersen, L. K., Lyytimäki, J., Normander, B., Hallin-Pihlatie, L., Bezák, P., Cil, A. and Varjopuro, R., et al. (2007). *Urban lifestyle and urban biodiversity*, ALTER-Net research report, University of Aarhus, Aarhus.

Pidgeon, A. M., Radeloff, V. C., Flather, C. H., Lepczyk, C. A., Clayton, M. K., Hawbaker, T. J. and Hammer, R. B.

(2007). Associations of forest bird species richness with housing and landscape patterns across the USA. *Ecological Applications*, **17**, 1989–2010.

Piper, S. D. and Catterall, C. P. (2003). A particular case and a general pattern: hyperaggressive behaviour by one species may mediate avifaunal decreases in fragmented Australian forests. *Oikos*, **101**, 602–614.

Reis, E., Lopez-Iborra, G. M. and Pinheiro, R. T. (2012). Changes in bird species richness through different levels of urbanization: Implications for biodiversity conservation and garden design in Central Brazil. *Landscape and Urban Planning*, **107**, 31–42.

Reisen, W. K., Takahashi, R. M., Carroll, B. D., and Quiring, R. (2008). Delinquent mortgages, neglected swimming pools, and West Nile Virus, California. *Emerging Infectious Diseases*, **14**, 1747–1749.

Richmond, S., Nol, E. and Burke, D. (2011). Avian nest success, mammalian nest predator abundance, and invertebrate prey availability in a fragmented landscape. *Canadian Journal of Zoology-Revue Canadienne De Zoologie*, **89**, 517–528.

Robertson, B., Kriska, G., Horvath, V. and Horvath, G. (2010). Glass buildings as bird feeders: urban birds expliot insects trapped by polarized light pollution. *Acta Zoologica Academiae Scientiarum Hungaricae*, **56**, 283–292.

Robinson, S. K. and Wilcove, D. S. (1994). Forest fragmentation in the temperate zone and its effects on migratory songbirds. *Bird Conservation International*, **4**, 233–249.

Rodewald, A. D. (2012). Evaluating factors that influence avian community response to urbanization. *Studies in Avian Biology*, **45**, 71–92.

Rodewald, A. D. and Shustack, D. P. (2008a). Urban flight: understanding individual and population-level responses of Nearctic-Neotropical migratory birds to urbanization. *Journal of Animal Ecology*, **77**, 83–91.

Rodewald, A. D. and Shustack, D. P. (2008b). Consumer resource matching in urbanising landscapes: Are synanthropic species over-matching? *Ecology*, **89**, 515–521.

Rosenberg, K. V., Terrill, S. B. and Rosenberg, G. H. (1987). Value of suburban habitats to desert riparian birds. *Wilson Bulletin*, **99**, 642–654.

Rosselli, L. and Stiles, F. G. (2012). Wetland habitats of the Sabana de Bogota Andean Highland Plateau and their birds. *Aquatic Conservation-Marine and Freshwater Ecosystems*, **22**, 303–317.

Rutz, C. (2008). The establishment of an urban bird population. *Journal of Animal Ecology*, **77**, 1008–1019.

Sandström, U. G., Angelstam, P. and Mikusinski, G. (2006). Ecological diversity of birds in relation to the structure of urban greenspace. *Landscape and Urban Planning*, **77**, 39–53.

Santos, C. D., Miranda, A. C., Granadeiro, J. P., Lourenco, P. M., Saraiva, S. and Palmeirim, J. M. (2010). Effects of artificial illumination on the nocturnal foraging of waders. *Acta Oecologica-International Journal of Ecology*, **36**, 166–172.

Saunders, D. A., Hobbs, R. J. and Margules, C. R. (1991). Biological consequences of ecosystem fragmentation—a review. *Conservation Biology*, **5**, 18–32.

Shanahan, D. F., Miller, C., Possingham, H. P. and Fuller, R. A. (2011a). The influence of patch area and connectivity on avian communities in urban revegetation. *Biological Conservation*, **144**, 722–729.

Shanahan, D. F., Possingham, H. and Riginos, C. (2011b). Models based on individual level movement predict spatial patterns of genetic relatedness for two Australian forest birds. *Landscape Ecology*, **26**, 137–148.

Shochat, E., Lerman, S. B., Anderies, J. M., Warren, P. S., Faeth, S. H. and Nilon, C. H. (2010). Invasion, competition, and biodiversity loss in urban ecosystems. *Bioscience*, **60**, 199–208.

Shochat, E., Lerman, S. B., Katti, M. and Lewis, D. B. (2004). Linking optimal foraging behavior to bird community structure in an urban-desert landscape: Field experiments with artificial food patches. *American Naturalist*, **164**, 232–243.

Shochat, E., Warren, P. S., Faeth, S. H., Mcintyre, N. E. and Hope, D. (2006). From patterns to emerging processes in mechanistic urban ecology. *Trends in Ecology and Evolution*, **21**, 186–191.

Sims, V., Evans, K. L., Newson, S. E., Tratalos, J. A. and Gaston, K. J. (2008). Avian assemblage structure and domestic cat densities in urban environments. *Diversity and Distributions*, **14**, 387–399.

Slabbekoorn, H. and Peet, M. (2003). Birds sing at a higher pitch in urban noise. *Nature*, **424**, 267.

Smith, R. M., Warren, P. H., Thompson, K. and Gaston, K. J. (2006). Urban domestic gardens (VI): environmental correlates of invertebrate species richness. *Biodiversity and Conservation*, **15**, 2415–2438.

Strohbach, M., Haase, D. and Kabisch, N. (2009). Birds and the city: urban biodiversity, land use, and socioeconomics. *Ecology and Society*, **14**, Art no. 31.

Sushinsky, J.R., Rhodes, J.R., Possingham, H.P., Gill, T.K. and Fuller, R.A. (2013). How should we grow cities to minimize their biodiversity impacts? *Global Change Biology*, **19**, 401–410.

Takami, Y.C., Koshio, C., Ishii, M., Fujii, H., Hidaka, T. Shimizu, I. (2004). Genetic diversity and structure of urban populations of Pieris butterflies assessed using amplified fragment length polymorphism. *Molecular Ecology* **13**, 245–258.

Thompson, K., Austin, K. C., Smith, R. M., Warren, P. H., Angold, P. G. and Gaston, K. J. (2003). Urban domestic gardens (I): Putting small-scale plant diversity in context. *Journal of Vegetation Science*, **14**, 71–78.

Tratalos, J., Fuller, R. A., Evans, K. L., Davies, R. G., Newson, S. E., Greenwood, J. J. D. and Gaston, K. J.

(2007). Bird densities are asociated with household densities. *Global Change Biology*, **13**, 1685–1695.

Tremblay, M. A. and St Clair, C. C. (2011). Permeability of a heterogeneous urban landscape to the movements of forest songbirds. *Journal of Applied Ecology*, **48**, 679–688.

Warren, P. S., Harlan, S., Boone, C., Lerman, S. B., Shochat, E. and Kinzig, A. P. (2010). Urban ecology and human social organisation. In K. J. Gaston, ed. *Urban Ecology*, pp. 172–201. Cambridge University Press, Cambridge.

Warren, P. S., Katti, M., Ermann, M. and Brazel, A. (2006). Urban bioacoustics: it's not just noise. *Animal Behaviour*, **71**, 491–502.

Wood, W. E. and Yezerinac, S. M. (2006). Song sparrow (*Melospiza melodia*) song varies with urban noise. *Auk*, **123**, 650–659.

Yeh, P. J. (2004). Rapid evolution of a sexually selected trait following population establishment in a novel habitat. *Evolution* **58**, 166–172.

CHAPTER 2

The impact of artificial light on avian ecology

Kamiel Spoelstra and Marcel E. Visser

2.1 Introduction

The advance of urbanization and the increase in the standard of living involves a progressive, worldwide expansion in nocturnal illumination. The presence of artificial light extends to all areas with human activities; two-thirds of the world's population are exposed to light levels above the threshold set for light pollution, which includes almost the entire population in North America and Europe (Cinzano et al., 2001). Areas all over the planet are affected: terrestrial habitats are illuminated by street lights and by a variety of lights for the illumination of structures, and lighted vehicles; marine habitats are disturbed by light from oil rigs, wind turbines, ships and light houses. Natural habitat is not only polluted by direct light (glare), but—in most cases—indirectly by reflection (sky glow) of light.

Ecological light pollution is defined as light which disrupts ecosystems (Longcore & Rich, 2004). This is induced by immediate changes in species composition as a result of deterrence or attraction of different species. Photoperiod is an important cue for many bird species and this makes light pollution potentially an important ecological trap (Schlaepfer et al., 2002): photoperiod may no longer be a reliable cue in the presence of artificial light. In illuminated areas, the response to day length may lead to non-adaptive behaviour. Moreover, even if artificial light has no direct effect, species may be affected by more complex mechanisms, such as cascading effects in the food chain. For example, insects attracted by light are no longer available for predators which hunt in darkness, or insects may not reproduce any longer (Perkin et al., 2011). Consequently, their larvae are no longer available as a food source for predatory species.

Effects of artificial night lighting have already been reported for most animal classes, and may involve substantial changes in behaviour and physiology (Navara & Nelson, 2007; Rich & Longcore, 2006). Birds feature prominently among the species affected. This is not surprising, given that most avian species have excellent light perception. Artificial light is known to affect foraging, migration, orientation, daily timing of behaviour (e.g. Jones & Francis, 2003; Poot et al., 2008; Dwyer et al., 2012), and potentially influences seasonal timing and stress physiology. Birds are therefore an important species group for the study of the impact of artificial light on the ecosystem.

In this chapter, we will discuss the information currently available on how light affects birds in different life-history stages on short time scales and we will identify potential long-term effects. We will discuss possibilities for mitigation of impact of ecological light pollution on birds and will outline experiments needed to further clarify the ecological impact of night light on birds.

2.2 Known effects of artificial light

Probably the best documented response of birds to artificial light is attraction during migration at night. Brightly lit spots in dark environments, such as oil rigs at sea, can disrupt orientation in night-migrating diurnal birds, especially when the sky

Avian Urban Ecology. Edited by Diego Gil and Henrik Brumm

is overcast (e.g. Jones & Francis, 2003; Poot et al., 2008). This form of disorientation often leads to death because of exhaustion, which ensues from a dramatic increase in flight time, but also directly by collision with the lighted structure. Direct mortality by light has also been reported from island breeding petrels, in which significant numbers of fledglings die as a result of predation and exhaustion after attraction to artificial light (Le Corre et al., 2002). Less dramatic effects are observed in habitats where light exposure has a more permanent character, and light is much more spatially distributed. Under these conditions light does not act as an attraction or deterrence but potentially has different negative consequences.

One of the best known effects of night light is an increase in nocturnal activity in birds living in illuminated habitats (Derrickson, 1988; Dwyer et al., 2012; Santos et al., 2010). Timing of daily activity is also affected—a frequently observed effect of artificial light is that birds start their dawn song earlier in the morning (Bergen & Abs, 1997; Kempenaers et al., 2010; Miller, 2006). Interestingly, species which start their dawn song earlier in the morning under natural conditions are affected much more by exposure to artificial light than birds starting dawn song later in the morning (e.g. blue tits *Cyanistes caeruleus*; Kempenaers et al., 2010, see Figure 2.1). This is probably related to interspecific variation in eye size, reflecting visual capability at low light intensities (Thomas et al., 2002). These effects are in line with the observation that birds can change their timing of song activity under natural light level fluctuations. For example, Leopold & Eynon (1961) already showed that birds sang earlier in the morning during moonlit nights compared to mornings with a cloud covered sky (see also Miller, 2006). The early onset of dawn song has potentially negative consequences such as depletion of energy levels and exhaustion (Kempenaers et al., 2010; Longcore & Rich, 2004), and the attraction of predators (Miller, 2006; Santos et al., 2010) but may also have positive effects (see Section 2.3, Consequences for fitness and population numbers). Miller (2006) showed that light also affected the timing of singing at dusk (birds stopped singing later when there was more light), but the effect at dawn was more pronounced. Continuation of song activity by artificial light after dusk has been reported in urban blackbirds (*Turdus merula*; Stephan,

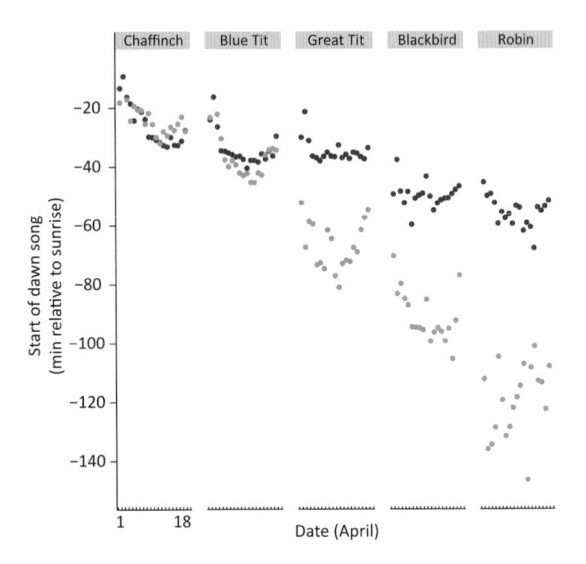

Figure 2.1 Start of dawn song of five common European passerine species, relative to sunrise between April 1–18. Dark dots indicate the onset of dawn song in dark territories, light dots indicate dawn song onset in illuminated territories. Species which start dawn song relatively early under normal—not illuminated—circumstances tend to advance dawn song much more by light than species that start relatively late in the morning. Reprinted from Kempenaers et al. (2010), Artificial night lighting affects dawn song, extra-pair siring success, and lay date in songbirds, *Current Biology* **20**(19) p. 1736, Copyright (2010), with permission from Elsevier.

1999), at the moment when forest blackbirds would have stopped all song activity. However the effect of artificial night lighting on dusk song has not been studied in detail.

Effects of artificial light on seasonal timing, such as mistiming of yearly reproduction, molt and migration, are seemingly logical (Navara & Nelson, 2007), but in fact there is very little evidence for such effects. For many bird species, especially those living in temperate zones, precise timing of reproduction, moult and migration is important, because there is often only a short 'optimal' time window for these activities (Dawson et al., 2001; Gwinner, 1999; Schaper et al., 2012). Indeed, seasonal timing can strongly impact fitness (Kempenaers et al., 2010; Poesel et al., 2006; Thomas et al., 2001; Verboven & Visser, 1998). Birds use the change in day length as a prime cue to anticipate optimal future conditions, for example in the timing of the onset of gonadal growth long before the start of egg laying (Dawson, 2008; Gwinner, 1999). Artificial light may lead to the false perception of longer days earlier in the season, by lengthening the light phase. This seems realistic, because, for example, the threshold for the onset of gonadal growth and body fattening in buntings (Emberizidae) is 10 Lux for white light (Kumar et al., 1992), which is comparable to the light level found at an illuminated parking lot (Rich & Longcore, 2006). In the laboratory, this effect is clearly present and has been known for a long time and used to induce gonadal growth in numerous experiments (e.g. Dawson et al., 2001; Lambrechts et al., 1997; Rowan, 1925; te Marvelde et al., 2012). Day length and the daily modulation of light intensity vary strongly with latitude. The distribution range of many species covers these different latitudes and the response to day length in different life-history stages of a local population may be fine-tuned to changes in day length at a specific latitude (Silverin et al., 1993), or in some cases this response is even specific for a region (Lambrechts et al., 1997; see Figure 2.2). The variance in day length perception caused by the presence of artificial light may therefore be problematic, even though these effects are seemingly subtle compared to the natural variation

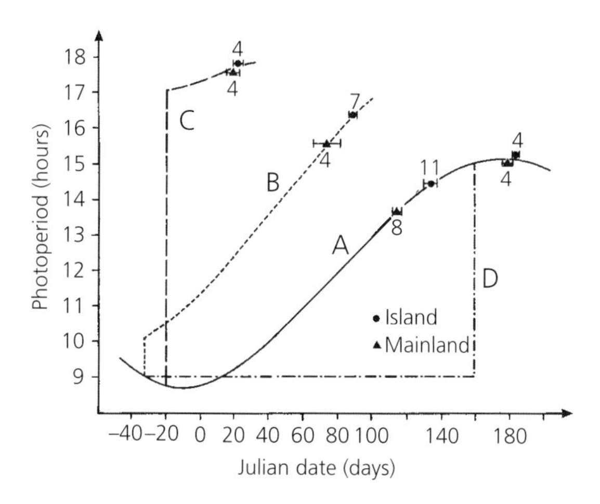

Figure 2.2 Onset of first clutches (average ± SE in Julian Dates; January 1 = 1; February 1 = 32) of captive blue tits from the mainland Europe (triangles) and from Corsica (dots), all breeding in aviaries in Montpellier, southern France. All birds were presented with one to four changes in photoperiod, one of which was the natural increase in photoperiod (solid line). Clearly, with additional artificial light birds will start to breed earlier, even when this increase is as early as January (treatment C). Figure from Lambrechts et al. (1997), Copyright 2014 National Academy of Sciences, USA.

over latitude to which populations have been able to adapt over a period of millennia.

Effects of misperception of day length on timing of egg laying as a result of artificial light may however be reduced by other environmental factors like temperature or food abundance. For example, when great tits (*Parus major*) were captured in their natural habitat before the start of the breeding season, exposed to a long day photoperiod in the laboratory and released back in the field, they showed advanced follicle growth, but eventually these birds did not differ in laying date in comparison to control birds (te Marvelde et al., 2012). This example shows that early exposure to light (long days) does not necessarily lead to actual changes in laying date under field conditions (although it may still be costly to develop gonads earlier than needed). The advancement of blue tits breeding in illuminated habitat reported by Kempenaers et al. (2010) may therefore be a direct effect of light, but it could also be an indirect effect mediated by male behaviour as male blue tits might have started singing earlier in the season under the influence of artificial night lighting, which may have stimulated females to lay earlier.

Apart from song activity, Relatively little is known about the impact of artificial light on other daily activity patterns of birds. There are, however, some examples of changes in these patterns. Indeed, wader species can successfully use illuminated estuarine and coastal wetlands at night. This different temporal utilization of habitat creates an artificial niche, which may change the level of competition between species (Dwyer et al., 2012; Hockin et al., 1992; Santos et al., 2010). Furthermore, several species of diurnal insectivorous birds have been reported to use nocturnal light to forage at night (Lebbin et al., 2007). Recently, an experimental field study by Titulaer et al. (2012) showed that great tits exposed to artificial light during the chick feeding stage strongly increased the frequency of food provisioning to the chicks during daytime (see Figure 2.3), but the underlying mechanism remains unclear.

There are thus clear effects of artificial light on birds. These effects, however, all play a role at the individual level and they are measured over a relative short time scale. Whether or not artificial light has fitness consequences or even effects on population numbers is still unknown.

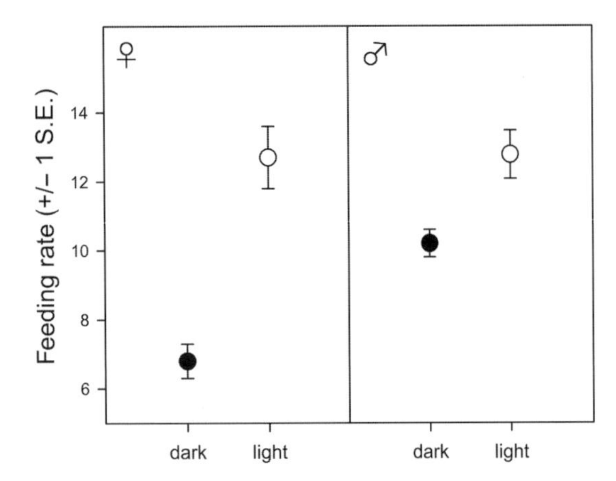

Figure 2.3 Feeding rate of great tits (*Parus major*) during the second half (9–16 day nestlings) of the chick rearing stage (in number of entries in the nest box per hour). Black circles: non-illuminated nest boxes; open circles: illuminated nest boxes. Error bars indicate 1 SE. Figure from Titulaer et al. (2012, slightly adapted figure under Creative Commons Attribution License).

2.3 Consequences for fitness and population numbers

Fitness consequences are likely to be present when there are effects of artificial light on one or more life history stages of birds. Photoperiod is an important cue for seasonal timing, and if birds alter for instance their timing due to artificial light they are likely to be less well adapted to their environment, i.e. have a lower fitness than their undisturbed counterparts (cf. Schlaepfer et al., 2002). There is, however, very little data on this. An example of fitness consequences of a change in daily timing is the paternity gain by male blue tits that occupy illuminated territories as reported by Kempenaers et al. (2010). Males that have a territory under street lights sired more extra-pair offspring in other nests than males that had a territory further away from the artificial light. The males near the light posts are potentially more attractive to females in adjacent, dark territories because of their early onset of dawn song. This is in line with earlier observations of female preference for early singing males (Poesel et al., 2006). However, the effect may also relate to differences in male quality and needs to be verified in an experimental setup.

De Molenaar et al. (2006) did a unique study on nesting behaviour of godwits (*Limosa l. limosa*) in a 230 ha grassland area crossed by a lamp lit highway. Road lights, and temporally placed light posts away from the road were experimentally switched off and on during two consecutive years, and timing of egg laying, nest site choice and nest predation was monitored. The results suggest that early arriving godwits choose nest sites at greater distance from the artificial light than late arriving birds, and that road lighting has a negative effect on the preference for breeding grounds. However, very little is known about the long-term consequences of exposure to artificial light on the population level of resident birds in a continuously lit terrestrial habitat. These effects may be latent and therefore much less visible. One reason for this gap in our knowledge is that there are very few other experimental studies comparing areas with and without artificial light (see 'call for experimental work'). Also, there is a severe lack of knowledge on the threshold values of light intensity at which light has no effect anymore in an ecological context. This makes it unclear which areas near a light source need to be studied to evaluate avian population consequences. Once we know about these values and consequently the magnitude of these effects, we can make predictions about the long-term effect these could have on the population. Whether these effects are indeed a serious threat cannot be foreseen a priori—studies on other effects have shown that negative effects, although causing significant mortality, might be compensated by the reproductive potential in some species (Arnold & Zink, 2011).

In short, we have very little knowledge about what the long-term consequences are, and it is essential that more work is done on this to assess the ecological consequences of artificial light.

2.4 What are the possibilities for mitigation?

Artificial light is often assumed to have negative ecological effects and over the past years there has been an increased effort to reduce these effects. However, current efforts for the mitigation of such effects are often inadequate (Montevecchi, 2006).

Possibilities for mitigation are thoroughly evaluated by Gaston et al. (2012) and include prevention, limitation of illumination during parts of the night, shielding, changing light intensity and change of spectral composition. Shielding of upward radiation can indeed significantly reduce impact of light on birds (Reed et al., 1985).

Reduction of negative effects of artificial illumination by changes in the spectrum has become a realistic option over the last few years with the current, global change to LED lighting. Probably the best known example of reduction of disruptive effects by light colour is the attraction and disorientation of migrating birds at night. When long wavelengths are removed from the light spectrum, migratory birds are much less disturbed by brightly illuminated spots (Gochfeld, 1973; Wiltschko & Wiltschko, 1995; Wiltschko et al., 1993). The application of green light could reduce the impact of brightly illuminated oil rigs amid a dark sea (Poot et al., 2008). These examples are very valuable but we do want to stress that they are still few and thus that generalisation is difficult.

In order to predict the response of different species of birds to light with a certain spectrum, it is essential to know which colours are actually perceived. Birds have excellent light perception—birds have been almost exclusively diurnal throughout their evolution and this may be the reason why they have retained a more complex colour vision compared to mammals (Bowmaker, 1980). Colour vision in most birds is mediated by four distinct single cone types, with peak sensitivities at long (~570 nm), medium (~500 nm), short (~470 nm), and very short (either around 365 nm or around 405 nm) wavelengths (Osorio & Vorobyev, 2008). In addition, the lens and cornea are transparent for wavelengths well below 400 nm and many birds have indeed good UV vision (Goldsmith, 1980; Govardovskiĭ & Zueva, 1977). Peak sensitivities vary between species because of different densities of highly variable, brightly coloured oil droplets within the receptor cells, which shift the peak cone sensitivity by filtering light (Bowmaker, 1980). In general, spectral sensitivity of each bird species is most likely driven by the habitat (light environment) in which the bird lives (e.g. Hart et al., 2001), but also by predation (inconspicuousness to predators, Håstad et al., 2005) and sexual selection (conspicuousness to conspecifics, e.g. Delhey et al., 2012).

The temporal organization of birds is intricate and highly dependent on light input. A complicating factor is that, unlike mammals, birds have extraocular perception of light by photoreceptors in the pineal gland and brain (Bellingham & Foster, 2002). Birds can be synchronized to an external light/dark cycle (entrained) by extraocular light (Cassone et al., 2009) and when the light pulses are of sufficient duration, extraocular light induces gonadal growth. In laboratory experiments, it has been shown that the spectral composition of light is important for different timing processes. For example, red light has a stronger effect on the induction of gonadal development than white light (Kumar & Rani, 1996; Lewis & Morris, 2000; Malik et al., 2002), which may be because it penetrates the skull relatively easily (Hartwig & Van Veen, 1979). Green light has little effect on gonadal development and on the induction of photorefractoriness (Rani & Kumar, 2000; Rani et al., 2001), but is much more effective in the synchronization of the circadian clock to the overt light/dark cycle (entrainment; Kumar & Rani, 1996; Malik et al., 2002).

In conclusion, the ecological effects of artificial light at night of different spectral composition are difficult to predict as birds have excellent vision with a very broad spectral sensitivity, and have extra-ocular light perception. Moreover, different potential ecological effects are under the influence of different light colours. Finally, the inter-species differences in spectral sensitivity are an additional confounding factor in predicting the effects for different species. This makes a 'one size fits all' approach to reduce the impact on birds impossible. Hence, it will depend on the circumstances which spectrum would be suitable, for instance whether the night light is in a breeding area or whether it is near a foraging area.

2.5 Call for experimental work, long-term study and effects of spectral composition

The ubiquity of light pollution, and the current possibilities of spectral adaptation by the application of LED technology, makes it of utmost importance to understand the impact of artificial light at different life-history stages of birds, and to be able to predict the long-term consequences on the population level. Immediate effects of artificial light, such as mortality of birds by disturbance of migration or by fledglings attracted by light can already be addressed relatively easy. However, in order to mitigate long-term effects of 'continuous' nocturnal light in populated areas, more research is needed. Detailed study of these effects should include the consequences of artificial light on (stress) physiology, circadian and seasonal timing, abundance and timing of prey species in illuminated areas, and effects on predators. As of today, most studies on the effects of artificial light are correlative. In order to discern effects of light from other factors, both anthropogenic and natural, there is a special need for experimental studies.

As a potential route of mitigation is the use of light with different wavelengths (Gaston et al., 2012), we also need to set up field experiments to study the effects different colours of light have on the various life-history stages in birds. Finally, and most challenging, we need to look at long-term effects, starting with the assessment of fitness consequences of artificial light. These should then be scaled up to population consequences, and related light intensity (how far from a light source effects are still present). These studies should include many species, as effects on prey species may cascade through the food chain.

In order to find answers to the questions raised above, a field experiment has recently started in the Netherlands where, in a large-scale monitoring project, the presence and density of many species and species groups are monitored at experimentally lit field sites. On each site (all at a forest edge) there are always four treatments in separate transects: light with a reduced red component (greenish), light with a reduced blue component (reddish), white light, and dark control. The spectra are chosen such that the light remains suitable for civil use. An assessment was made of the species present before the application of illumination, and over the course of three years the change in species and species composition is carefully monitored (see also www.lichtopnatuur.org). Experiments like these will answer questions on the ecological impact of artificial light of different wavelengths. However, many more of such experiments are needed, for instance in different habitats, to get a comprehensive overview.

Given the degree of light pollution—two-thirds of the world's population lives in light-polluted areas—it is amazing how little is known on the ecological effects of light pollution. While there are a number of good studies that show the effects of artificial light on individual behaviour, what is lacking are studies on the fitness consequences of the effects and on how population numbers are affected by light pollution. Only then we can fully assess the ecological consequences and take useful mitigation measures.

Acknowledgements

We thank Bart Kempenaers for his input in the manuscript and the scientific staff of the *Light on Nature* project for their collaboration. This research is supported by the Dutch Technology Foundation STW, which is the applied science division of NWO, and the Technology Programme of the Ministry of Economic Affairs.

References

Bellingham, J. and Foster, R. G. (2002). Opsins and mammalian photoentrainment. *Cell and Tissue Research*, **309**, 57–71.

Bergen, F. and Abs, M. (1997). Etho-ecological study of the singing activity of the Blue Tit (*Parus caeruleus*), Great Tit (*Parus major*) and Chaffinch (*Fringilla coelebs*). *Journal of Ornithology*, **138**(4), 451–467.

Bowmaker, J. K. (1980). Colour vision in birds and the role of oil droplets. *Trends in Neurosciences*, **3**(8), 196–199.

Cassone, V. M., Paulose, J., Whitfield-Rucker, M., and Peters, J. (2009). Time's arrow flies like a bird: two paradoxes for avian circadian biology. *General and Comparative Endocrinology*, **163**(1–2), 109–116.

Cinzano, P., Falchi, F., and Elvidge, C. D. (2001). The first World Atlas of the artificial night sky brightness. *Monthly Notices of the Royal Astronomical Society*, **328**(3), 689–707.

Dawson, A., King, V. M., Bentley, G. E., and Ball, G. F. (2001). Photoperiodic control of seasonality in birds. *Journal of Biological Rhythms*, **16**(4), 365–380.

Dawson, A. (2008). Control of the annual cycle in birds: endocrine constraints and plasticity in response to ecological variability. *Philosophical Transactions of the Royal Society of London. Series B, Biological Sciences*, **363**(1497), 1621–1633.

Delhey, K., Hall, M., Kingma, S. A., and Peters, A. (2012). Increased conspicuousness can explain the match between visual sensitivities and blue plumage colours in fairy-wrens. *Proceedings of the Royal Society B: Biological Sciences*, **280**(1750), doi: 10.1098/rspb.2012.1771.

Derrickson, K. C. (1988). Variation in repertoire presentation in northern mockingbirds. *The Condor*, **90**(3), 592–606.

Dwyer, R. G., Bearhop, S., Campbell, H. A., and Bryant, D. M. (2012). Shedding light on light: benefits of anthropogenic illumination to a nocturnally foraging shorebird. *Journal of Animal Ecology*, **82**(2), 478–485.

Gaston, K. J., Davies, T. W., Bennie, J., and Hopkins, J. (2012). Reducing the ecological consequences of night-time light pollution: options and developments. *Journal of Applied Ecology* **49**(6), 1256–1266.

Gochfeld, M. (1973). Confused nocturnal behavior of a flock of migrating yellow wagtails. *The Condor*, **75**(2), 252–253.

Goldsmith, T. H. (1980). Hummingbirds see near ultraviolet light. *Science*, **207**, 786–788.

Govardovskiĭ, V. I., and Zueva, L. V. (1977). Visual pigments of chicken and pigeon. *Vision Research*, **17**(4), 537–543.

Gwinner, E. (1999). Rigid and flexible adjustments to a periodic environment: Role of circadian and circannual programs. In N. J. Adams and R. H. Slotow, eds, Proc. XXII Int. Ornithol. Congr. Durban. BirdLife South Africa; Johannesburg, South Africa: 1999, pp. 2366–2378.

Hartwig, H. G., and Van Veen, T. (1979). Spectral Characteristics of Visible Radiation Penetrating into the Brain and Stimulating Extraretinal Photoreceptors. *Journal of Comparative Physiology A*, **130**(3), 277–282.

Hastad, O., Victorsson, J., and Odeen, A. (2005). Differences in color vision make passerines less conspicuous in the eyes of their predators, *Proceedings of the National Academy of Sciences USA* **102**(18), 6391–6394.

Hockin, D., Ounsted, M., Gorman, M., Hill, D., Keller, V., and Barker, M. A. (1992). Examination of the effects of disturbance on birds with reference to its importance in ecological assessments. *Journal of Environmental Management*, **36**(4), 253–286.

Jones, J., and Francis, C. M. (2003). The effects of light characteristics on avian mortality at lighthouses. *Journal of Avian Biology*, **34**(4), 328–333.

Kempenaers, B., Borgström, P., Loës, P., Schlicht, E., and Valcu, M. (2010). Artificial night lighting affects dawn song, extra-pair siring success, and lay date in songbirds. *Current Biology*, **20**(19), 1735–1739.

Kumar, V., Kumar, B. S., and Singh, B. P. (1992). Photostimulation of blackheadedbunting: subjective interpretation of day and night depends upon both photophase contrast and light intensity. *Physiology and Behavior*, **51**(6), 1213–1217.

Kumar, V., and Rani, S. (1996). Effects of wavelength and intensity of light in initiation of body fattening and gonadal growth in a migratory bunting under complete and skeleton photoperiods. *Physiology and Behavior*, **60**(2), 625–631.

Lambrechts, M., Blondel, J., Maistre, M., and Perret, P. (1997). A single response mechanism is responsible for evolutionary adaptive variation in a bird's laying date. *Proceedings of the National Academy of Sciences USA*, **94**(10), 5153–5155.

Le Corre, M., Ollivier, A., Ribes, S., and Jouventin, P. (2002). Light-induced mortality of petrels: a 4-year study from Réunion Island (Indian Ocean). *Biological Conservation*, **105**(1), 93–102.

Lebbin, D. J., Harvey, M. G., Lenz, T. C., Andersen, M.J., and Ellis, J. M. (2007). Nocturnal migrants foraging at night by artificial light. *The Wilson Journal of Ornithology* **119**(3), 506–508.

Leopold, A., and Eynon, A. E. (1961). Avian daybreak and evening song in relation to time and light intensity. *The Condor*, **63**(4), 269–293.

Lewis, P. D., and Morris, T. R. (2000). Reviews Poultry and coloured light. *World's Poultry Science Journal*, **56**(3), 189–207.

Longcore, T., and Rich, C. (2004). Ecological Light Pollution. *Frontiers in Ecology and the Environment*, **2**(4), 191.

Malik, S., Rani, S., and Kumar, V. (2002). The influence of light wavelength on phase-dependent responsiveness of the photoperiodic clock in migratory blackheaded bunting. *Biological Rhythm Research*, **33**(1), 65–73.

Miller, M. (2006). Apparent effects of light pollution on singing behavior of American robins. *The Condor*, **108**(1), 130–139.

Molenaar, J. G. De, Sanders, M. E., and Jonkers, D. A. (2006). Road lighting and grassland birds: local influence of road lighting on a black-tailed godwit population. In C. Rich and T. Longcore, eds., *Ecological Consequences of Artificial Night Lighting*, pp. 114–136. Island Press, Washington.

Montevecchi, W. A. (2006). Influences of artificial light on marine birds. In C. Rich and T. Longcore, eds., *Ecological Consequences of Artificial Night Lighting*, pp. 94–113. Island Press, Washington.

Navara, K. J., and Nelson, R. J. (2007). The dark side of light at night: physiological, epidemiological, and ecological consequences. *Journal of Pineal Research*, **43**(3), 215–24.

Osorio, D., and Vorobyev, M. (2008). A review of the evolution of animal colour vision and visual communication signals. *Vision Research*, **48**(20), 2041–2052.

Perkin, E. K., Hölker, F., Richardson, J. S., Sadler, J. P., Wolter, C., and Tockner, K. (2011). The influence of artificial light on stream and riparian ecosystems: questions, challenges, and perspectives. *Ecosphere*, **2**(11), 1–16.

Poesel, A., Kunc, H. P., Foerster, K., Johnsen, A., and Kempenaers, B. (2006). Early birds are sexy: male age, dawn song and extrapair paternity in blue tits, *Cyanistes* (formerly *Parus*) *caeruleus*. *Animal Behaviour*, **72**(3), 531–538.

Poot, H., Ens, B. J., Vries, H. De, Donners, M. A. H., Wernand, M. R., and Marquenie, J. M. (2008). Green Light for Nocturnally Migrating Birds. *Ecology and Society*, **13**(2), 47.

Rani, S, and Kumar, V. (2000). Phasic response of the photoperiodic clock to wavelength and intensity of light in the redheaded bunting, *Emberiza bruniceps*. *Physiology and behavior*, **69**(3), 277–283.

Rani, Sangeeta, Singh, S., Misra, M., and Kumar, V. (2001). The influence of light wavelength on reproductive photorefractoriness in migratory blackheaded bunting (*Emberiza melanocephala*). *Reproduction Nutrition Development*, **41**(4), 277–284.

Reed, J. R., Sincock, J. L., and Hailman, J. P. (1985). Light attraction in endangered procellariform birds: reduction by shielding upward radiation. *The Auk*, **102**(2), 377–383.

Rich, C., and Longcore, T. (2006). *Ecological Consequences of Artificial Night Lighting*. Island Press, Washington.

Rowan, W. (1925). Relation of light to bird migration and developmental changes. *Nature*, **115**, 494–495.

Santos, C. D., Miranda, A. C., Granadeiro, J. P., Lourenço, P. M., Saraiva, S., and Palmeirim, J. M. (2010). Effects of artificial illumination on the nocturnal foraging of waders. *Acta Oecologica*, **36**(2), 166–172.

Schaper, S. V, Dawson, A., Sharp, P. J., Caro, S. P., and Visser, M. E. (2012). Individual variation in avian reproductive physiology does not reliably predict variation in laying date. *General and Comparative Endocrinology*, **179**(1), 53–62.

Schlaepfer, M. A., Runge, M. C., and Sherman, P. W. (2002). Ecological and evolutionary traps, *TRENDS in Ecology and Evolution*, **17**(10), 474–480.

Silverin, B., Massa, R. and Stokkan, K. A. (1993). Photoperiodic adaptation to breeding at different latitudes in Great Tits. *General and Comparative Endocrinology*, **90**(1), 14–22.

Stephan, B. (1999). *Die Amsel*. Die Neue Brehm-Bücherei, Hohenwarsleben.

te Marvelde, L., Schaper, S. V, and Visser, M. E. (2012). A single long day triggers follicle growth in captive female great tits (*Parus major*) in winter but does not affect laying dates in the wild in spring. *PloS One*, **7**(4): e35617. doi:10.1371/journal.pone.0035617

Thomas, D. W., Blondel, J., Perret, P., Lambrechts, M. M., and Speakman, J. R. (2001). Energetic and fitness costs of mismatching resource supply and demand in seasonally breeding birds. *Science*, **291**(5513), 2598–2600.

Thomas, R. J., Székely, T., Cuthill, I. C., Harper, D. G. C., Newson, S. E., Frayling, T. D., and Wallis, P. D. (2002). Eye size in birds and the timing of song at dawn. *Proceedings of the Royal Society B*, **269**(1493), 831–837.

Titulaer, M., Spoelstra, K., Lange, C. Y. M. J. G., and Visser, M. E. (2012). Activity patterns during food provisioning are affected by artificial light in free living great tits (*Parus major*). *PloS One*, **7** (5): e37377. doi:10.1371/journal.pone.0037377.

Verboven, N., and Visser, M. E. (1998). Seasonal variation in local recruitment of great tits: the importance of being early. *Oikos*, **81**(3), 511–524.

Wiltschko, W., and Wiltschko, R. (1995). Migratory orientation of European Robins is affected by the wavelength of light as well as by a magnetic pulse. *Journal of Comparative Physiology A*, **177**(3), 363–369.

Wiltschko, W., Munro, U., Ford, H., and Wiltschko, R. (1993). Red light disrupts magnetic orientation of migratory birds. *Nature*, **364**, 525–527.

CHAPTER 3

Wild bird feeding (probably) affects avian urban ecology

Valentin Amrhein

3.1 Introduction

Why do people feed wild birds? It is intuitively clear that humans feel delighted by the presence of birds that they can manage to attract by offering ordinary seeds at a feeder. According to surveys on human motivations for feeding wild birds in Australia, some people may gain experiential knowledge from observing the birds in their gardens, or feed the birds in return for the massive habitat destruction caused by humans (Howard & Jones, 2004; Ishigame & Baxter, 2007). However, outside Australia, the largely unstudied motivations of people feeding birds are just one example for our gaps of knowledge with respect to bird feeding (Jones & Reynolds, 2008).

It is in sharp contrast to our limited knowledge on the habit that wild bird feeding is probably the most widespread and popular form of human–wildlife interaction throughout the world (Jones, 2011), and, at least in northern temperate regions, the largest wildlife management activity (Martinson & Flaspohler, 2003). In the UK, sufficient commercial wild bird foods are sold to support a hypothetical number of over 30 million great tits (*Parus major*; Robb et al., 2008a), which is many more than the 2 million pairs of great tits that are actually present (www.bto.org). Surveys have found that 64% of households provide supplementary food for birds in the UK (Davies et al., 2012), and 43% in the USA (Martinson & Flaspohler, 2003). In suburban and rural environments of Australia, estimated household feeding rates range from 36% to 48% (Ishigame & Baxter, 2007). Although BirdLife Australia does not encourage supplementary feeding of wild birds (Bird Observation and Conservation Australia, 2010), northern organizations such as the British Trust for Ornithology (BTO), the Royal Society for the Protection of Birds (RSPB) and the Cornell Laboratory of Ornithology now recommend feeding birds for promoting nature conservation (Jones, 2011). In Germany, a renowned ornithologist recently wrote a popular book on bird feeding that sold 50,000 copies in the first 1.5 years, in which he advocates feeding the birds year-round and on a massive scale (Berthold & Mohr, 2008). Indeed, wild bird feeding is not only here to stay (Jones, 2011), it also seems to be increasing, at least in the UK (Chamberlain et al., 2005).

Today, in many countries, the huge effort in providing supplementary food for birds may be one of the largest human influences on bird populations, in addition to habitat loss and change, human-induced climate change and hunting. As I will show in this chapter, we now have a sound basis of knowledge about the impacts that supplementary feeding can have on birds. However, as I will also show, our knowledge mostly comes from small-scale experiments that researchers did in natural and rural habitats. Surprisingly little is known on the impact of feeding birds in our urban gardens and backyards. However, humans are influencing urban food supply for birds not only directly by providing feeders, but also via waste treatment and by creating, changing, or destroying urban or natural habitats and food sources in our cities (Chace & Walsh, 2006). Because such additional human influences are usually weaker in rural landscapes, the effects of feeding wild birds are likely

Avian Urban Ecology. Edited by Diego Gil and Henrik Brumm

to differ between natural study sites and urban environments.

In this chapter, I will highlight important findings on the effects of food supplementation on avian ecology. Food supply clearly influences bird numbers both in winter and during the breeding season (Martin, 1987; Newton, 1998), and feeding short-lived passerines can alter almost every aspect of their ecology, from reproductive parameters to behaviour and distribution (Martinez-Abrain & Oro, 2010). Because providing supplementary food is so easily done, manipulating food supply was a method of choice in many areas of research. Throughout, I will distinguish between results obtained from rural or natural study sites and from studies in urban habitat. At several places, I will point to study questions and methods that I think would be worth considering. Finally, I will discuss possible reasons why there is so little research on bird feeding in cities and give examples what could be done to change that.

3.2 Body condition and survival

The most obvious reason why people provide supplementary food for birds in winter is probably that they hope to enhance body condition of the birds to help them survive.

As expected, wild bird feeding appears to affect body condition. In Ohio, supplementary food improved the nutritional condition of wintering woodland birds, as indicated by the faster daily growth rates of feathers of provisioned birds than of unprovisioned birds (Grubb & Cimprich, 1990). Australian magpies *Cracticus tibicen* receiving supplementary food had a higher body mass, and when supplied with sausages, they also had higher plasma cholesterol levels (Ishigame et al., 2006). It seems that birds may often gain higher body mass when receiving supplementary food in natural habitats (Boutin, 1990), but this may not always be found to a similar degree in urban habitats: when Liker et al. (2008) kept house sparrows in aviaries under *ad libitum* food supply, urban sparrows had a consistently lower body mass than sparrows from rural origin, which the authors explain by habitat differences in nestling development or by adaptive divergence of sparrow populations. This is clearly calling for more studies comparing how wild bird feeding affects body condition in urban and rural areas.

In one of the first reviews on food supplementation experiments under field conditions, Boutin (1990) states that survival over the winter and subsequent breeding density has been the focus of most food addition studies. Survival rates were improved in all of the six reviewed cases where survival was measured (Boutin, 1990). For example, black-capped chickadees *Poecile atricapillus* with access to winter food had much higher over-winter survival rates than birds on control sites (69% vs. 37%; Brittingham & Temple, 1988). In all six studies, however, supplementary food was provided in rural and natural forests and not in urban areas. In one of the few studies in urban landscapes, Egan and Brittingham (1994) found that black-capped chickadees had higher monthly survival rates (94%) in a suburban area with mature trees and bird feeders throughout than in mature forests without supplementary food provision (81%). However, the authors suggest that feeders probably influenced movements of the birds and that greater apparent survival was measured rather than actual survival; thus, the birds may have been more closely attached to a study site when food was provided, and feeding could have reduced emigration rather than have enhanced survival. As is often the case, it is difficult to distinguish between emigration from a study site and actual death in food supplementation studies, unless the degree of dispersal is estimated and accounted for; this could be done by capturing at multiple neighbouring plots (Marshall et al., 2004) or by using radio telemetry (Sandercock, 2006).

I suggest studying winter dispersal for estimating true survival in urban habitats in which supplementary food is provided, and to compare this with control sites without bird feeders in similar urban habitat and in rural landscapes. Because urban habitats are often highly heterogeneous, it would be particularly advisable to randomly select fed and unfed urban sites and then to swap these over between years to try and control for habitat differences. Dispersal and true survival could be estimated by capturing individuals in several gardens within an urban area. To account for the degree of dispersal that is likely to be a function of the distance between study plots, distances between capture sites should be varied, but capture sites should be within potential reach for dispersing subjects. It should also be possible to study the movements of supplemented birds in urban landscapes by using radio telemetry in winter.

3.3 Density and distribution

Immigration, emigration, and survival determine how effects of winter feeding carry over to influence bird density during the next breeding season. Newton (1998) reviewed studies that experimentally manipulated food supply to investigate changes in bird numbers. He found 15 cases in which extra winter food provision was followed by an increase in subsequent breeding density by a factor ranging from 1.2 to 2.4, while in 11 cases, there was no clear increase. Although those numbers include some results that are not strictly independent because they are from different species investigated during the same studies, it appears that winter feeding can enhance breeding density by increasing survival or immigration, or by reducing emigration. A mechanism leading to higher bird numbers could be that supplementary feeding influences breeding density by affecting territorial behaviour. For example, if males reduce the size of their territories when supplementary food is provided (Enoksson & Nilsson, 1983), density of territories could be increased, which again calls for studies combining supplementary feeding with radio telemetry.

Often, winter food supplementation affects breeding numbers only when natural food supply is poor; for example, van Balen (1980) found that in great tits, supplementary feeding almost doubled the number of breeding pairs in years of poor beech *Fagus silvatica* crop, but had little impact in years with good beech crop. Generally, of course, many factors will contribute to variation in the effects of food supplementation found between studies. For example, long-lived species that often inhabit the same breeding territories for several years may show less sudden increases in density after supplementary feeding than short-lived species (Newton, 1998); also, studies using short time periods of experimental food provision are likely to yield different outcomes than studies providing supplementary food for longer periods (Harrison et al., 2010; Newton, 1998; Saggese et al., 2011). Results are clearly most valuable if they come from long-term studies evaluating potential effects of inter-year variance in environmental conditions (Schoech, 2009).

The studies reviewed by Newton (1998) are generally from natural habitats. In urban areas, post-winter densities of Carolina wrens *Thryothorus ludovicianus* were predicted by the presence of bird feeders (Job & Bednekoff, 2011). Data from the BTO Garden Bird Feeding Survey showed that several species are positively associated with the number of feeding stations provided by survey participants (Chamberlain et al., 2005). In the city of Sheffield, the density of feeding households was positively related to bird abundance (Figure 3.1), but had no apparent effect on overall species richness; however, in an individual garden, wild bird feeding will almost always increase the range of species and the number of individuals (Fuller et al., 2008). The studies cited above are all correlational, and, so far, no experiment manipulating food supply for investigating effects on bird density and distribution has been reported from an urban environment.

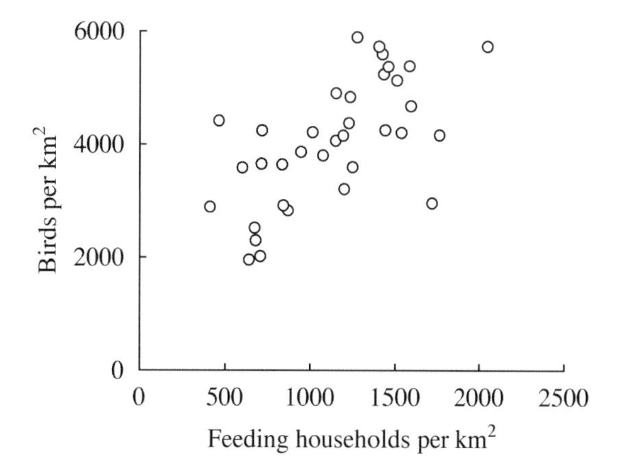

Figure 3.1 Relationship between bird abundance and levels of bird feeding in 35 neighbourhood types, summarized from 160 1-km squares with 442 bird survey points spread across urban Sheffield (UK). Points are mean values per km^2, calculated from the survey points falling within each neighbourhood type. Abundance is the mean number of individuals of all bird species per km^2 calculated using distance sampling. The density of households providing food for wild birds was a strong predictor of bird abundance in the urban landscape after accounting for the effect of cover by greenspace. Data are from Fuller et al. (2008), with kind permission by the authors.

On a larger scale, wild bird feeding may be partially responsible for northward range expansions, for example of the northern cardinal *Cardinalis cardinalis* and the American goldfinch *Carduelis tristis* in the USA (Robb et al., 2008b). In Europe, a rapid evolution of a new migratory divide has been observed in blackcaps *Sylvia atricapilla* that since the 1960s have established winter quarters in Britain, 1000 to 1500 km north of their traditional wintering areas. Among other factors such as climate warming, improved wintering conditions in Britain due to wild bird feeding may have contributed to this development (Bearhop et al., 2005, Berthold et al., 1992).

3.4 Productivity

In his review, Boutin (1990) noted that few studies have examined the effect of food supplementation on reproductive success in birds. Today, however, reproductive performance is probably the most widely researched topic in studies on food supplementation, as indicated by the 59 studies reviewed by Robb et al. (2008b) that provide data on how food supplementation influences the dates of egg laying.

In 58% of those 59 studies, feeding led to significantly earlier laying dates (Robb et al., 2008b). An increase of either clutch size, hatching success, chick growth rate, or fledging success due supplementary food was found in 44% to 64% of studies, and while between 36% and 55% of studies found no significant effect of supplementary feeding on a particular measure of breeding success, only three studies found negative effects on one of the measures (Robb et al., 2008b). It thus seems well established that providing supplementary food often enhances productivity of birds.

One recent study producing mixed evidence was by Harrison et al. (2010), who found that while supplementary feeding advanced laying in blue tits *Cyanistes caeruleus* and great tits, brood size at hatching was reduced by about half a chick in both species. The authors speculate that the provided peanut cakes may have led to smaller clutch sizes due to their high contents of fat rather than of protein; however, they note a striking similarity of their results from a natural woodland to urban habitat, in which often food is provided throughout the year, and in which both species of tits usually lay smaller clutches than in natural habitats (Chamberlain et al., 2009). Unlike other studies that often provided food for relatively short periods of time, Harrison et al. (2010) supplied bird food continuously from several weeks prelaying until hatching of the young, thus more closely mimicing how food is provided during the breeding season in an increasing number of urban gardens in which birds are fed year-round (Jones & Reynolds, 2008). However, the authors note that also other factors in addition to supplementary feeding may influence reproductive parameters of urban birds, and that although logistically challenging, further research in urban garden habitats would clearly be of value (Harrison et al., 2010). Indeed, the only way to study the effects of wild bird feeding on productivity while controlling for additional urban influences is probably to compare birds that do have access and that do not have access to bird feeders in otherwise similar urban landscapes.

Again, most studies on productivity have been conducted in natural habitats so far, and we are generally lacking information on how food supplementation affects reproductive performance of birds breeding in our gardens. This is exemplified by a brief survey of the 56 studies that Robb et al. (2008b) list in the electronic Appendix of their review. Of those 56 studies dealing with various impacts of supplementary feeding on breeding success, only four were carried out in urban habitat, and one was in suburban habitat. The only study on a non-corvid urban passerine was by Crossner (1977) on European starlings *Sturnus vulgaris* breeding in nestboxes on a university campus. The other urban studies were, again on a university campus and in other urban habitat, by Hochachka and Boag (1987) and Dhindsa and Boag (1990) on the black-billed magpie *Pica pica* and by Richner (1992) on the carrion crow *Corvus corone corone*, and, in suburban habitat, by Schoech et al. (2004) on the Florida scrub-jay *Aphelocoma coerulescens*. The other 49 studies were done in natural habitat outside residential areas, and two studies were on zebra finches *Taeniopygia guttata* in the laboratory. Thus, most of our knowledge on the influence of wild bird feeding on reproductive performance of city birds comes from outside cities, or from corvid species within cities. A subtle preference by researchers for corvids such as magpies and crows may be explained by the often easily visible nests alongside public roads. In natural habitats, many if not most published experiments using supplementary food provision were done on tits in forest nestbox-plots (Newton, 1998), and we have yet no information on how urban birds breeding in nestboxes are dealing with the bird feeders that are often found in close proximity to their nestboxes.

Further, most studies on the influence of artificial food supply on reproductive performance provided supplementary food during the breeding season. Although city birds are now often fed well into the breeding season or even year-round (Jones & Reynolds, 2008), most wild bird feeding is still being done in midwinter (Chamberlain et al., 2005). Only two studies seem to have investigated how food supplementation that is restricted to the winter season carries over to influence reproductive performance during the breeding season: Robb et al. (2008a) found that feeding that stopped 6 weeks prior to laying advanced laying dates and increased fledging success in the blue tit. Plummer et al. (2013) reported that provision of fat during winter resulted in smaller relative yolk mass in larger eggs and reduced yolk carotenoid concentrations in early breeding blue tits. This suggests that at least when the nutritional composition of provisioned foods is reduced to vegetable fat, carry-over effects on breeding performance may not always be beneficial (Plummer et al., 2013).

3.5 Song and territorial behaviour

A number of studies showed that supplementary feeding enhances territory defence activity. Ydenberg (1984) placed feeding tables in the territories of great tits and then simulated territorial intrusions using song playback and a stuffed mount of a male. He found that males with access to a feeding table attacked the model more than males who received no extra food. Ydenberg and Krebs (1987) then suggested that supplementary food allowed the birds to invest more time in territorial activity by allowing them to meet their food requirements more rapidly. Also, foraging temporally conflicts with singing (Gil & Gahr, 2002), and song output of male songbirds is likely to be constrained by energy reserves (Berg et al., 2005). Accordingly, eight of nine studies that were reviewed by Thomas (1999) found that food-supplemented males increased song output.

A particularly sensitive time of singing is during the dawn chorus in the hour before sunrise. In some species, dawn singing seems to be important for territory defence (Amrhein et al., 2004; Slagsvold et al., 1994), and studies on the winter wren *Troglodytes troglodytes* found that the number of songs sung before sunrise reflect past territorial intrusions by conspecific males more clearly than singing after sunrise (Amrhein & Erne, 2006; Erne & Amrhein, 2008).

Studies providing supplementary food for 1 to 5 days found an earlier start of dawn singing or a higher dawn song output in the blackbird *Turdus merula* (Cuthill & Macdonald, 1990), the silvereye *Zosterops lateralis* (Barnett & Briskie, 2007), and the black-capped chickadee (Grava et al., 2009). However, wild birds are usually fed over much longer periods, which will influence other variables that in turn will affect song output in complex ways (Thomas, 1999). For example, wild bird feeding may attract conspecific males into the territory of a supplemented male (Berg et al., 2005; Cuthill & Macdonald, 1990; Davies & Houston, 1981; Tobias, 1997), and Ydenberg (1984) and Tamm (1985) found that higher intruder pressure due to supplementary feeding leads to more chasing and attacking, but not to more singing or display activity by the territory owner.

We studied how a longer period of feeding would affect dawn singing of great tits in an extensive nestbox-plot near the city of Oslo (Saggese et al., 2011). Supplementary food was provided for 16 or 17 days *ad libitum* within the territories, starting about 5 weeks before the first egg was laid in a given territory. We found that by the end of the feeding period, supplemented males started to sing later relative to sunrise than control males (second observation session in Figure 3.2). Further, the later start of dawn singing carried over to a third observation session, 17 days after supplementary feeding had ended (Figure 3.2). This suggests that supplementary feeding had long-term effects on behaviour beyond the end of food provision (for

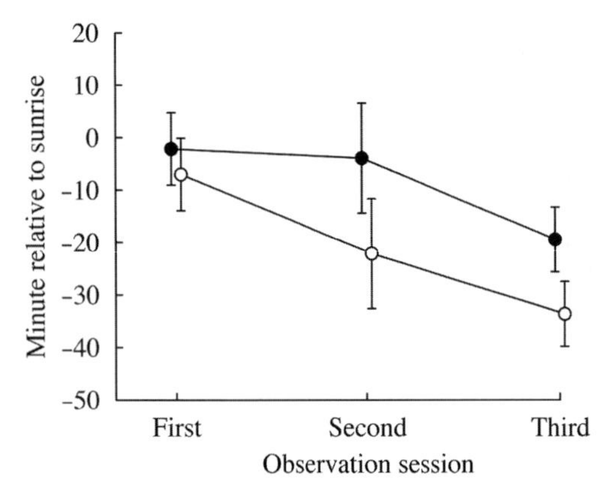

Figure 3.2 Mean ± SE start of dawn singing in minutes relative to sunrise, in 14 food-supplemented male great tits (filled circles) and in 14 control males (open circles). There were 16–17 days between the first and the second, and 17 days between the second and the third observation sessions. From immediately after the first observation session until after the second observation session, supplemented males were supplied with sunflower seeds and wild bird fat balls in their territories. Individual observation sessions in which a male did not sing or was not found in its territory were deleted, so that sample sizes for the first, second and third observation sessions were $N = 14$, 10, and 12 for supplemented males and $N = 12$, 9, and 11 for control males. Data are from Saggese et al. (2011).

details, see Saggese et al., 2011). Our results contrast with those from studies using shorter periods of food supplementation (Barnett & Briskie, 2007; Cuthill & Macdonald, 1990; Grava et al., 2009) in that we found a later, not an earlier start of dawn singing in supplemented birds. Similarly to our study, however, Clarkson (2007) found that 2 weeks after long-term food supplementation had ended, supplemented prothonotary warblers *Protonotaria citrea* sang less frequently at dawn in comparison with control birds.

These examples suggest that to investigate realistic effects of wild bird feeding as practised in urban gardens, food needs to be regularly provided for longer periods of time, which may lead to results that differ from what we know from short-term supplementation studies. Needless to say that none of the studies cited in this paragraph investigated territorial and singing behaviour within urban residential areas.

In general, future research could focus on how supplementary feeding influences distribution and reproduction via effects on social interactions. Further, although studies have shown that birds feeding at feeders may not bear a higher risk of predation, wild bird feeding could affect distribution and behaviour of both predator and prey species (Robb et al., 2008b; Saggese et al., 2011). We certainly need more food supplementation studies that take a behavioural point of view, because the mechanisms behind changes in survival, distribution, and productivity are far from being understood (Jones, 2011). Because resident birds may form pairs already in winter (e.g., Lemmon et al., 1997), changes in singing and territorial behaviour that are caused by food supplementation within the territory of a male during winter likely affect pair bonding and later reproductive performance.

3.6 Conclusion

In the urban backyards and gardens of the world, there are millions of bird feeders alongside millions of nestboxes. From what we know about the effects of supplementary feeding in rural and natural areas, it seems clear that wild bird feeding heavily influences multiple aspects of avian urban ecology. To date, however, we have little published information on the influence of wild bird feeding on reproductive performance of city birds. Further, there are hardly any experimental studies on how manipulated food supply in our cities affects survival, dispersal, distribution, and behaviour of birds. Why is that so? Of course, the problem about private gardens is that they are private. But if we want to learn how wild bird feeding affects avian urban ecology, we must learn to avoid the natural settings and seek out garden habitats.

We certainly know a lot about how supplementary feeding affects different aspects of avian ecology from many well-designed experiments performed in natural and seminatural study sites. However, in our cities, birds have greater access to anthropogenic foods of all sorts, while access to insects and other natural food resources may be reduced (Chace & Walsh, 2006). Environmental cues and species assemblages are so different in urban and rural areas that for many bird species, the effects of wild bird feeding probably differ between urban, suburban, and rural areas. However, even for such a basic statement, it is hard to find any reference.

It is clear that studies using experimental manipulation of food supply in urban areas will pose a much greater logistical challenge than studies in easily accessible rural nestbox-plots where food supplement availability is relatively easily controlled, and where habitat may be broadly homogeneous. In contrast, urban habitats seem to be characterized by small-scale variation in environmental factors that include differences in vegetation and natural food resources from garden to garden.

However, why not make use of this small-scale variation, by investigating several smaller study plots rather than a large one, perhaps with the help of voluntary nest recorders and bird ringers in city gardens? It should be possible to convince interested garden owners to put up new bird feeders, for the sake of a scientific experiment. It may also be possible to convince already keen bird feeders to stop feeding in their gardens, and maybe in some neighbouring gardens, at least for restricted periods of time. Using such local food deprivation or supplementation, future studies could compare multiple scattered urban study plots rather than investigating one continuous area as has usually been done in experiments from non-urban study sites. On such local urban study plots, one could then differentially vary, for example, the quantity of food supply or the lengths of food provision periods, to highlight multiple research questions such as the contested issue whether wild birds should be fed only in winter or year-round (Berthold & Mohr, 2008; Jones & Reynolds, 2008).

To study how wild bird feeding affects avian urban ecology, most of the research discussed in this chapter should be replicated in urban environments. Conducting research on bird feeding ecology in our cities is probably not for the faint-hearted (Jones, 2011); but it is urgently needed and a promising new field for adventurous scientists.

Acknowledgements

I thank Diego Gil, Darryl Jones, Tobias Roth, and an anonymous referee for comments on the manuscript. This work was funded by the Swiss Association Pro Petite Camargue Alsacienne and the Fondation de Bienfaisance Jeanne Lovioz.

References

Amrhein, V. and Erne, N. (2006). Dawn singing reflects past territorial challenges in the winter wren. *Animal Behaviour*, **71**, 1075–1080.

Amrhein, V., Kunc, H. P., and Naguib, M. (2004). Non-territorial nightingales prospect territories during the dawn chorus. *Proceedings of the Royal Society of London, Series B* (Suppl.), **271**, S167–S169.

Barnett, C. A. and Briskie, J. V. (2007). Energetic state and the performance of dawn chorus in silvereyes (*Zosterops lateralis*). *Behavioral Ecology and Sociobiology*, **61**, 579–587.

Bearhop, S., Fiedler, W., Furness, R. W., Votier, S. C., Waldron, S., Newton, J., Bowen, G. J., Berthold, P., and Farnsworth, K. (2005). Assortative mating as a mechanism for rapid evolution of a migratory divide. *Science*, **310**, 502–504.

Berg, M. L., Beintema, N. H., Welbergen, J. A. and Komdeur, J. (2005). Singing as a handicap: the effects of food availability and weather on song output in the Australian reed warbler *Acrocephalus australis*. *Journal of Avian Biology*, **36**, 102–109.

Berthold, P. and Mohr, G. (2008). *Vögel füttern—aber richtig*. Kosmos Verlag, Stuttgart.

Berthold, P., Helbig, A. J., Mohr, G., and Querner, U. (1992). Rapid microevolution of migratory behaviour in a wild bird species. *Nature*, **360**, 668–670.

Bird Observation and Conservation Australia (2010). Policy paper: Supplementary feeding of wild birds. Birdlife Australia (http://birdlife.org.au/).

Boutin, S. (1990). Food supplementation experiments with terrestrial vertebrates: patterns, problems, and the future. *Canadian Journal of Zoology*, **68**, 203–220.

Brittingham, M. C. and Temple, S.A. (1988). Impacts of supplemental feeding on survival rates of black-capped chickadees. *Ecology*, **69**, 581–589.

Chace, J. F. and Walsh, J. J. (2006). Urban effects on native avifauna: a review. *Landscape and Urban Planning*, **74**, 46–69.

Chamberlain, D. E., Vickery, J. A., Glue, D. E., Robinson, R. A., Conway, G. J., Woodburn, R. J. W., and Cannon, A. R. (2005). Annual and seasonal trends in the use of garden feeders by birds in winter. *Ibis*, **147**, 563–575.

Chamberlain, D. E., Cannon, A. R., Toms, M. P., Leech, D. I., Hatchwell, B. J., and Gaston, K. J. (2009). Avian productivity in urban landscapes: a review and meta-analysis. *Ibis*, **151**, 1–18.

Clarkson, C. E. (2007). Food supplementation, territory establishment, and song in the prothonotary warbler. *Wilson Journal of Ornithology*, **119**, 342–349.

Crossner, K. A. (1977). Natural selection and clutch size in the European starling. *Ecology*, **58**, 885–892.

Cuthill, I. C. and Macdonald, W. A. (1990). Experimental manipulation of the dawn and dusk chorus in the blackbird *Turdus merula*. *Behavioral Ecology and Sociobiology*, **26**, 209–216.

Davies, N. B. and Houston, A. I. (1981). Owners and satellites: the economics of territory defence in the pied wagtail, *Motacilla alba*. *Journal of Animal Ecology*, **50**, 157–180.

Davies, Z. G., Fuller, R. A., Dallimer, M., Loram, A., and Gaston, K. J. (2012). Household factors influencing participation in bird feeding activity: a national scale analysis. *PLoS ONE*, **7**, e39692.

Dhindsa, M. S. and Boag, D. A. (1990). The effect of food supplementation on the reproductive success of black-billed magpies *Pica pica*. *Ibis*, **132**, 595–602.

Egan, E. S. and Brittingham, M. C. (1994). Winter survival rates of a southern population of black-capped chickadees. *Wilson Bulletin*, **106**, 514–521.

Enoksson, B. and Nilsson, S. G. (1983). Territory size and population density in relation to food supply in the nuthatch *Sitta europaea* (Aves). *Journal of Animal Ecology*, **52**, 927–935.

Erne, N. and Amrhein, V. (2008). Long-term influence of simulated territorial intrusions on dawn and dusk singing in the winter wren: spring versus autumn. *Journal of Ornithology*, **149**, 479–486.

Fuller, R. A., Warren, P. H., Armsworth, P. R., Barbosa, O., and Gaston, K.J. (2008). Garden bird feeding predicts the structure of urban avian assemblages. *Diversity and Distributions*, **14**, 131–137.

Gil, D. and Gahr, M. (2002). The honesty of bird song: multiple constraints for multiple traits. *Trends in Ecology and Evolution*, **17**, 133–141.

Grava, T., Grava, A., and Otter, K.A. (2009). Supplemental feeding and dawn singing in black-capped chickadees. *Condor*, **111**, 560–564.

Grubb, T. C. and Cimprich, D. A. (1990). Supplementary food improves the nutritional condition of wintering woodland birds: evidence from ptilochronology. *Ornis Scandinavica*, **21**, 277–281.

Harrison, T. J. E., Smith, J. A., Martin, G. R., Chamberlain, D. E., Bearhop, S., Robb, G. N., and Reynolds, S. J. (2010). Does food supplementation really enhance productivity of breeding birds? *Oecologia*, **164**, 311–320.

Hochachka, W. M and Boag, D. A. (1987). Food shortage for breeding black-billed magpies (*Pica pica*)–an experiment using supplemental food. *Canadian Journal of Zoology*, **65**, 1270–1274.

Howard, P. and Jones, D. N. (2004). A qualitative study of wildlife feeding in south-east Queensland. In S. K. Burger and D. Lunney, eds, *Urban Wildlife: more than meets the eye*, pp. 55–62. Transactions of the Royal Zoological Society of New South Wales, Sydney.

Ishigame, G. and Baxter, G.S. (2007). Practice and attitudes of suburban and rural dwellers to feeding wild birds in Southeast Queensland, Australia. *Ornithological Science*, **6**, 11–19.

Ishigame, G., Baxter, G. S., and Lisle, A. T. (2006). Effects of artificial foods on the blood chemistry of the Australian magpie. *Austral Ecology*, **31**, 199–207.

Job, J. and Bednekoff, P. A. (2011). Wrens on the edge: feeders predict carolina wren *Thryothorus ludovicianus* abundance at the northern edge of their range. *Journal of Avian Biology*, **42**, 16–21.

Jones, D. N. (2011). An appetite for connection: why we need to understand the effect and value of feeding wild birds. *Emu*, **111**, i–vii.

Jones, D. N. and Reynolds, S. J. (2008). Feeding birds in our towns and cities: a global research opportunity. *Journal of Avian Biology*, **39**, 265–271.

Lemmon, D., Withiam, M. L., and Barkan, C. P. L. (1997). Mate protection and winter pair-bonds in black-capped chickadees. *Condor*, **99**, 424–433.

Liker, A., Papp, Z., Bókony, V., and Lendvai, A. Z. (2008). Lean birds in the city: body size and condition of house sparrows along the urbanization gradient. *Journal of Animal Ecology*, **77**, 789–795.

Marshall, M. R., Diefenbach, D. R., Wood, L. A., and Cooper, R. J. (2004). Annual survival estimation of migratory songbirds confounded by incomplete breeding site-fidelity: study designs that may help. *Animal Biodiversity and Conservation*, **27.1**, 59–72.

Martin, T. E. (1987). Food as a limit on breeding birds: a life history perspective. *Annual Review of Ecology, Evolution, and Systematics*, **18**, 453–487.

Martínez-Abraín, A. and Oro, D. (2010). Applied conservation services of the evolutionary theory. *Evolutionary Ecology*, **24**, 1381–1392.

Martinson, T. J. and Flaspohler, D. J. (2003). Winter bird feeding and localized predation on simulated bark-dwelling arthropods. *Wildlife Society Bulletin*, **31**, 510–516.

Newton, I. (1998). *Population Limitation in Birds*. Academic Press, London.

Plummer, K. E., Bearhop, S., Leech, D. I., Chamberlain, D. E., and Blount, J. D. (2013). Fat provisioning in winter impairs egg production during the following spring: a landscape-scale study of blue tits. *Journal of Animal Ecology*, **82**, 673–682.

Richner, H. (1992). The effect of extra food on fitness in breeding carrion crows. *Ecology*, **73**, 330–335.

Robb, G. N., McDonald, R. A., Chamberlain, D. E., Reynolds, S. J., Harrison, T. J. E., and Bearhop, S. (2008a). Winter feeding of birds increases productivity in the subsequent breeding season. *Biology Letters*, **4**, 220–223.

Robb, G. N., McDonald, R. A., Chamberlain, D. E., and Bearhop, S. (2008b). Food for thought: supplementary feeding as a driver of ecological change in avian populations. *Frontiers in Ecology and the Environment*, **6**, 476–484.

Saggese, K., Korner-Nievergelt, F., Slagsvold, T., and Amrhein, V. (2011). Wild bird feeding delays start of dawn singing in the great tit. *Animal Behaviour*, **81**, 361–365.

Sandercock, B. K. (2006). Estimation of demographic parameters from live-encounter data: a summary review. *Journal of Wildlife Management*, **70**, 1504–1520.

Schoech, S. J. (2009). Food supplementation experiments: a tool to reveal mechanisms that mediate timing of reproduction. *Integrative and Comparative Biology*, **49**, 480–492.

Schoech, S. J., Bowman, R., and Reynolds, S.J. (2004). Food supplementation and possible mechanisms underlying early breeding in the florida scrub-jay (*Aphelocoma coerulescens*). *Hormones and Behavior*, **46**, 565–573.

Slagsvold, T., Dale, S., and Sætre, G.-P. (1994). Dawn singing in the great tit (*Parus major*): mate attraction, mate guarding, or territorial defence? *Behaviour*, **131**, 115–138.

Tamm, S. (1985). Breeding territory quality and agonistic behavior: effects of energy availability and intruder pressure in hummingbirds. *Behavioral Ecology and Sociobiology*, **16**, 203–207.

Thomas, R. J. (1999). Two tests of a stochastic dynamic programming model of daily singing routines in birds. *Animal Behaviour*, **57**, 277–284.

Tobias, J. (1997). Food availability as a determinant of pairing behaviour in the European robin. *Journal of Animal Ecology*, **66**, 629–639.

Van Balen, J. H. (1980). Population fluctuations of the great tit and feeding conditions in winter. *Ardea*, **68**, 143–164.

Ydenberg, R. C. (1984). The conflict between feeding and territorial defence in the great tit. *Behavioral Ecology and Sociobiology*, **15**, 103–108.

Ydenberg, R. C. and Krebs, J. R. (1987). The tradeoff between territorial defense and foraging in the great tit (*Parus major*). *American Zoologist*, **27**, 337–346.

PART 2

Behaviour and Physiology

CHAPTER 4

Attention, habituation, and antipredator behaviour: implications for urban birds

Daniel T. Blumstein

4.1 Antipredator behaviour as a key to understanding human impacts

The presence of humans may have a profound effect on the distribution and abundance of animals, including birds (e.g. Marzluff et al., 2001). Why, for instance, when we develop hiking trails does avian biodiversity change (Kangas et al., 2010; Miller et al., 2001)? What explains variation in the tolerance that individuals, populations and species' may have towards humans? And, does habituation play an important role in this? In addition, humans introduce noise to environments and the cities are noisy places (e.g. Kight & Swaddle, 2011). Why do anthropogenic sounds seemingly influence some species more or differently than others and how might this happen (e.g. Francis et al., 2011a)? Given a fundamental understanding of the mechanisms involved in such distraction we may be able to develop novel ways to manage noise and other anthropogenic stimuli so they do not have negative effects on populations of animals.

Predation can have a really bad effect on one's direct fitness! Antipredator behaviour includes those features or phenotypic traits that animals do to reduce the probability of being *detected* by a predator, *attacked* by a predator, or *killed* by a predator (Caro, 2005). It includes adaptations to detect predators (identify them, antipredator vigilance), escape from predators (flee them, use refugia), and communicate about them (Lima & Dill, 1990).

Thus, studying antipredator behaviour is important to understand if we want to manage human impacts because predation is a strong selective force that influences habitat selection and population persistence (Blumstein & Fernández-Juricic, 2010). For instance, marmots (*Marmota* spp.) persist in areas of good visibility and protective rocks (Blumstein et al., 2006). There are also indirect effects of predation risk on populations and communities. Fear—what I will define here as the perceived risk of predation—alone may influence where animals go and what they do and thus fear may structure communities. A recent fascinating paper showed that birds hearing the sounds of predators were less able to feed their young and this decline in feeding was directly responsible for lower reproductive success (Zanette et al., 2011). Thus, simply the presence of predators, or stimuli perceived as predators, may be costly.

A key to understanding how humans impact animals is to view people as predators (Frid & Dill, 2002), as they have been during much of their entire evolutionary history. Predators approach animals, animals flee and hide from them and we can capitalize on these flight responses to gain fundamental insights into how species perceive humans (Møller, 2010; Stankowich & Blumstein, 2005). I will review several studies below.

4.2 Explaining variation in disturbance susceptibility

I want to start by asking a deceptively simple question: why do some species tolerate disturbance

Avian Urban Ecology. Edited by Diego Gil and Henrik Brumm

while others do not? I, and others (e.g. Møller, 2008, 2009a, 2010; 2012; Fernández-Juricic et al., 2009) have used flight initiation distance (FID), the distance at which an individual flees an approaching person, to quantify this. We know that flighty birds suffer from greater risk of raptor predation (Møller et al., 2008), and I assume that it is useful to understand the effects of urbanization because flighty species may be more vulnerable to anthropogenic disturbance (see also Møller, 2008). I will focus here on one key insight that I developed from my studies that created a comparative data set with about 10,000 flushes on >300 species of birds.

Before I do so, however, I would like to acknowledge three key researchers who have adopted similar methods and techniques to study urbanization effects on birds. Joanna Burger conducted a number of pioneering studies using human disturbance on birds, some of which used FID to study avian responses to human disturbance. Among other things, key findings have been to show that the type of approach (direct or tangential) may influence FID (Burger & Gochfeld, 1981, but see Heil et al., 2007), variation in human activity levels are associated with variation in foraging behaviour (Burger & Gochfeld, 1991a), and birds apparently habituate in areas where there are many visitors (Burger & Gochfeld, 1991b). Esteban Fernández-Juricic has also conducted many urbanization studies using FID (and other measures) to draw inferences about human exposure. He too has discovered that birds in urban parks seemingly have habituated (birds tolerate close approaches; a phenomena that could also arise from park populations being comprised of immigrants with little fear of humans) and that visitation in highly visited parks may not actually be disturbing (Fernández-Juricic et al., 2001a), and that alert distance (the distance that birds first detect an approaching threat) may indeed be a superior measure to FID when quantifying disturbance (Fernández-Juricic et al., 2001b). Anders Møller has recently conducted a number of studies that have demonstrated (among other things): that urbanization initially reduces both the mean and variation in FID because intolerant populations decline, before mean and variation increase with increased population size from those animals that tolerated the initial bout of urbanization (Møller, 2010); that urban species (Møller, 2009a) and populations (Møller, 2008) tolerate closer approaches from humans; and that differential FIDs of predators and their prey make urban areas predator-free refugia for some prey (Møller, 2012).

A key assumption of using flight initiation distance to study vulnerability of birds to humans is that we must assume that flight initiation distance is a species-specific trait. A number of years ago, we measured FID of shorebirds at a variety of locations in and around Botany Bay in Sydney, Australia (Blumstein et al., 2003). Botany Bay was a good place to conduct a study of how birds respond to humans because the bay includes areas that are protected and have few visitors, public beaches with many visitors, and private properties with relatively fewer visitors. We asked whether and how species responded to approaching humans at these different sites with different exposures to humans. We expected that species would respond differently as a function of their exposure to humans, but if there were species-specific responses, we expected that flighty species would generally be flighty. We tested this by looking at both the main effects of species and site, and importantly the interaction between species in sight on explaining variation in FID. We found that there was no significant interaction between species and site implying that those species with larger FIDs had larger FIDs at whatever site they were studied. This is illustrated in Figure 4.1 where you can see that few lines cross. From this, we concluded that flighty species are flighty, and this species-specific nature of FID has stimulated a lot of research.

I have focused a number of analyses on the 150 species with >10 independent data points, and therefore a good estimate of FID (Blumstein, 2006). There are other ways to study data sets with variation in sample sizes. For instance Møller (e.g. Møller, 2008, 2009a, 2010) has weighted each species' value by sample size. I shall review a series of phylogenetically based analyses that I conducted that ask what are the natural history and life history correlates of flightiness. In all cases, analyses were conducted using phylogenetically independent contrast values to control for the expected (and documented) similarity among relatives due to common phylogenetic descent.

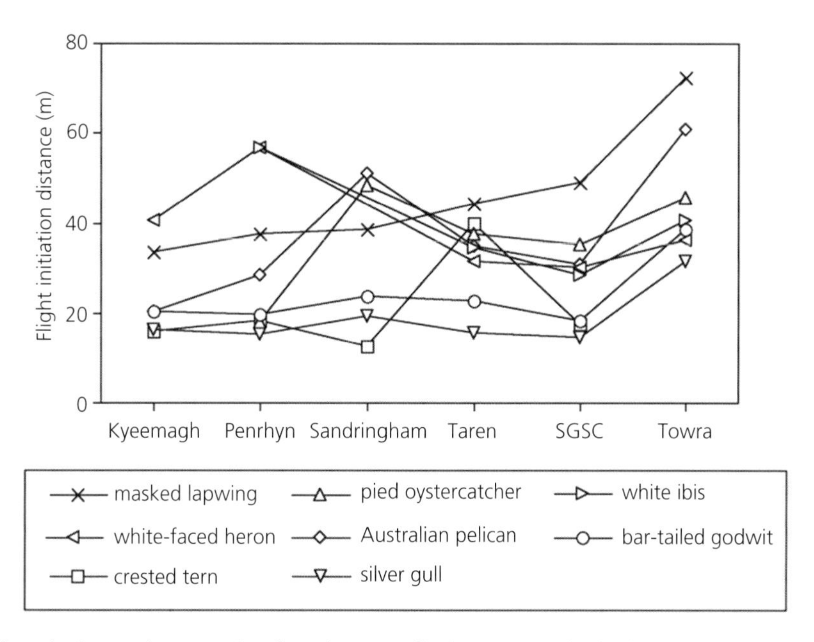

Figure 4.1 The relationship between location and FID for eight species of birds. You can see that few lines cross suggesting that flighty species are typically flighty; indeed, while there were significant effects of site and species, there were no significant interactions between site and species. Modified from Blumstein et al. (2003).

Larger body-sized birds detected approaching humans at greater distances and most of the variation in first response distance is explained by body size (Blumstein et al., 2005; Figure 4.2). Body size explains most of the explained variation in flight initiation distance as well (Blumstein, 2006). Body size is an important life-history trait that has a lot of predictive ability (Bennett & Owens, 2002)—specifically with respect to antipredator behaviour. However, there is some variation that requires further exploration.

One might expect that relative eye size influences the ability of birds to respond to threats. This ultimately is because eye size is associated with higher visual acuity (Kiltie, 2000), and higher visual acuity is associated with a greater ability to resolve objects from farther away (Land & Nilsson, 2002). One analysis suggests that after controlling for body size, eye size does not explain variation in FID (Blumstein et al., 2004), while another analysis suggests that it does (Møller & Erritzøe, 2010). Another couple of studies measured visual acuity of passerines (considering both eye size and the density of retinal ganglion cells) and found that species with higher visual acuity were able to detect a predator model from farther away (Dolan & Fernández-Juricic, 2010; Fernández-Juricic & Kowalski, 2011). My suspicion is that birds have large eyes for several reasons, and there may be a relatively small effect of eye size on vulnerability to humans. Note, however, the effect reported by Møller and Erritzøe (2010) accounted for 11% of the variance, which is not an insignificant amount. However, in other analyses (Blumstein, 2006) I found that flightiness coevolved with capturing live prey (as did Møller & Erritzøe, 2010), and being a cooperative breeder. Thus, perhaps having eyes that detect movement (which ultimately is about how the retina functions), not eye size, per se, is associated with flightiness.

And, after controlling for body mass and starting distance (the distance that the observer started an experimental approach to a bird), statistically, birds that reproduce at an older age are more flighty (also see Møller & Garamszegi, 2012). I examined a variety of other life history traits but found that most did not have a large effect. For instance, there was no effect of clutch size, no effect of the number of days young were fed, no effect of longevity (although

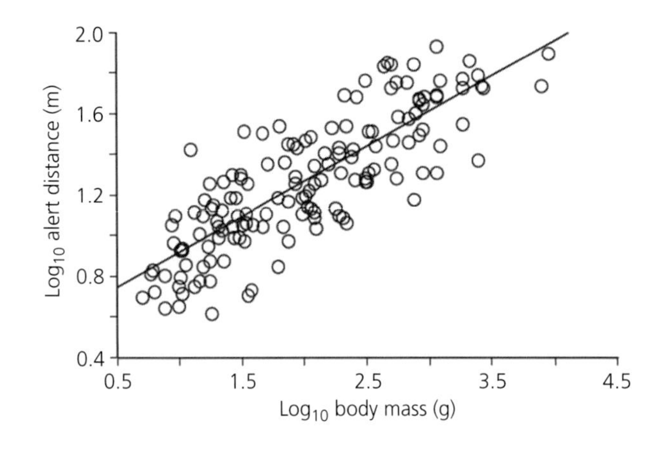

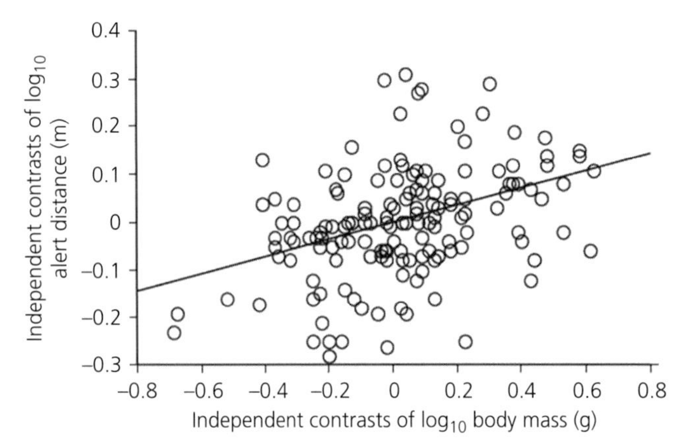

Figure 4.2 Larger body-sized birds first detect approaching humans at a greater distance whether quantified with species values or by calculating phylogenetically independent contrast values. Modified from Blumstein et al. (2005).

these are notoriously difficult to estimate), and no effect of habitat openness (Blumstein, 2006; but see Møller & Erritzøe, 2010 who report an effect of habitat openness). These results were somewhat surprising. However, being social, which I quantified by noting whether a species was a cooperative breeder or not, was associated with enhanced flightiness.

In a recent analysis, with different data, Møller (2009b) found that after controlling for body size, birds with greater basal metabolic rates have greater FID estimates. This is an important finding because it suggests that risk taking is part of a life-history syndrome of traits. And this life-history variation has important implications for how species respond to humans.

Thus, while life history variation is an important factor in explaining the evolution of species-specific differences in flight initiation distance, body size is the single biggest predictor yet identified. And, recent work by Møller (2012) suggests that this has important consequences for coexistence of smaller prey and larger predators in urban areas. I conclude from these analyses that body size has a profound effect on species vulnerability, and suggest that, without any other knowledge, body size alone may be a useful surrogate for vulnerability; large species are more vulnerable to human disturbance than smaller species (Bennett & Owens, 2002).

4.3 The natural history of habituation

4.3.1 Habituation and the geography of fear

Psychologists have formally studied habituation and its doppelganger, sensitization, for over 100 years. Yet even the ancients knew that animals may habituate to repeated exposure. Thompson

(2009) quoted an Aesop's Fable about the fox and the lion:

> A fox who had never yet seen a lion, when he fell in with him for the first time in the forest was so frightened that he was near dying with fear. On his meeting with him for the second time, he was still much alarmed, but not to the same extent as at first. On seeing him the third time, he so increased in boldness that he went up to him and commenced a familiar conversation with him.

Habituation is seen when a response declines over repeated exposures to a particular stimulus. By contrast, sensitization is seen when the response is enhanced with repeated exposure to stimuli. While this is well known, what is not really well understood is what I will refer to as 'the natural history' of these phenomena. For instance, under what conditions do animals habituate, and under what conditions do animals sensitize? What are the life history and natural history correlates or predictors of habituation or sensitization in different individuals, populations and species? I will first discuss some insights from studying ungulates, and then about several studies of birds, including one that I have not previously reported.

Günther's dik-diks (*Madoqua guentheri*) are small, monogamous, African ungulates that are eaten by about 36 species of mammals and birds—assessing risk to them is *essential*. Many species respond to the sounds of their predators and take evasive action. We (Coleman et al., 2008) capitalized on this expected response to predator sounds and broadcast jackal calls (a potential predator) and non-alarming bird song (a benign sound) within 0.5 km of human habitation, and >0.5 km of human habitation. We found that unhabituated dik-diks were unable to discriminate between the sounds of predators and benign sounds. This is important because ecotourists at more pristine places are likely to disturb animals and interfere with their risk assessment. What was interesting was the observation of this 0.5 km discrimination ability threshold.

To follow up on this, and in a study of how mule deer (*Odocoileus hemionus*, a North American ungulate) respond to marmot alarm calls (a form of interspecific communication) we (Carasco & Blumstein, 2012) found that deer discriminated marmots and white-crowned sparrow (*Zonotrichia leucophrys*) song when they were relatively close to people, *again within 0.5 km of houses*, while they failed to discriminate farther away, suggesting that there was some sort of a ceiling effect.

These two studies raise an interesting question that is ripe for study: is 0.5 km a 'magic number' for ungulates? What about other taxa?

Working along the beaches of Santa Monica, California, we (Webb & Blumstein, 2005) found that there were quantitatively different patterns of human visitation on either side of a very popular pier. Interestingly, FID to humans varied on either side of the pier (Figure 4.3). On the side with reasonably constant visitation, FID was reasonably constant. On the side with decreasing visitation, gulls were more flighty when there were fewer people. This pattern is evident over several kilometres and it illustrates the scale and pattern of human impacts on wildlife.

However, the scale of human disturbance can be much shorter. Working in two southern California wetlands, we (Ikuta & Blumstein, 2003) found that the presence of a fence that separated an area where ecotourists were common from an area where there were few visitors was sufficient to explain variation in avian FID. Indeed, species responded in a similar way to their responses in a nearby, protected wetland with very few human visitors. And, Fernández-Juricic et al. (2009) found that other wetland birds had shorter FID in areas with greater visitation.

I think that these studies suggest that we need to know much more about the spatial ecology of fear. How generalizable is the 0.5 km threshold for making biologically important discriminations? Are there different thresholds in highly urbanized areas? Over what distances do humans influence FID—a biologically important antipredator behaviour? And, what factors explain variation in the scale of interference.

An important caveat, and opportunity for future research, is that to really understand whether habituation, differential recruitment, or local adaptation to human disturbance is occurring, one must study marked individuals; something that is rarely done with studies of avian FID. My group has shown that marmot FID is repeatable, and that some individuals may habituate (Runyan & Blumstein, 2004; M. Petelle et al., in review). Recent work with birds

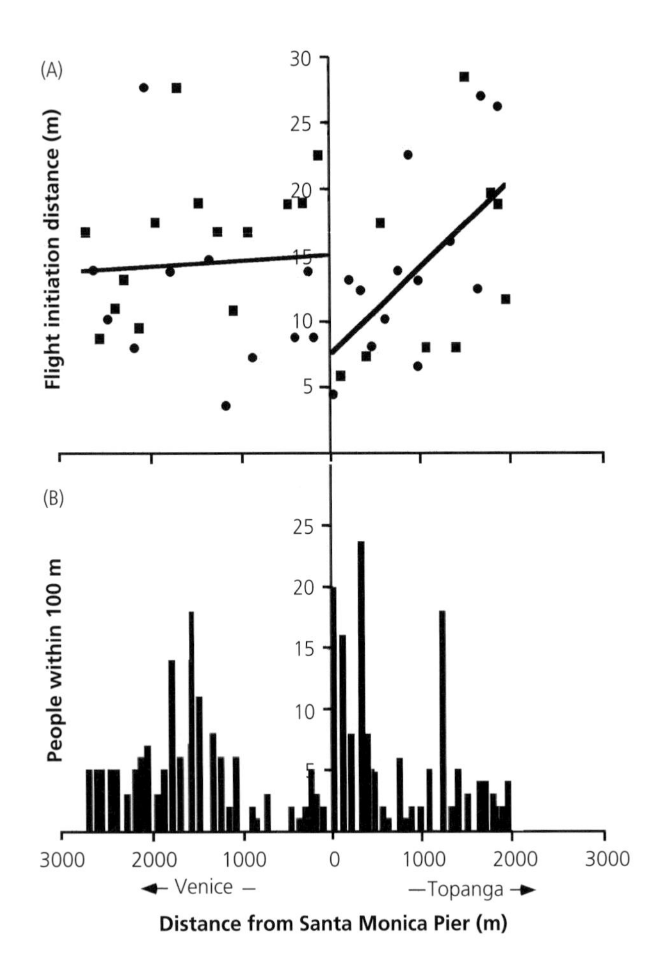

Figure 4.3 Flight initiation distance of Western gulls varies rapidly with distance on a side of a popular pier where visitation drops off compared to the side where visitation is more constant. Modified from Webb and Blumstein (2005).

has shown that FID may be significantly repeatable (Carrete & Tella, 2010), and other work (e.g. Møller 2008) suggests that local adaption to human disturbance is possible in the long run.

4.3.2 Sensitization and the contiguous habitat hypothesis

But not all species habituate. My work at both very patchy wetlands and very contiguous chaparral habitat in California led me to develop a novel hypothesis. While I suspected that increasing human visitation would typically habituate birds, this is not what I found.

Rina Fernandez, an undergraduate student working in my lab, visited six different sites weekly over 10 weeks. Sites were trails and fire roads in the Santa Monica Mountains (outside Los Angeles, California) that had different degrees of human visitation that she quantified by counting the number of pedestrians she encountered. Two sites, Albertson Motorway and Zuma Canyon, were classified as high impact based on the number of pedestrians encountered during censuses while Lower Chesboro, Palo Comodo Social Trail, Sage Hill, Morrison Ranch Road, and the Zuma Loop trail were considered lower impact areas. Thus, we had some replication for our two levels of impact.

While at the sites, birds were experimentally approached and FID was estimated. Of 49 species studied, we focused on 14 species with ≥4 observations per impact level (the majority of species had many more observations and California towees (*Melozone crissalis*) had a total of 208). We expected to see that there would be significant differences whereby increased human disturbance would be

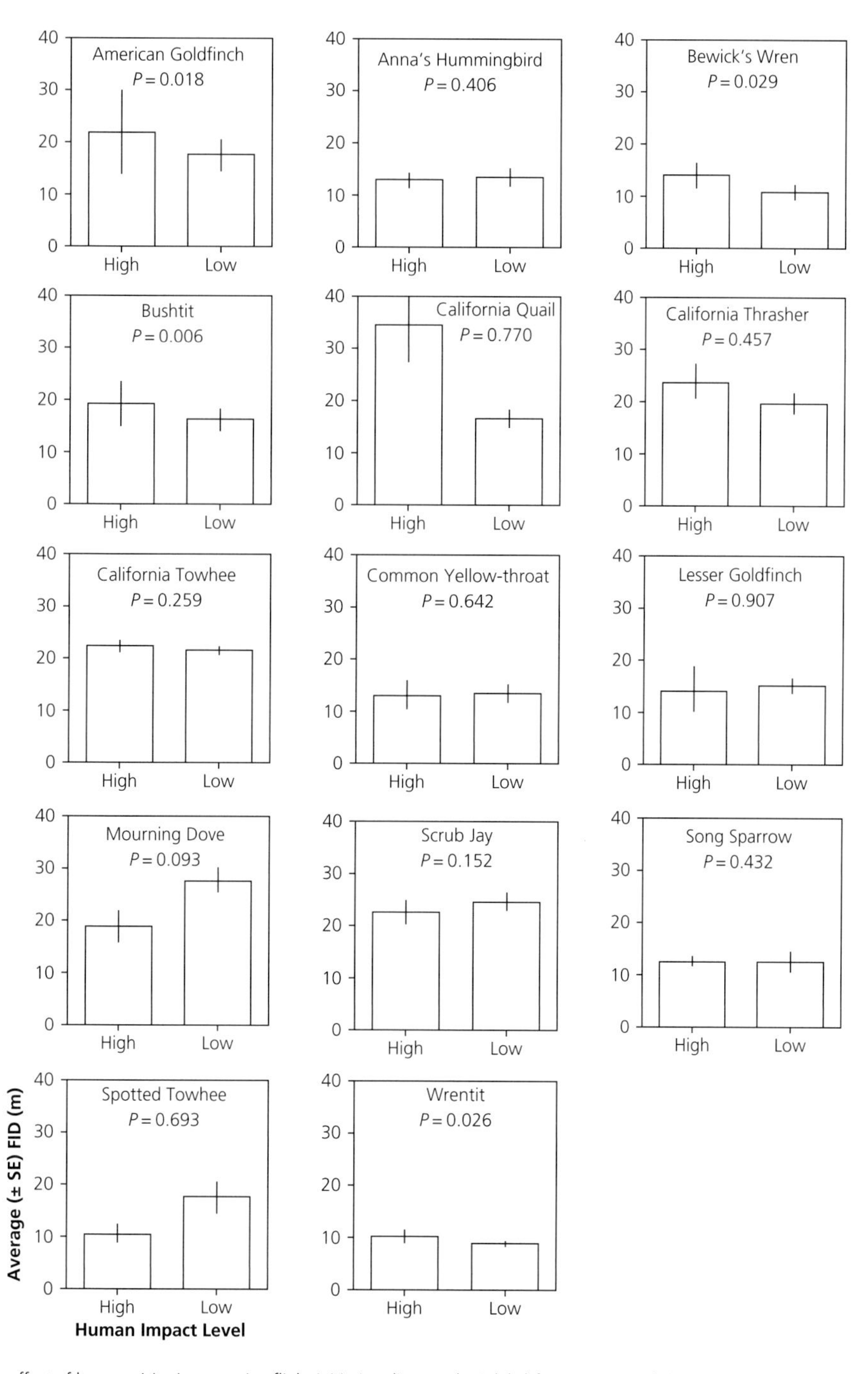

Figure 4.4 The effect of human visitation on avian flight initiation distance (± SE) (m) for 14 species of chaparral birds. In only four species, American goldfinch, Bewick's wrens, bushtits, and wrentits did human visitation have any impact on FID. Results suggest that these three species were sensitized to human visitation in that their FID was greater with increased human visitation. *P*-values are from a general linear model after explaining significant variation accounted for by the distance the observer began walking towards the focal bird.

associated with decreased FID. *We found the opposite.* Indeed only 4 of these 14 species of chaparral birds had significantly different FID estimates as a function of high or low human disturbance levels. And, for the four species that varied, they sensitized, rather than habituated (Figure 4.4).

By contrast, all of the species of shorebirds habituated to increased visitation in the Ikuta and Blumstein (2003) study. A number of factors might influence this difference including (but not limited to) the degree to which species are territorial, group living, etc. However, perhaps there is something about the options that individuals have that explains this difference.

I would like to suggest a novel hypothesis to explain these patterns: 'the contiguous habitat hypothesis'. The contiguous habitat hypothesis states that 'species living in limited habitats (e.g. wetland fragments), may be more likely to habituate than those living in more contiguous habitats (e.g. chaparral)'. If this is generally true, it seems to provide a mechanistic model that may explain species vulnerability to urbanization.

We know from many studies that the spatial scale of habituation can be stark (e.g. with a fence), or gradual and occur over several kilometres. Animals that have no options other than to remain in a small, constrained habitat type (as exemplified by wetland birds in Southern California) should be more likely to habituate. Alternatively, there has already been some species-sorting on these wetlands that has led to the loss of species that are unable to habituate. By contrast, species that have more contiguous habitat may actually be more vulnerable to human disturbance because it is likely that the first response for many species is to sensitize to increased human disturbance.

Gill et al. (2001) noted that we must be careful when assume that a behavioural response indicates that there has been a population consequence, and, importantly, the lack of response indicates no population consequence. They also emphasized that the availability of alternative habitat was likely a key feature in determining how human disturbance was reflected in observed differences. The contiguous habitat hypothesis differs in emphasis: I assume that habituating to humans allows coexistence of those populations and species that are able to habituate. Thus, the contiguous habitat hypothesis may help explain an equilibrium condition. We need to know more about how individuals and individual species respond across a range of human disturbances and those researchers who adopt a reaction norm approach will likely make fundamental discoveries.

The contiguous habitat hypothesis requires proper testing and I hope that by articulating it here, other data sets will be examined to see if and how it explains variation in habituation/sensitization with increased human disturbance and if, ultimately, populations and species that tolerate disturbance because of limited options, are more likely to persist in areas with humans.

4.4 Noise and its varied effects

Let us change topics a bit and think about the effects of anthropogenic noise, which is present in the seas and on land. Noise may mask biologically important signals and is associated with changes in signal structure (reviewed in Chapter 6). Noise is associated with physiological stress (Campo et al., 2005). Noise is associated with a decline in avian reproductive success (reviewed in Chapter 7), with changes in population distributions (Bayne et al., 2008; Bejder et al., 2006; Reijen et al., 1998), and noise may directly harm animals (Popper & Hastings, 2009). Avian community structure (Francis et al., 2009, 2011b; Slabbekoorn & Halfwerk, 2009), and important species interactions—like pollination and seed dispersal (Francis et al., 2012), can be influenced by anthropogenic noise. Noise may also provide distracting stimuli that interfere with biologically important assessments. I will focus first on masking and then discuss how noise may be distracting.

4.4.1 Noise masks important signals

While the importance of the acoustic environment on signal structure has long been recognized, it has only been since Slabbekoorn and Peet (2003) that conservation and behavioural biologists have been particularly sensitive to the idea that urban noise can influence the structure of bird song. Birds sing at higher frequencies (Parris & Schneider, 2009) and at greater amplitudes (Brumm, 2004) to sing over background noise. Those that naturally sing above the lower-frequency noise may be less likely to be impacted less by urban noise than those that sing at higher frequencies. However, other than Francis et al. (2011a) who focused on well compressor noise, I am not aware of a systematic test of this hypothesis in urban ecosystems. Indeed, if we look at how species avoid biological noise in their environment, we see a number of potential adaptations to avoid noise (Kirschel et al., 2009) that may be used by birds encountering anthropogenic noise.

The initial Slabbekoorn and Peet result has created a cottage industry in studying birds' responses to urban noise (as of March 2012, the article has been cited 290 times). But, I believe, that this cottage industry has largely emphasized bottom-up perceptual processes (Miller & Bee, 2012), in that researchers have gone out and focused on quantifying the noise spectra and the frequencies produced in noisy areas and more rural and presumably more quiet areas. This is fine, but there are a variety of top-down cognitive processes (Miller & Bee, 2012) that have been neglected. Below I will discuss attentional processes.

4.4.2 Noise competes for limited attention

Stimuli, in any modality, that do not provide information about biologically important features of the environment, may distract individuals from making biologically important assessments (Chan & Blumstein, 2011). I shall first explain this in the context of signal detection theory (Green & Swets, 1966) and then discuss how attention may also be modified.

Animals must discriminate useful and informative stimuli from those that are not useful or informative. To make such discrimination, individuals must set a decision threshold which will, inevitably have two types of error: false positives mean that individuals respond to non-informative stimuli and false negatives means that individuals miss responding to informative stimuli. Signal detection theory is a statistical framework to understand and quantify this tradeoff (Wiley, 2006).

Using signal detection theory we can view stimuli as having certain potentially overlapping characteristics. For instance, both predators and non-predators move through the air, and to discriminate a predator from a non-predator some threshold about how to respond to these stimuli must be set (Figure 4.5).

The more perceivable stimuli in the environment are, the greater the risk of an error. And, the relative cost of the two sorts of errors might cause several things to happen. First, a predetermined threshold might be unchanged. This will cause individuals to have more false positive responses and if these responses are costly, it will cause a waste of both time and energy. Of course, in response to these costly responses, thresholds may change which would reduce the false positive responses but at the cost of more false negative responses! Thus, animals may simply not respond to the presence of real predators and this too may have a fitness cost.

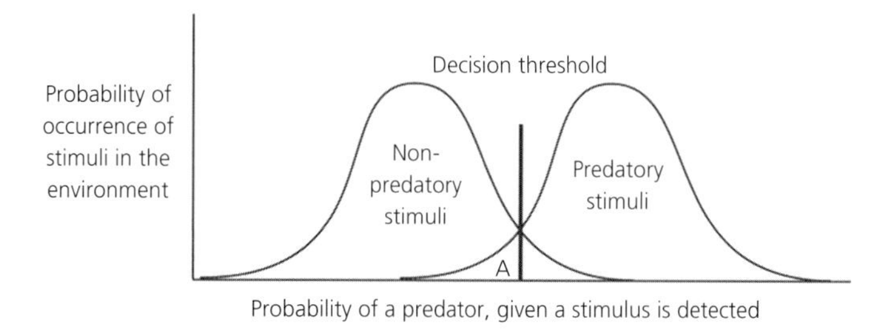

Figure 4.5 Fundamental signal detection problem that all species face when trying to detect predators. Predatory and non-predator stimuli overlap in characteristics (e.g. raptors and leaves both move through the air but only raptors are dangerous). To differentiate between them, animals must set a response threshold which inevitably trades-off certainty with error. Area A, to the left of the decision threshold, illustrates missed detections because predatory stimulus characteristics fully overlap non-predator stimulus characteristics. If these errors are costly, then we expect natural selection will shift the decision threshold further to the left.

Another view of competing stimuli is how they may act on attention. Attention is viewed as a limiting resource: individuals must allocate it among biologically important tasks such as looking for suitable habitat, food, mates, and detecting predators. Given these competing demands, extraneous stimuli can capture valuable attention and distract it from tasks at hand. I think that to understand how humans impact wildlife and to manage these impacts we must develop mechanistic hypotheses (Blumstein & Fernández-Juricic, 2010).

Chan et al. (2010) developed the 'distracted prey hypothesis' that states that: 'Any stimulus an animal can perceive is capable of distracting it by reallocating part of its finite attention and thus preventing it from responding to an approaching threat'.

The main idea of the distracted prey hypothesis is that attention is finite (e.g. Dukas, 2004), and that attention can be compromised by stimuli in multiple modalities. We know that birds, mammals, reptiles, and invertebrates all use attention to focus on relevant stimuli. Thus, and as discussed above, cities create many extraneous stimuli that may distract animals.

While studies of hermit crabs (Chan et al., 2010) and lizards (Huang et al., 2011) found that boat motor noise and camera sounds (respectively) distracted these species, we need studies focusing directly on birds to better understand how anthropogenic stimuli can distract birds. Thus, there are many research opportunities that may generate novel insights that will better allow us to manage urbanization and its deleterious consequences on birds. And, it is essential to try to properly quantify fitness consequences of limited attention, a potential challenge to much urbanization work.

4.5 Conclusions

Ultimately, human coexistence with wildlife in urban settings will require active management. I have focused largely on my own studies to suggest that by focusing on antipredator behaviour, we can gain valuable mechanistic insights into how birds are likely to respond to urbanization. Mechanistic studies are particularly important because if we know the mechanism, we should be able to manipulate the impacts (Blumstein & Fernández-Juricic, 2010). For instance, if we know that prey are distracted, and that this distraction increases predation risk, then we can better understand and manage causes of decline for threatened or endangered species. If we know that species on contiguous habitat are more sensitive to increased human visitation, vulnerable species will be managed by reducing visitation. And, if we know that large body-sized birds are generally going to be more vulnerable to disturbance, we know that larger patches are required to contain them.

The future will determine whether such management interventions work and I (Blumstein, 2007, 2013) strongly advocate the use of evidence-based evaluation (Sutherland et al., 2004) because this will be the most efficient way to test these management recommendations. Finally, I believe these studies suggest that there is a very fruitful collaboration between behavioral biologists and wildlife managers and I hope that this review helps stimulate these collaborations.

Acknowledgements

For discussion about these ideas, I thank Alvin Chan, Esteban Fernández-Juricic, Anders Møller, Nicole Munoz, and Ted Stankowich. Rina Fernandez, Lori Ikuta, Nick Webb, and Claire Zugmeyer were key collaborators in the study of the natural history of habituation. My work on avian antipredator behaviour has been supported by the Australian Research Council, Macquarie University, and the University of California Los Angeles. While writing this review, I was supported by both NSF-IDBR-0754247, and DEB-1119660. I thank Diego Gill, Esteban Fernández-Juricic, and Anders Møller for excellent and constructive comments on a previous version of this chapter.

References

Bayne, E. M., Habib, L., and Boutin, S. (2008). Impacts of chronic anthropogenic noise from energy-sector activity on abundance of songbirds in the Boreal forest. *Conservation Biology*, **22**, 1186–1193.

Bejder, L., Samuels, A., Whitehead, H., Gales, N., Mann, J., Connor, R., Heithaus, M., Watson-Capps, J., Flaherty, C., and Krutzen, M. (2006). Decline in relative abundance of bottlenosed dolphins exposed to long-term disturbance. *Conservation Biology*, **20**, 1791–1798.

Bennett, P. M. and Owens, P. F. (2002). *Evolutionary Ecology of Birds: Life Histories, Mating Systems, and Extinction*. Oxford University Press, Oxford.

Blumstein, D. T. (2006). Developing an evolutionary ecology of fear: how life history and natural history traits affect disturbance tolerance in birds. *Animal Behaviour*, **71**, 389–399.

Blumstein, D. T. (2007). Darwinian decision-making: putting the adaptive into adaptive management. *Conservation Biology*, **21**, 552–553.

Blumstein, D. T. (2013). Why we *really* don't care about the evidence in evidence-based decision-making in conservation (and how to change this). In M. Bekoff, ed., *Ignoring Nature: Animal Losses and What We Must Do About Them*. University of Chicago Press, Chicago, 103–112.

Blumstein, D. T. and Fernández-Juricic, E. (2010). *A Primer of Conservation Behavior.* Sinauer Associates, Sunderland, MA.

Blumstein, D. T., Anthony, L. L., Harcourt, R. G., and Ross, G. (2003). Testing a key assumption of wildlife buffer zones: is flight initiation distance a species-specific trait? *Biological Conservation*, **110**, 97–100.

Blumstein, D. T., Fernández-Juricic, E., Ledee, O., Larsen, E., Rodriguez-Prieto, I., and Zugmeyer, C. (2004). Avian risk assessment: effects of perching height and detectability. *Ethology*, **110**, 273–285.

Blumstein, D. T., Fernández-Juricic, E., Zollner, P. A. and Garity, S. C. (2005). Inter-specific variation in avian responses to human disturbance. *Journal of Applied Ecology*, **42**, 943–953.

Blumstein, D. T., Ozgul, A., Yovovitch, V., Van Vuren, D. H., and Armitage, K. B. (2006). Effect of predation risk on the presence and persistence of yellow-bellied marmot (*Marmota flaviventris*) colonies. *Journal of Zoology London*, **270**, 132–138.

Brumm, H. (2004). The impact of environmental noise on song amplitude in a territorial bird. *Journal of Applied Ecology*, **73**, 434–440.

Burger, J. and Gochfeld, M. (1981). Discrimination of the threat of direct versus tangential approach to the nest by incubating herring and great black-backed gulls. *Journal of Comparative Physiology and Psychology*, **95**, 676–684.

Burger, J. and Gochfeld, M. (1991a). Human activity influence and diurnal and nocturnal foraging of sanderlings (*Calidris alba*). *Condor*, **93**, 259–265.

Burger, J. and Gochfeld, M. (1991b). Human distance and birds: tolerance and response distances of resident and migrant species in India. *Environmental Conservation*, **18**, 158–165.

Campo, J. L., Gil, M. G., and Davila, S. G. (2005). Effects of specific noise and music on stress and fear levels of laying hens of several breeds. *Applied Animal Behaviour Science*, **91**, 75–84.

Caro, T. (2005). *Antipredator Defenses in Birds and Mammals*, University of Chicago Press, Chicago.

Carrasco, M. F. and Blumstein, D. T. (2012). Mule deer (*Odocoileus hemionus*) respond to yellow-bellied marmot (*Marmota flaviventris*) alarm calls. *Ethology*, **118**, 243–250.

Carrete, M. and Tella, J. L. (2010). Individual consistency in flight initiation distances in burrowing owls: a new hypothesis on disturbance-induced habitat selection. *Biology Letters*, **6**, 167–170.

Chan, A. A. Y.-H. and Blumstein, D. T. (2011). Attention, noise, and implications for wildlife conservation and management. *Applied Animal Behaviour Science*, **131**, 1–7.

Chan, A. A. Y.-H., Giraldo-Perez, P., Smith, S., and Blumstein, D. T. (2010). Anthropogenic noise affects risk assessment and attention: the distracted prey hypothesis. *Biology Letters*, **6**, 458–461.

Coleman, A., Richardson, D., Schechter, R., and Blumstein, D. T. (2008). Does habituation to humans influence predator discrimination in Gunther's dik-diks (*Madoqua guentheri*)? *Biology Letters*, **4**, 250–252.

Dolan, T. and Fernández-Juricic, E. (2010). Retinal ganglion cell topography of five species of ground foraging birds. *Brain, Behavior and Evolution*, **75**, 111–121.

Dukas, R. (2004). Causes and consequences of limited attention. *Brain, Behavior and Evolution*, **63**, 197–210.

Fernández-Juricic, E. and Kowalski, V. (2011). Where does a flock end from an information perspective? A comparative experiment with live and robotic birds. *Behavioral Ecology*, **22**, 1304–1311

Fernández-Juricic, E., Jimenez, M. D., and Lucas, E. (2001a). Bird tolerance to human disturbance in urban parks of Madrid (Spain): management implications. In Marzluff, J. M., Bowman, R., and Donnelly, R., eds., *Avian Ecology and Conservation in an Urbanizing World*, pp. 259–274. Kluwer Academic Publishers, Norwell, MA.

Fernández-Juricic, E., Jimenez, M. D., and Lucas, E. (2001b). Alert distance as an alternative measure of bird tolerance to human disturbance. Implications for park design. *Environmental Conservation*, **28**, 263–269.

Fernández-Juricic, E., Zahn, E. F., Parker, T., and Stankowich, T. (2009). California's endangered Belding's savannah sparrow (*Passerculus sandwichensis beldingi*): tolerance of predation disturbance. *Avian Conservation and Ecology—Écologie et conservation des oiseaux*, **4** (2), 1.

Francis, C. D., Ortega, C. P., and Cruz, A. (2009). Noise pollution changes avian communities and species interactions. *Current Biology*, **19**, 1415–1419.

Francis, C. D., Ortega, C. P., and Cruz, A. (2011a). Vocal frequency change reflects different responses to anthropogenic noise in two suboscine tryant flycatchers. *Proceedings of the Royal Society B*, **278**, 2025–2031.

Francis, C. D., Paritsis, J., Ortega, C. P., and Cruz, A. (2011b). Landscape patterns of avian habitat use and nest success are affected by chronic gas well compressor noise. *Landscape Ecology*, **26**, 1269–1280.

Francis, C. D., Klein, N. J., Ortega, C. P., and Cruz, A. (2012). Noise pollution alters ecological services: enhanced pollination and disrupted seed dispersal. *Proceedings of the Royal Society B*.

Frid, A. and Dill, L. M. (2002). Human-caused disturbance stimuli as a form of predation risk. *Conservation Ecology*, **6**, 11.

Gill, J. A., Norris, K., and Sutherland, W. J. (2001). Why behavioural responses may not reflect the population consequences of human disturbance. *Biological Conservation*, **97**, 265–268.

Green, D. M. and Swets, J. A. (1966). *Signal Detection Theory and Psychophysics*. Wiley, New York.

Halfwerk, W., Hollemann, L., Lessells, K., and Slabbekoorn, H. (2011). Negative impact of traffic noise on avian reproductive success. *Journal of Applied Ecology*, **48**, 210–219.

Heil, L., Fernández-Juricic, E., Renison, D., Nguyen, V., Cingolani, A. M., and Blumstein, D. T. (2007). Avian responses to tourism in the biogeographically isolated high Córdoba Mountains, Argentina. *Biodiversity and Conservation*, **16**, 1009–1026.

Huang, B., Lubarsky, K., Teng, T., and Blumstein, D. T. (2011). Take only pictures, leave only . . . fear? The effects of photography on the West Indian anole, *Anolis cristatellus*. *Current Zoology*, **57**, 77–82.

Ikuta, L. A. and Blumstein, D. T. (2003). Do fences protect birds from human disturbance? *Biological Conservation*, 112, 447–452.

Kangas, K., Luoto, M., Ihantola, A., Tomppo, E., and Siikamaki, P. (2010). Recreation-induced changes in boreal communities in protected areas. *Ecological Applications*, **20**, 1775–1786.

Kight, C. R. and Swaddle, J. P. (2011). How and why environmental noise impacts animals: an integrative, mechanistic review. *Ecology Letters*, **14**, 1052–1061.

Kiltie, R. A. (2000). Scaling of visual acuity with body size in mammals and birds. *Functional Ecology*, **14**, 226–234.

Kirschel, A. N. G., Blumstein, D. T., Cohen, R. F., Buermann, W., Smith, T. B., and Slabbekoorn, H. (2009). Birdsong tuned to the environment: green hylia song varies with elevation, tree cover, and noise. *Behavioral Ecology*, **20**, 1089–1095.

Land, M. F. and Nilsson, D.-E. (2002). *Animal Eyes*. Oxford University Press, Oxford.

Lima, S. L. and Dill, L. M. (1990). Behavioral decisions made under the risk of predation: a review and prospectus. *Canadian Journal of Zoology*, **68**, 619–640.

Marzluff, J. M., Bowman, R., and Donnelly, R., eds. (2001). *Avian Ecology and Conservation in an Urbanizing World*. Kluwer Academic Publishers, Norwell, MA.

Miller, C. T. and Bee, M. A. (2012). Receiver psychology turns 20: is it time for a broader approach? *Animal Behaviour*, **83**, 331–343.

Miller, S. G., Knight, R. L., and Miller, C. K. (2001). Wildlife responses to pedestrians and dogs. *Wildlife Society Bulletin*, **29**, 124–132.

Møller, A. P. (2008). Flight distance of urban birds, predation, and selection for urban life. *Behavioral Ecology Sociobiology*, **63**, 63–75.

Møller, A. P. (2009a). Successful city dwellers: a comparative study of the ecological characteristics of urban birds in the Western Palearctic. *Oecologia*, **159**, 849–858.

Møller, A. P. (2009b). Basal metabolic rate and risk-taking behaviour in birds. *Journal of Evolutionary Biology*, **22**, 2420–2429.

Møller, A. P. (2010). Interspecific variation in fear responses predicts urbanization in birds. *Behavioral Ecology*, **21**, 365–371.

Møller, A. P. (2012). Urban areas as refuges from predators and flight distance of prey. *Behavioral Ecology*

Møller, A. P. and Erritzøe, J. (2010). Flight distance and eye size in birds. *Ethology*, **116**, 458–465.

Møller, A. P. and Garamszegi, L. Z. (2012). Between individual variation in risk-taking behavior and its life history consequences. *Behavioral Ecology*, doi: 10.1093/beheco/ars040.

Møller, A. P., Nielsen, J. T., and Garamszegi, L. Z. (2008). Risk taking by singing males. *Behavioral Ecology*, **19**, 41–53.

Parris, K. M. and Schneider, A. (2009). Impacts of traffic noise and traffic volume on birds of roadside habitats. *Ecology and Society*, **14**(1), 29.

Petelle, M.B., McCoy, D., Alejandro, V., and D.T. Blumstein. Development of boldness and docility in yellow-bellied marmots. *Animal Behaviour*, in review.

Popper, A. N. and Hastings, M. C. (2009). The effects of human-generated sound on fish. *Integrative Zoology*, **4**, 43–52.

Reijen, R., Foppen, R., Braak, C. T., and Thissen, J. (1998). The effects of car traffic and breeding bird populations in woodland. III. Reduction of density in relation to the proximity of main roads. *Journal of Applied Ecology*, **32**, 187–202.

Runyan, A. and Blumstein, D. T. (2004). Do individual differences influence flight initiation distance? *Journal of Wildlife Management*, **68**, 1124–1129.

Slabbekoorn, H. and Halfwerk, W. (2009). Behavioural ecology: noise annoys at the community level. *Current Biology*, **19**, R693–R695.

Slabbekoorn, H. and Peet, M. (2003). Birds sing at a higher pitch in urban noise. *Nature*, **424**, 267.

Stankowich, T. and Blumstein, D. T. (2005). Fear in animals: a meta-analysis and review of risk assessment. *Proceedings of the Royal Society B*, **272**, 2627–2634.

Sutherland, W. J., Pullin, A. S., Dolman, P. M., and Knight, T. M. (2004). The need for evidence-based conservation. *Trends in Evolution and Ecology*, **19**, 305–308.

Thompson, R. F. (2009). Habituation: a history. *Neurobiology of Learning and Memory*, **92**, 127–134.

Webb, N. and Blumstein, D. T. (2005). Variation in human disturbance differentially affects predation risk assessment in western gulls. *Condor*, **107**, 178–181.

Wiley, R. H. (2006). Signal detection and animal communication. *Advances in the Study of Behavior*, **36**, 217–247.

Zanette, L. Y., White, A. F., Allen, M. C., and Clinchey, M. (2011), Perceived predation risk reduces the number of offspring songbirds produce per year. *Science*, **334**, 1398–1401.

CHAPTER 5

Behavioural and ecological predictors of urbanization

Anders Pape Møller

5.1 Introduction

Geographical urbanization is the process of conversion of rural habitats into urban ones. This conversion now happens at an unprecedented rate as evidenced by some countries such as the Netherlands now having more than 15% of the entire land surface being covered by urban habitats (European Commission, 2006; Schneider et al., 2009). By 2008 more than half of all humans were living in urban areas (Handwerk, 2008), and this trend is predicted to increase in the coming years as the total human population size grows further (United Nations, 2007). In contrast, biological urbanization is the process by which rural species of animals invade, become established and expand their populations in urban areas, similar to any biological invasion process. In the following I will only use the term urbanization to refer to the phenomenon of adaptation of living organisms to urban habitats. Numerous species of animals and plants now have their main populations in man-made habitats with population densities often being orders of magnitude higher than in the ancestral rural habitats. Given a rapidly changing world the successful urbanization of birds and other animals can be considered a predictor of future success as urban areas expand and ancestral habitats are lost to conversion to urban habitats, agriculture and forestry.

Humans have in the Middle East lived in cities for more than 10,000 years, which in evolutionary terms is a very short period. For many animals this constitutes considerably less than 10,000 generations, making the study of urbanization an interesting investigation in micro-evolution. For example, house sparrows *Passer domesticus* have been associated with cities and humans since the emergence of agriculture 10,000 years ago (Sætre et al., 2012; Summers-Smith, 1963), and there are subfossil barn swallow *Hirundo rustica* remains from a flint mine from the older Stone Age in Denmark (Møller, 1992). More recently Conrad Gesner wrote in his *Vogelbuch* published in 1669 that house sparrows, starlings *Sturnus vulgaris*, barn swallows, and many other birds already then bred in close association with humans. Such close associations between humans and birds have arisen from tremendous selective advantages in terms of protection against predators, as evidenced by nest predation rates for neighboring populations breeding inside buildings or outside on average being 0.8% and 22.1% across 11 species of birds, or more than a 28-fold difference (Møller, 2010b). More recently, as cities started to grow, new species of birds invaded urban habitats. An example of this more recent wave of urban invasion in Europe is the blackbird *Turdus merula* that was already established as a breeding bird in Rome in Italy at the beginning of the 19th century (Bonaparte, 1828). Even more recently numerous other species have invaded urban habitats and established populations that often have densities that exceed the ancestral densities by more than two orders of magnitude. For example, that is the case for the blackbird and several other species (Møller et al., 2012; Stephan, 1990). Therefore, the frequency distribution of the timing of urban invasions is strongly left-skewed with many more recent cases than ancient cases of urbanization. We can exclude

Avian Urban Ecology. Edited by Diego Gil and Henrik Brumm

the possibility that this is a simple effect of increased observer activity because recently urbanized species have long and invariable flight distances, song posts relatively low in the vegetation and not higher population density in urban compared to rural habitats (Møller, 2010a; Møller et al., 2012).

Urban habitats differ from nearby rural habitats in terms of disturbance regimes, light conditions, habitat characteristics and distribution, anthropogenic food, and climate (Alberti, 2005; Gilbert, 1989; Miller, 2005; Rebele, 1994; Turner et al., 2004). Because the phenotype of urban animals differs from that of conspecifics in nearby rural habitats, and because the composition of the community of urban animals differs from that of the rural community, these differences will also affect the urbanization process.

Which are the factors that contribute to urbanization? The literature on urban birds is full of suggestions about variables that account for successful invasion of urban habitats. These range from association with specific habitats such as rocks, cliffs and mountains that would predispose birds to becoming associated with buildings (Klausnitzer, 1989), the use of holes as nest sites (Tomialojc, 1970), residency being favoured over migration (Pulliainen, 1963), omnivory being favoured over species with more specialized diets (Dickman & Doncaster, 1987), and species with broad environmental tolerance being preadapted to urban habitats (Bonier et al., 2007). More recently, Møller (2009) investigated a total of 12 predictors of urbanization in all species of birds from the Western Palearctic and found significant differences in mean values for urbanized species and closely related rural species belonging to 39 pairs of such sister taxa. In an analysis of nine variables (the remaining three variables had too many missing values to merit inclusion) these variables all remained significant predictors of urbanization. Among these variables, response to predators (as shown by force required to remove feathers, flight distance when approached by a human observer), life history (adult survival rate, annual fecundity), dispersal (maximum dispersal distance, number of subspecies), ecological success (population size, total breeding range), and body mass accounted for significant variation in urbanization. Likewise, Evans et al. (2010) analysed the independent effects of ground nesting, plant and invertebrate diet, fecundity, migration, dispersal, niche position and relative brain mass as predictors of population density in urban habitats or ratio of density in urban to rural habitats. Although Evans et al. did not find a significant phylogenetic signal in the predictor variables in their study, and therefore did not conduct a phylogenetic analysis; the absence of a phylogenetic signal in a trait has no relevance for a phylogenetic signal in a regression model (Revell, 2010).

The urbanization process can be broken up into its component parts (Figure 5.1) made up of the problem of (1) arriving in urban areas in sufficient numbers, (2) becoming adjusted to the novel environmental conditions, or already having characteristics that predispose a species to succeed, and (3) expanding the population in the novel urban environment. Each of these three steps of the invasion process is likely to be influenced by several different factors. The first step deals with dispersal and density-dependent dispersal of urban colonizers. The second step concerns the ability to survive in urban habitats: (i) with its peculiar availability of food and nutrients, (ii) ability to cope with predators associated with humans such as cats and dogs, (iii) ability to cope with proximity of humans, (iv) cognitive ability that facilitates colonization of novel environments, and (v) life history that facilitates survival and reproduction. I will briefly review each of these components of the urbanization process.

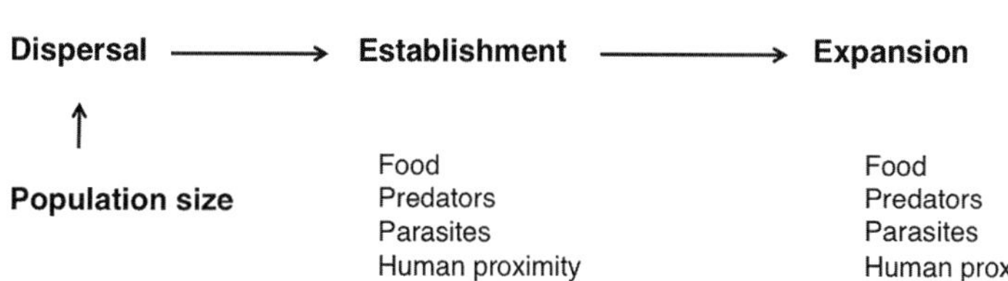

Figure 5.1 Schematic presentation of urbanization from the initial immigration into urban areas followed by establishment and expansion. The associations between different biological phenomena and these three steps of the invasion process are indicated by arrows.

While studies of the biological invasion process of urbanization are superficially simple because there are so many cities, the scientific literature is replete with studies conducted on only a single species in one single city and in a single nearby rural area. Such studies based on a single replicate of urbanization contrasted with a single replicate of ancestral rural habitat constitute a prime example of pseudoreplication (Hurlbert 1984). In cases where the phenotypes of urban and rural populations are contrasted, we are investigating the recent change in phenotype *after* divergence between such pairs of populations, and they do indeed constitute statistically independent observations. These can be urban and rural populations of the same species or urban and rural populations of different species (only one species is urban and the closely related species is rural), or they can be samples obtained at different locations along a gradient from rural to urban areas, as in the study of corticosterone in tree sparrows *Passer montanus* by Zhang et al. (2011). However, even observations of single species share their common phylogenetic history with closely related species that also tend to be either rural or urban, but not both, and rigorous phylogenetic analyses that distinguish convergent evolution from similarity due to common phylogenetic descent must therefore be adopted. Among the 526 species of birds from the Western Palearctic there is a highly significant effect of genus on whether a species is classified as urbanized or not (sensu Møller, 2009), accounting for more than 50% of the variance ($F = 1.46$, d.f. = 219, 306, $r^2 = 0.51$, $P = 0.0012$). Therefore, there are good reasons for conducting analyses of urbanization in a rigorous phylogenetic context.

The objectives of this chapter were to assess: (1) the extent to which different indices of urbanization reflected the same underlying biological phenomenon; and (2) the extent to which different life history and ecological factors accounted for urbanization when comparing populations or communities of birds.

5.2 Definition of urbanization

Urbanization needs a rigorous definition to allow for rigorous analyses. Previously used definitions of biological urbanization include: (1) year when a population became urbanized as judged by direct observation (Møller, 2008); (2) difference in breeding population density between urban and rural habitats, with relatively higher density in urban habitats reflecting relatively higher success in urban areas and hence more ancient urbanization (Møller et al., 2012); (3) at least one urban population with higher population density than that of nearby rural populations (Møller, 2009); and (4) whether a bird species has been found breeding in city centres (Croci et al., 2008).

A correlation matrix for these four different variables is shown in Table 5.1. Different measures of urbanization may vary in their information content. Of the six correlation coefficients reported in Table 5.1 based on more than 80 species of birds from the Western Palearctic (depending on the comparison), all but one were strong, explaining 25–34% of the variance (a strong effect sensu Cohen, 1988). Analyses based on Kendall rank order correlations provided similar results. These correlations suggest that no single index provided complete information on urbanization, and that the use of multiple indicators may provide a better grasp of different features of urbanization. Evans et al. (2010) reported correlations exceeding 0.85 among four measures of breeding population density or ratio of urban to rural population density for 88 species of birds in the UK.

Principal component analysis on the correlations for the four variables for 82 species of birds produced a first principal component with an eigenvalue of 2.27 that accounted for 57% of the variance, which is a considerable improvement over the 25% variance explained by most correlations reported in Table 5.1. Loadings by the four variables were all greater than 0.43, suggesting that all variables contributed significantly to the first principal

Table 5.1 Correlation matrix between four different indices of urbanization of birds.

	Breeds in urban centre	Year urbanized	Difference in density
Urbanized species	0.58***	0.55***	0.54***
Breeds in urban centre		0.51***	0.28*
Year urbanized			−0.53***

*$P < 0.05$, *** $P < 0.0001$.

component in the same direction. All other principal components had eigenvalues smaller than 0.89 that thus could not be considered meaningful to include.

5.3 Factors associated with urbanization

I obtained a total of 94 effect sizes of the relationship between urbanization and ecological characteristics from an exhaustive search of the literature using Web of Science and Google Scholar as sources (Appendix 5.1). I was unable to obtain effect sizes from Croci et al. (2008) because test statistics were not reported. All meta-analyses were made with MetaWin using an unstructured random effects model with effect sizes weighted by sample size (Rosenberg et al., 2000). I evaluated the strength of relationships between urbanization and predictor variables using Pearson's product moment correlation coefficient as a standardized estimate of effect size, adopting the suggestion by Cohen (1988) that a correlation of $r = 0.10$, accounting for 1% of the variance is small, a correlation of $r = 0.30$, accounting for 9% of the variance intermediate, and a correlation of $r = 0.50$, accounting for 25% of the large variance. Møller and Jennions (2002) showed in a meta-analysis of meta-analyses in biology that on average main effects accounted for 5–7% of the variance, thus accounting for an intermediate effect. Results should preferably be replicated if we should have confidence in the generality of research findings (Kelly, 2006). Therefore, I tested for consistency in findings among studies based on Pearson's coefficients.

5.3.1 Predictors of urbanization

Dispersal is important by reflecting the problems of arriving in urban areas in sufficient numbers. Species with high dispersal abilities should more readily colonize urban habitats (Møller, 2009), although once arrived in urban habitats there may subsequently be selection for reduced dispersal (Cheptou et al., 2008). The mean effect size did not reach statistical significance and overall was small (Figure 5.2A; Table 5.2, Appendix 5.1). While Møller (2009) used maximum distance of established breeding populations from the European mainland as a measure of dispersal, Evans et al. (2010) relied on distances moved by ringed birds without information on whether these individuals actually reproduced. Hence, such indirect measures do not necessarily represent effective dispersal in the genetic sense of the term. Dispersal may be further promoted by high population density, with seven effect sizes showing associations with urbanization (Figure 5.2B; Appendix 5.1). Dispersal is well known for being density dependent (Clobert et al., 2001). Again, range size is generally positively associated with greater density and larger global population sizes (Brown, 1995).

Urban habitats are peculiar with respect to availability of food and nutrients. Urban habitats often have abundant food for a longer part of the year than nearby rural habitats and a greater diversity of potential food items that may favor an omnivorous diet (Bonier et al., 2007; Dickman & Doncaster, 1987; Gilbert, 1989). Food may be more readily available in the proximity of humans during winter thereby facilitating urbanization (Pulliainen, 1963). Thus, there are studies suggesting that diet generalism, a granivorous diet or a non-insectivorous diet may facilitate urbanization, with the six effect sizes having a mean close to zero (Table 5.2). Møller et al. (2009) reported that urban and rural birds differed in their liver contents of antioxidants like vitamin E and carotenoids. This can have significant consequences for urbanization (Appendix 5.1). A deficit of carotenoids in urban habitats is also shown for carotenoid-based plumage of a few species (Hõrak et al., 2001; Isaksson et al., 2005; Slagsvold & Lifjeld, 1985). A lack of antioxidants is consistent with problems of normal brain development in urban habitats as reflected by left-skewed frequency distributions of brain size (Møller & Erritzøe, 2012). Such effects of deficiency of antioxidants have already been shown experimentally for zebra finches *Taeniopygia guttata* (Bonaparte et al., 2011).

Urban habitats may be colonized because of a reduction in the abundance of predators. Both urbanization and domestication are affected by the ability to cope with the proximity of domesticated predators such as dogs and cats (Møller, 2012). Indeed, domestic cats can have significant effects on the distribution and behaviour of birds, but also on their reproductive behaviour and life history decisions (e.g. Baker et al., 2008; Møller et al., 2010d).

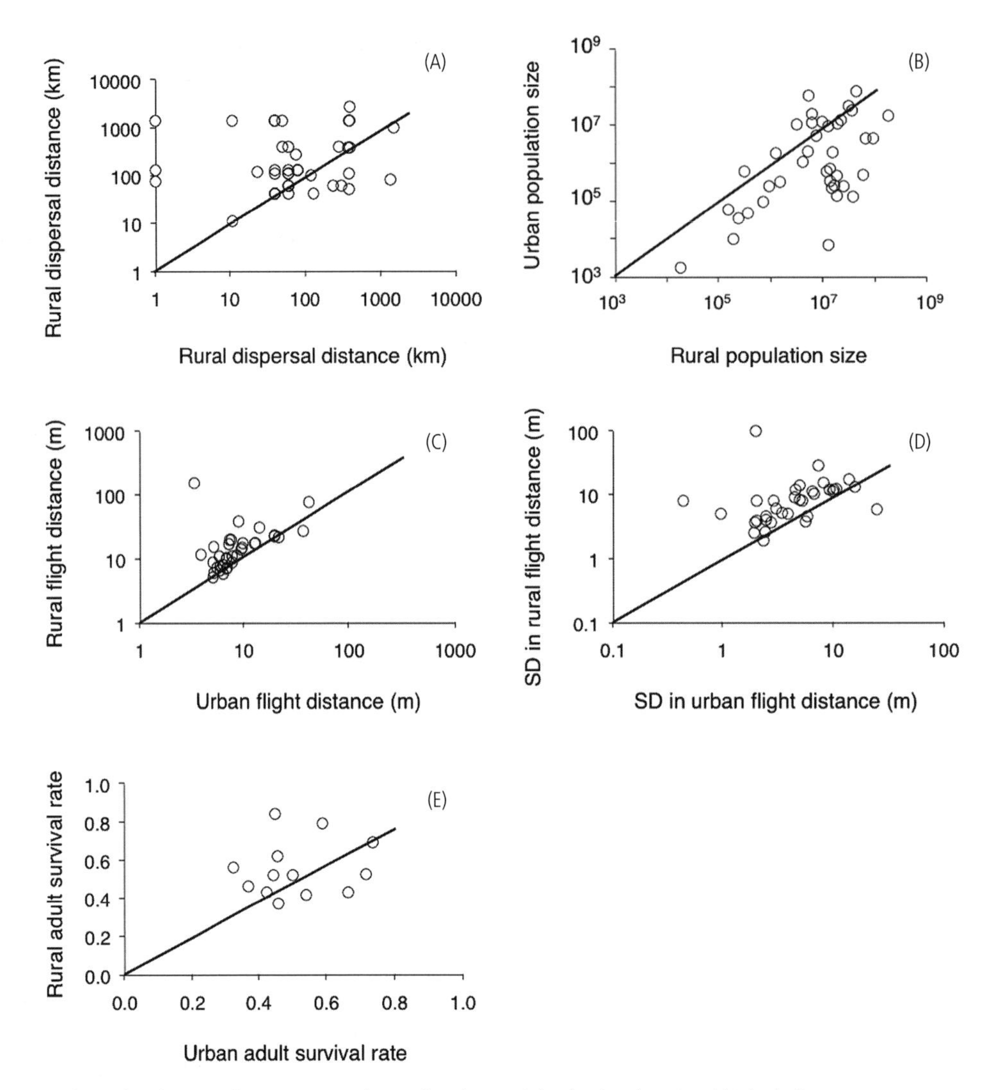

Figure 5.2 Relationships between character size in urban and rural pairs of closely related species of birds. The lines represent Y = X. (A) Dispersal distance (km). (B) Population size in the Western Palearctic. (C) Mean flight distance (m). (D) Standard deviation (SD) in flight distance (m). (E) Adult survival rate. Mainly adapted from Møller (2009).

Thus antipredator adaptations may facilitate urbanization as shown for differences in behavioural responses to proximity of potential predators such as humans (Møller & Ibañez-Alamo, 2012). Responses range from short flight distances to an approaching human over loss of feathers to facilitate escape, tonic immobility to feign death, and direct attempts to escape through emission of fear screams, wriggle behaviour, biting and scratching, to alarm calls that may attract the attention of conspecifics and heterospecifics (Møller & Ibañez-Alamo, 2012). There were 26 effect sizes for antipredator responses with an overall mean effect of large magnitude (Table 5.2, Appendix 5.1). This category had together with geographic range the strongest mean effect of all categories (Table 5.2).

Urbanization may relate to the ability to cope with proximity of humans. Humans constitute a specific class of potential predators for birds, and birds respond to the approach of humans by flight

Table 5.2 Mean effect size (Pearson's product moment correlation coefficient), 95% bootstrap C. I. for different categories of variables predicting urbanization. Mean weighted effect sizes labelled with * are significant at $P < 0.05$. See Appendix 5.1 for original data.

Category	Mean effect size	95% bootstrap CI	Sample size
Predation	0.472*	0.352, 0.588	29
Geographic range and density	0.455*	0.325, 0.568	7
Body size	−0.289	−0.687, 0.118	5
Sociality	0.231	−0.074, 0.391	3
Brain and cognition	0.199*	0.044, 0.382	17
Dispersal	0.196	−0.138, 0.448	3
Habitat	0.191*	−0.043, 0.345	5
Sexual selection	0.146	−0.082, 0.332	5
Antioxidants	0.086	−0.157, 0.308	6
Migration	0.072	−0.226, 0.062	2
Life history	0.062	−0.166, 0.335	8
Diet	−0.019	−0.199, 0.173	6
Parasitism	0.004	−0.590, 0.696	3

(Cooke, 1980). Urban birds cope with the proximity of humans better than nearby rural conspecifics (Cooke, 1980; Møller, 2008; Figure 5.2C). This is not due to habituation because individuals are highly consistent in their flight behaviour among days and even among years (Møller & Garamszegi 2012). Birds that have successfully invaded urban environments already had shorter and more variable flight distances in their ancestral rural habitat than unsuccessful conspecifics, followed by a further loss in mean and variance in flight distance upon urbanization (Carrete & Tella, 2011; Møller, 2010a; Møller and Garamszegi 2012). Once a species had become established in urban habitats, there was a gradual increase in intra-specific variation in flight distance over time again consistent with gradual adaptation rather than habituation as an explanation for such temporal change in behaviour (Møller, 2010a; Møller and Garamszegi 2012). Variance in flight distance showed intermediate to large effects (Figure 5.2D; Appendix 5.1).

Brains and cognitive ability may facilitate urbanization. Cognitive abilities have been hypothesized to facilitate urbanization because this process basically concerns the ability to cope with a novel environment (Kark et al., 2007). There are 15 effect size estimates that differ widely in magnitude (Appendix 5.1). The mean effect size was small to intermediate (Table 5.2). In particular the effect size reported by Maklakov et al. (2011) was more than ten standard errors above the mean. In contrast, the by far largest study based on brain mass for over 4,000 individuals only showed a small effect of 0.08 for the relationship between brain mass and urbanization (Møller & Erritzøe, 2012). Interestingly, variance in brain mass (Figure 5.3A) and skewness in brain mass (Figure 5.3B) showed much stronger effect sizes of 0.26 and 0.28 than mean brain mass for the same large-scale study (Møller & Erritzøe, 2012). This suggests that a population subset with cognitive deficiencies is associated with urbanization rather than mean cognitive ability as such.

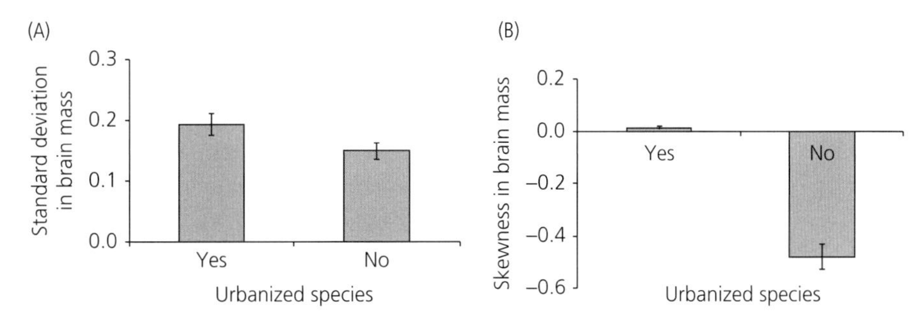

Figure 5.3 (A) Mean (SE) standard deviation in brain mass (g) and (B) mean (SE) skewness in brain mass in bird species that did or did not become urbanized. Adapted from Møller and Erritzøe (2012).

Urbanization may be facilitated by a life history that facilitates survival and reproduction. Proximity of humans and their domestic animals such as cats and dogs will select for changes in the life history such as early start of reproduction, many reproductive events, small investment in each single offspring and elevated adult survival rate compared to the situation with a low density of potential predators (Møller, 2012). Indeed bird species that have become urbanized have higher fecundity and higher adult survival rate than sister species that have not become urbanized (Figure 5.2E; Møller, 2009). Mean effect size was small and not significant (Table 2, Appendix 1).

There were effect size estimates available for several other categories, although most of these were small. Two effect sizes for migration were both small (Table 5.2, Appendix 5.1). Three effect sizes for parasitism were small (Table 5.2, Appendix 5.1). Five effect sizes for sexual selection were small (Table 5.2, Appendix 5.1). The strongest effect was for song post height in response to presence of cats in urban environments (Appendix 5.1). Finally, three effect size estimates for sociality were small (Table 5.2, Appendix 5.1).

5.3.2 Ranking of effect sizes

Average effect size weighted by sample size was of intermediate magnitude (Pearson $r = 0.219$, 95% bootstrap confidence intervals (CI) = 0.140, 0.294) and statistically highly heterogeneous (Q_{total} = 151.34, d.f. = 93, $P = 0.00013$). Rosenthal's fail-safe number as an estimate of the number of null results required to eliminate this mean effect was 1477 unpublished null results. Mean effect sizes for the 13 different categories are shown in Table 5.2. Four of these categories had 95% CI that differed from zero. The strongest effect was for predation that accounted for 22% of the variance, followed by geographic range accounting for 21% of the variance, and predation and brain size and cognition that each accounted for 4% of the variance. Therefore, two categories had strong effects and two small to intermediate effects. There was significant heterogeneity among categories (Q_{total} = 155.91, d.f. = 93, $P = 0.00005$). Rosenthal's fail-safe number was 1525. Mean weighted effect size estimates for intraspecific and interspecific estimates were marginally different (Q_{total} = 140.25, d.f. = 91, $P = 0.034$) with the estimate for interspecific studies being 0.151 (95% CI = 0.049, 0.247) and for intraspecific studies 0.330 (95% CI = 0.211, 0.451). Thus there was a tendency for larger effects in intraspecific than in interspecific studies as expected from the more powerful paired design of intraspecific studies. Rosenthal's fail-safe number was in this case 1298.

5.3.3 Partial effects after controlling for other variables

Many of the variables associated with successful urbanization of birds are correlated among each other, implying that partial correlations are the appropriate measure of association. Møller (2009) and Evans et al. (2010) reported such independent partial effects after accounting for the correlations among predictor variables.

Although all effects reviewed here were quantified using a single metric, these effect sizes are not necessarily directly comparable. For example, an effect size based on a pairwise comparison of sister taxa is much more likely to reflect the true underlying effect size because the potentially confounding effects of other variables is automatically controlled statistically. Sister taxa are by definition very similar in most respects because they share most of their evolutionary past, making it unlikely that a large number of such comparisons will be confounded by other factors (Møller & Birkhead, 1992).

5.3.4 Repeatability among cities

Ideally analyses of effect sizes from different cities along a latitudinal gradient would be very informative because they would provide replication while environmental conditions change. Møller et al. (2012) have shown that for both year of urbanization and difference in breeding population density between urban and nearby rural habitats there is a significant small repeatability, implying that the same species tended to show similar patterns of urbanization across geographic scales. In contrast, there was no significant effect of latitude for these two measures of urbanization.

5.4 Discussion

5.4.1 Different indices of urbanization

Despite considerable interest in the ecology and evolutionary biology of urbanization there has been little methodological investigation of the variables used as indicators of extent of urbanization. Here I analysed four different indicators of the extent of urbanization; whether at least one population had higher density in urban than in rural habitats (Møller, 2009), whether populations breeding in city centres existed (Croci et al., 2008), the difference in population density between urban and nearby rural habitats (Evans et al., 2010; Møller et al., 2012), and the estimated year of urbanization as assessed by amateur ornithologists (Møller, 2008). These four variables generally showed positive correlations exceeding 0.50, thus accounting for strong effects of consistency, with a single exception for the correlation of 0.28 between difference in population density and whether a species was breeding in city centres (Table 5.1). However, almost half of the variance remained unexplained, suggesting that the use of multiple indicators of urbanization may be warranted.

Evans et al. (2010) have previously analysed the correlation between four urbanization indices based on population density or the ratio of urban to rural density. Although these indices showed correlations greater than 0.85, they were all basically based on the same variables and thus cannot be considered independent indices. Because their data have not been published, it is impossible to make comparisons with the data analysed here.

One of the few examples of an extensive analysis of urbanization across large geographical scales is reported by Møller et al. (2012), who analysed sources of variation for estimated year of urbanization and difference in breeding population density for pairs of urban and rural sites across a latitudinal gradient from Granada in Southern Spain to Rovaniemi in Northern Finland. Both variables showed significant, but small repeatabilities among cities, implying that the same species tended to have similar values across cities. That was the case independent of whether anciently urbanized species that have been urbanized for hundreds of years, such as house sparrow and starling, were excluded from the analyses. There were no clear latitudinal trends in these two indices of urbanization, suggesting that differences in climate or other variables varying in a consistent manner along latitudinal gradients are unlikely to play a major role in urbanization.

5.4.2 Indicators of urbanization

Numerous factors have been suggested to play a role in urbanization, and only three studies have investigated the role of more than a single variable (Carrete & Tella, 2011; Evans et al., 2010; Møller, 2009). Even these studies are not comparable because the first relied on sister taxa comparisons, the second did not find evidence of phylogenetic signal in the data despite other studies of urbanization having found such phylogenetic effects (above I reported that genus accounted for more than 50% of the variance in whether a species was classified as urbanized or not), and the third study relied on inclusion of order and family as random factors in statistical models. The risk of effects of unidentified confounding variables is the greatest for the second and to some extent the third study, while the first automatically controls for such confounding variables because sister taxa that already resemble each other due to a shared evolutionary past are unlikely to consistently be confounded by other variables (Møller & Birkhead, 1992).

If we assume that variables with larger effect sizes are likely of the greatest importance, we can rank mean effect sizes of the different categories of predictor variables (Table 5.2). This exercise suggests that predation and geographic range are important because they all explain more than 10% of the variance. These variables are followed by brain size and cognition and habitats that only explain 4% of the variance. The remaining categories all explain less than 1% of the variance. Therefore, interspecific interactions (predation) and large populations and geographic ranges are likely to have played a major role in urbanization.

Effects are often far from clear-cut as shown by studies of brain size and urbanization in birds. Kark et al. (2007) found no significant effect in a small sample of species not analysed in a phylogenetic context. Evans et al. (2010) did not find a significant effect of brain mass on population density in urban habitats or ratio of density in urban to rural habitats among 88 species of British birds. In contrast, Carrete & Tella (2011) showed recently that bird species with relatively larger brains for their body size were more likely to successfully invade urban environments in Argentina. Maklakov et al. (2011) reported that families of birds with relatively larger brains had more urbanized species. All these analyses relied on mean brain mass calculated from cranial volume of skulls. Møller et al. (2012) analysed data on brain mass and used both mean, variance and skewness in brain mass based on the assumption that brain mass would facilitate invasion of urban habitats, variance in brain mass would reflect a greater diversity of behavioural phenotypes, and that left skewness in brain mass would reflect a fraction of individuals with poor brain development for environmental reasons including poor diet and poor availability of antioxidants (Møller et al. 2010c, 2011). There was no significant effect of mean brain mass, but species with larger variance in brain mass and a smaller degree of left skew in brain mass were more likely to become urbanized. In addition, analyses of urban and rural populations of the same species showed that the degree of left skew increased from the ancestral state in rural habitats as species became urbanized. Thus urban habitats and the diet derived from these habitats especially in terms of antioxidant contents (Møller et al., 2009) are likely to be the direct cause of poor brain development. In fact Bonaparte et al. (2011) have shown experimentally for zebra finches that a protein-poor diet with low antioxidant levels caused stunted brain development. The study by Møller et al. (2012) illustrates the fact that mean brain mass may not be the actual factor behind a certain correlation, and that closer scrutiny of the effects of variation and skew may prove rewarding. Just to show that this is not a single example Møller (2010a) showed for flight distance that species that became urbanized originally had a high variance in flight distance in the ancestral rural habitat, and this variation was reduced upon urbanization, followed by a subsequent increase in variance in flight distance. In other words, the diversity of behavioural phenotypes had a significant bearing on initial invasion, subsequent establishment and final expansion. Therefore, the 'effect' of mean values on urbanization, even when intuitively obvious, is sometimes far from clear.

5.4.3 Future studies

Most of the studies analysed here were from the temperate zone in the northern hemisphere with just a few effects reported from other continents. In particular, it would be interesting to have studies of the effects of urbanization in tropical cities because tropical species are generally considered to be specialists compared to non-tropical sibling species. If specialization is an impediment to successful invasion of urban areas (Julliard et al., 2006), specialized tropical species should become urbanized with greater difficulty than temperate species.

Multiple factors are contributing to urbanization, as already shown by several studies (e.g. Croci et al., 2008; Evans et al., 2010; Møller, 2009). The analyses reported here raise questions about the ancestral transition(s) for successful urbanization, and which are the ecological traits that predispose species to successful invasion of urban habitats.

There are also important conservation implications of urbanization. Growth of cities and increasing conversion of rural habitat are generally assumed to result in loss of biodiversity. However, urbanization may also create new biodiversity because of high rates of genetic divergence among urban populations (Baratti et al., 2009; Björklund et al., 2010; Evans et al. 2009a; Fulgione et al., 2000; Rutkowski et al., 2005). Thus conservation biologists may consider how to make urbanization work in favor of greater biodiversity rather than loss of biodiversity.

Acknowledgements

D. Gil kindly invited me to contribute this chapter.

References

Alberti, M. (2005). The effects of urban patterns on ecosystem function. *International Regional Science Reviews*, **28**, 168–92.

Baker, P. J., Molony, S. E., Stone, E., Cuthill, I. C., and Harris, S. (2008). Cats about town: is predation by free-ranging pet cats *Felis catus* likely to affect urban bird populations? *Ibis*, **150**, 86–99.

Baratti, M., Cordaro, M., Dessi-Fulgheri, F., Vannini, M., and Fratini, S. (2009). Molecular and ecological characterization of urban populations of the mallard (*Anas platyrhynchos*) in Italy. *Italian Journal of Zoology*, **76**, 330–9.

Björklund, M., Ruiz, I., and Senar, J.C. (2010). Genetic differentiation in the urbanhabitat: The great tits (*Parus major*) of the parks of Barcelona city. *Biological Journal of the Linnean Society*, **99**, 9–19.

Bonaparte, C.-L. (1828). *Ornithologie comparé de Rome et de Philadelphie*. C. L. Bonaparte, Rome.

Bonaparte, K.M., Riffle-Yokoi, C., and Burley, N.T. (2011). Getting a head start: Diet, sub-adult growth, and associative learning in a seed-eating passerine. *PLoS One*, **6**(9), e23775.

Bonier, F., Martin, P. R., and Wingfield, J. C. (2007). Urban birds have broader environmental tolerance. *Biology Letters*, **3**, 670–3.

Brown, J.H. (1995). *Macroecology*. Chicago University Press, Chicago, IL.

Carrete, M. and Tella, J. L. (2010). Individual consistency in flight initiation distances in burrowing owls: a new hypothesis on disturbance-induced habitat selection. *Biology Letters*, **23**, 167–170.

Carrete, M. and Tella, J. L. (2011). Inter-individual variability in fear of humans and relative brain size of the species are related to contemporary urban invasion in birds. *PLoS One*, **6**(4), e18859.

Cheptou, P.-O., Carrue, O., Rouifed, S., and Cantarel, A. (2008). Rapid evolution of seed dispersal in an urban environment in the weed *Crepis sancta*. *Proceedings of the National Academy of Sciences USA*, **109**, 3796–9.

Clobert, J., Nichols, J. D., Danchin, E., and Dhondt, A., eds. (2001). *Dispersal*. Oxford University Press, Oxford, UK.

Cohen, J. (1988). *Statistical Power Analysis for the Behavioral Sciences*. 2nd edn. Lawrence Erlbaum, Hillsdale, NJ.

Cooke, A.S. (1980). Observations on how close certain passerine species will tolerate an approaching human in rural and suburban areas. *Biological Conservation*, **18**, 85–8.

Croci, S., Butet, A. and Clergeau, P. (2008). Does urbanization filter birds on the basis of their biological traits? *Condor*, **110**, 223–240.

Dickman, C.R. and Doncaster, C.P. (1987). The ecology of small mammals in urban habitats. I: Populations in a patchy environment. *Journal of Animal Ecology*, **56**, 629–40.

European Commission. (2006). *Urban Sprawl in Europe*. European Environmental Agency, Copenhagen, Denmark.

Evans, K. L., Gaston, K. J., Frantz, A. C., Simeoni, M., Sharp, S. P., McGowan, A., Dawson, D. A., Walasz, K., Partecke, J., Burke, T., and Hatchwell, B.J. (2009). Independent colonization of multiple urban centres by a formerly forest specialist bird species. *Proceedings of the Royal Society of London B*, **276**, 2403–10.

Evans, K. L., Chamberlain, D. E., Hatchwell, B. J., Gregory, R. D., and Gaston, K.J. (2010). What makes an urban bird?*Global Change Biology*, **17**, 32–44.

Fulgione, D., Rippa, D., Procaccini, G., and Milone, M. (2000). Urbanisation and the genetic structure of *Passer italiae* (Viellot 1817) populations in the south of Italy. *Ethology Ecology & Evolution*, **12**, 123–30.

Gesner, C. (1669). *Vollkommenes Vogelbuch*. Schlütersche Verlagsanstalt, Hannover, Germany, reprint 1981.

Gilbert, O.L. (1989). *The Ecology of Urban Habitats*. Chapman & Hall, London, UK.

Handwerk, D. (2008). Half of humanity will live in cities by year's end. http://news.nationalgeographic.com/news/2008/03/080313-cities.html (accessed June 17, 2013).

Hõrak, P., Ots, I., Vellau, H., Spottiswoode, C. and Møller, A.P. (2001). Carotenoid-based plumage coloration reflects hemoparasite infection and local survival in breeding great tits. *Oecologia*, **126**, 166–73.

Hurlbert, S. H. (1984). Pseudoreplication and the design of ecological filed experiments. *Ecological Monographs*, **54**, 187–211.

Isaksson C., Ornborg, J., Stephensen, E. and Andersson, S. (2005). Plasma glutathione and carotenoid coloration as potential biomarkers of environmental stress in great tits. *EcoHealth*, **2**, 136–138.

Julliard, R., Clavel, J., Devictor, V., Jiguet, F., and Couvet, D. (2006). Spatial segregation of specialists and generalists in bird communities. *Ecology Letters*, **9**, 1237–44.

Kark, S., Iwaniuk, A., Schalimtzek, A., and Banker, E. (2007). Living in the city: can anyone become an 'urban exploiter'? *Journal of Biogeography*, **34**, 638–51.

Kelly, C. D. (2006). Replicating empirical research in behavioral ecology: how and why it should be done but rarely ever is. *Quarterly Review of Biology*, **81**, 221–36.

Klausnitzer, B. (1989). *Verstädterung von Tieren*. Neue Brehm-Bücherei, Wittenberg-Lutherstadt, Germany.

Maklakov, A. A., Immler, S., Gonzalez-Voyer, A., Rönn, J., and Kolm, N. (2011). Brains and the city: Big-brained passerine birds succeed in urban environments. *Biology Letters*, **7**, 730–2.

Miller, J. R. (2005). Biodiversity conservation and the extinction of experience. *Trends in Ecology and Evolution*, **20**, 430–4.

Møller, A. P. (1992). The hirundines. *Natur og Museum*, **31**, 1–32.

Møller, A. P. (2008). Flight distance of urban birds, predation and selection for urban life. *Behavioral Ecology & Sociobiology*, **63**, 63–75.

Møller, A. P. (2009). Successful city dwellers: A comparative study of the ecological characteristics of urban birds in the Western Palearctic. *Oecologia*, **159**, 849–58.

Møller, A. P. (2010a). Interspecific variation in fear responses predicts urbanization in birds. *Behavioral Ecology*, **21**, 365–71.

Møller, A. P. (2010b). The fitness benefit of association with humans: Elevated success of birds breeding indoors. *Behavioral Ecology*, **21**, 913–18.

Møller, A. P. (2011). Song post height in relation to predator diversity and urbanization. *Ethology*, **117**, 529–38.

Møller, A. P. (2012). Reproduction. In B. Wong and U. Candolin, eds, *Behavioural Responses to a Changing World: Mechanisms and consequences*, pp. 106–118. Oxford University Press, Oxford, UK.

Møller, A. P. and Birkhead, T. R. (1992). A pairwise comparative method as illustrated by copulation frequency in birds. *American Naturalist*, **139**, 644–56.

Møller, A. P. and Erritzøe, J. (2012). The mean is not everything: Moment statistics for brain size and innovative behavior in birds. Unpublished manuscript.

Møller, A. P. and Ibáñez-Álamo, J. D. (2012). Capture behavior of urban birds provides evidence of predation being involved in urbanization. *Animal Behaviour*, **84**, 341–348.

Møller, A. P. and Jennions, M. D. (2002). How much variance can be explained by ecologists and evolutionary biologists? *Oecologia*, **132**, 492–500.

Møller, A. P., Erritzøe, J., and Karadas, F. (2010c). Levels of antioxidants in rural and urban birds and their consequences. *Oecologia*, **163**, 35–45.

Møller, A. P., Erritzøe, J., and Nielsen, J.T. (2010d). Causes of interspecific variation in susceptibility to cat predation on birds. *Chinese Birds*, **1**, 97–111.

Møller, A. P., Bonisoli-Alquati, A., Rudolfsen, G., and Mousseau, T.A. (2011). Chernobyl birds have smaller brains. *PLoS One* **6**(2), e16862.

Møller, A.P., Diaz, M., Flensted-Jensen, E., Grim, T., Ibáñez-Álamo, J.D., Jokimäki, J., Mänd, R., Markó, G., and Tryjanowski, P. (2012). High urban population density of birds reflects their timing of urbanization. *Oecologia*, **170**, 867–875.

Pulliainen, E. (1963). On the history, ecology and ethology of the mallards (*Anas platyrhynchos*) overwintering in Finland. *Ornis Fennica*, **40**, 45–66.

Rebele, F. (1994). Urban ecology and special features of urban ecosystems. *Global Ecology and Biogeography Letters*, **4**, 173–87.

Revell, L.J. 2010. Phylogenetic signal and linear regression on species data. *Methods in Ecology and Evolution*, **1**, 319–329.

Rosenberg, M.S., Adams, D.C., and Gurevitch, J. (2000). *MetaWin: Statistical software for meta-analysis*. Version 2.1. Sinauer Associates, Sunderland, MA.

Rutkowski, R., Rejt, L., Gryczynska-Siematkowska, A., and Jagolkowska, P. (2005). Urbanization gradient and genetic variability of birds: Example of kestrels in Warsaw. *Berkut*, **14**, 130–6.

Sætre, G.-P., Ruyahi, S., Aliabadian, M.et al.(2012) Single origin of human commensalism in the house sparrow. *Journal of Evolutionary Biology*, **35**, 788–96.

Schneider, A., Friedl, M. A., and Potere, D. (2009). A new map of global urban extent from MODIS satellite data. *Environmental Research Letters*, 4044003 (11 pp.)., doi:10.1088/1748-9326/4/4/044003.

Slagsvold, T. and Lifjeld, J. T. (1985). Variation in plumage color of the great tit *Parus major* in relation to habitat, season and food. *Journal of Zoology*, **206**, 321–8.

Stephan, B. (1990). *Die Amsel*. Neue Brehm-Bücherei, Wittenberg-Lutherstadt, Germany.

Summers-Smith, D. (1963). *The House Sparrow*. Collins, London, UK.

Tomialojc, L. (1970). Quantitative studies on the synanthropic avifauna of Legnica town and its environs. *Acta Ornithologica*, **12**, 293–392

Turner, W.R., Nakamura, T., and Dinetti, M. (2004). Global urbanization and the separation of humans from nature. *Bioscience*, **54**, 585–90.

United Nations. (2007). *World Urbanization Prospects*: The 2007 revisions. United Nations, New York.

Zhang, S., Lei, F., Liu, S., Li, D., Chen, C., and Wang, P. (2011). Variation in baseline corticosterone levels of tree sparrow (*Passer montanus*) populations along an urban gradient. *Journal of Ornithology*, **152**, 801–6.

Appendix 5.1 Effect sizes for factors accounting for urbanization used in the meta-analyses. See the chapter for sources and additional information.

Character	Category	Test statistic	Value	*P*	Effect size	*z*	*N*	Intraspecific study	Reference
Carotenoids in blood	Antioxidants	*F*	6.01	0.016	0.224	0.228	125	0	Møller et al. (2009)
Vitamin E in blood	Antioxidants	*F*	5.01	0.027	0.200	0.203	125	0	Møller et al. (2009)
Carotenoids in blood	Antioxidants	*t*	2.47	0.019	−0.385	−0.406	36	1	Møller et al. (2009)
Vitamin E in blood	Antioxidants	*t*	2.11	0.0006	−0.336	−0.349	36	1	Møller et al. (2009)
Carotenoids and time since urbanization	Antioxidants	*F*	3.27	0.081	0.314	0.324	33	1	Møller et al. (2009)
Vitamin E and time since urbanization	Antioxidants	*F*	6.2	0.019	0.432	0.463	33	1	Møller et al. (2009)
Body mass	Body size	χ^2	2.33	0.13	0.220	0.224	48	0	Møller (2010a)
Body mass	Body size	χ^2	657.38	<0.0001	−0.390	−0.411	152	0	Møller & Erritzøe unpublished data
Body mass and time since urbanization	Body size	*F*	4.46	0.037	0.206	0.209	102	1	Møller & Erritzøe unpublished data
Body mass	Body size	*t*	9.99		−0.909	−1.521	22	0	Maklakov et al. (2011)
Body mass	Body size	*F*	4.05	0.05	−0.314	−0.325	39	0	Carrete & Tella (2009)
Innovation rate	Brain and cognition	*F*	5.57	0.038	0.336	0.350	12	0	Møller (2009)
Brain mass	Brain and cognition	χ^2	26.49	<0.0001	0.078	0.078	152	0	Møller & Erritzøe unpublished data
Variance in brain mass	Brain and cognition	χ^2	295.14	<0.0001	0.261	0.267	152	0	Møller & Erritzøe unpublished data
Skewness in brain mass	Brain and cognition	χ^2	318.28	<0.0001	−0.271	−0.278	152	0	Møller & Erritzøe unpublished data
Mean brain mass	Brain and cognition	*t*	1.31	0.19	0.051	0.051	70	1	Møller & Erritzøe unpublished data
Variance in brain mass	Brain and cognition	*t*	1.45	0.15	0.079	0.079	70	1	Møller & Erritzøe unpublished data
Skewness in brain mass	Brain and cognition	*t*	2.8	0.0067	0.173	0.175	70	1	Møller & Erritzøe unpublished data
Brain mass and time since urbanziation	Brain and cognition	*F*	0.02	0.89	0.014	0.014	102	1	Møller & Erritzøe unpublished data
Variance in brain mass and time since urbanization	Brain and cognition	*F*	2.01	0.16	−0.140	−0.141	102	1	Møller & Erritzøe unpublished data
Skewness in brain mass and time since urbanization	Brain and cognition	*F*	21.97	<0.0001	0.423	0.451	102	1	Møller & Erritzøe unpublished data
Brain size	Brain and cognition	*t*	9.44		0.900	1.470	22	0	Maklakov et al. (2011)
Brain volume	Brain and cognition	*t*	1.73	0.11	0.420	0.447	40	0	Kark et al. (2007)
Brain mass	Brain and cognition	*F*	0.03	0.86	0.071	0.071	9	0	Carrete & Tella (2009)
Brain mass	Brain and cognition	*F*	4.01	0.07	0.486	0.530	15	0	Carrete & Tella (2009)
Brain mass	Brain and cognition				0.067	0.067	88	0	Evans et al. (2010)

continued

Appendix 5.1 *Continued*

Character	Category	Test statistic	Value	*P*	Effect size	*z*	*N*	Intraspecific study	Reference
Granivorous	Diet	χ^2	2.54	<0.20	0.252	0.258	40	0	Kark et al. (2007)
Insectivorous	Diet	χ^2	3.97	<0.05	−0.315	−0.326	40	0	Kark et al. (2007)
Omnivorous	Diet	χ^2	0.65	<0.10	0.127	0.128	40	0	Kark et al. (2007)
Diet generalism	Diet	*F*	2.07	0.16	0.250	0.256	33	0	Carrete & Tella (2009)
Plant diet	Diet				−0.291	−0.300	88	0	Evans et al. (2010)
Invertebrate diet	Diet				−0.060	−0.060	88	0	Evans et al. (2010)
Mainland distance	Dispersal	*F*	9.55	0.0037	0.448	0.482	39	0	Møller (2009)
No. subspecies	Dispersal	*F*	3.85	0.043	0.303	0.313	39	0	Møller (2009)
Dispersal	Dispersal				−0.137	−0.138	88	0	Evans et al. (2010)
Total breeding range	Geographic range	*F*	6.2	0.017	0.375	0.394	39	0	Møller (2009)
Western Palearctic breeding range	Geographic range	*F*	17.89	0.0001	0.566	0.642	39	0	Møller (2009)
Population size	Geographic range	*F*	29.27	<0.0001	0.660	0.793	39	0	Møller (2009)
Population density	Geographic range	*F*	12.11	0.0013	0.492	0.539	39	0	Møller (2009)
Rural abundance	Geographic range	*F*	3.6	0.07	0.298	0.307	39	0	Carrete & Tella (2009)
Rural density	Geographic range	*t*	2.9	0.004	0.175	0.177	250	1	Møller et al. (2012)
Rural density	Geographic range	*t*	115.71	<0.0001	0.570	0.648	250	1	Møller et al. (2012)
Cavity nester	Habitat	χ^2	0.01	<1	0.016	0.016	40	0	Kark et al. (2007)
Habitat generalism	Habitat	*F*	3.25	0.08	0.308	0.318	33	0	Carrete & Tella (2009)
Environmental tolerance	Habitat	*F*	0.08	0.78	0.046	0.046	39	0	Carrete & Tella (2009)
Ground nesting	Habitat				0.084	0.084	88	0	Evans et al. (2010)
Niche position	Habitat				0.437	0.469	88	0	Evans et al. (2010)
Annual fecundity	Life history	*F*	4.24	0.047	0.321	0.333	38	0	Møller (2009)
Adult survival	Life history	*F*	7.29	0.017	0.585	0.670	15	0	Møller (2009)
Altricial	Life history	χ^2	1.27	<1	−0.178	−0.180	40	0	Kark et al. (2007)
No. clutches per year	Life history	*t*	−0.86	0.40	0.204	0.207	40	0	Kark et al. (2007)
Clutch size	Life history	*t*	1.67	0.11	−0.375	−0.395	40	0	Kark et al. (2007)
Incubation period	Life history	*t*	1.24	0.24	−0.315	−0.326	40	0	Kark et al. (2007)
No. fledglings	Life history	*t*	0.06	0.96	−0.016	−0.016	40	0	Kark et al. (2007)
Age ratio	Life history	*t*	3.3	0.008	0.720	0.908	9		Evans et al. (2009b)
Migration	Migration	χ^2	2.04	<0.20	−0.226	−0.230	40	0	Kark et al. (2007)
Migration	Migration				0.062	0.062	88	0	Evans et al. (2010)
Mass of bursa of Fabricius	Parasitism	*F*	9.36	0.012	0.696	0.860	11	0	Møller (2009)
Malaria prevalence	Parasitism	*r*	0.26		−0.260	−0.266	12	0	Evans et al. (2009b)
Tick prevalence	Parasitism	*r*	0.59		−0.590	−0.678	9	0	Evans et al. (2009b)
Flight distance	Predation	*F*	12.1	<0.0001	0.955	1.886	15	1	Møller (2009)

continued

Appendix 5.1 *Continued*

Character	Category	Test statistic	Value	*P*	Effect size	*z*	*N*	Intraspecific study	Reference
Residual force required to remove rump feathers	Predation	*F*	10.24	0.015	0.771	1.023	9	0	Møller (2009)
Mean flight distance	Predation	χ^2	1.21	0.27	0.159	0.160	48	1	Møller (2010)
Variance in flight distance	Predation	χ^2	14.38	0.0002	0.547	0.615	48	1	Møller (2010)
Variance in flight distance and time since urbanization	Predation	*F*	4.44	0.041	0.300	0.310	48	0	Møller (2010)
Variance in flight distance and difference in density	Predation	*F*	4.51	0.04	0.332	0.345	48	0	Møller (2010)
Flight distance	Predation	*z*	–4.02	<0.0001	0.877	1.364	20	1	Carrete & Tella (2009)
Variance in flight distance	Predation	*F*	6.05	0.049	0.709	0.884	9	0	Carrete & Tella (2009)
Mean flight distance	Predation	*F*	0.23	0.63	0.097	0.098	26	0	Carrete & Tella (2009)
Variance in flight distance	Predation	*F*	18.31	<0.01	0.658	0.789	26	0	Carrete & Tella (2009)
Flight distance	Predation	*t*	8.82	<0.0001	0.803	1.106	44	1	Møller (2008)
Flight distance	Predation	*F*	323.43	<0.0001	0.382	0.402	2030	0	Møller (2008)
Flight distance and time since urbanization	Predation	*F*	7.57	0.009	0.422	0.450	39	1	Møller (2008)
Flight distance and proportion of individuals in urban areas	Predation	*F*	0.68	0.41	0.138	0.139	39	1	Møller (2008)
Alarm call and time of urbanization	Predation	*F*	0.41	0.33	–0.175	–0.177	15	1	Møller & Ibañez-Alamo (2012)
Biting and time of urbanization	Predation	*F*	6.27	0.025	0.570	0.648	15	1	Møller & Ibañez-Alamo (2012)
Wriggle and time of urbanization	Predation	*F*	6.89	0.057	0.589	0.675	15	1	Møller & Ibañez-Alamo (2012)
Fear scream and time of urbanization	Predation	*F*	0.001	0.83	0.009	0.009	15	1	Møller & Ibañez-Alamo (2012)
Feather loss and time of urbanization	Predation	*F*	1.22	0.1	0.293	0.302	15	1	Møller & Ibañez-Alamo (2012)
Tonic immobility and time of urbanization	Predation	*F*	0.001	0.73	–0.009	–0.009	15	1	Møller & Ibañez-Alamo (2012)
Alarm call	Predation	*F*	4.29	<0.0001	0.484	0.529	17	1	Møller & Ibañez-Alamo (2012)
Biting	Predation	*F*	4.07	<0.0001	0.475	0.516	17	1	Møller & Ibañez-Alamo (2012)
Fear scream	Predation	*F*	4.42	<0.0001	0.490	0.536	17	1	Møller & Ibañez-Alamo (2012)

continued

Appendix 5.1 *Continued*

Character	Category	Test statistic	Value	*P*	Effect size	*z*	*N*	Intraspecific study	Reference
Feather loss	Predation	*F*	3.21	<0.0001	0.432	0.462	17	1	Møller & Ibañez-Alamo (2012)
Tonic immobility	Predation	*F*	2.42	0.0014	0.384	0.405	17	1	Møller & Ibañez-Alamo (2012)
Wriggle	Predation	*F*	3.31	<0.0001	0.437	0.469	17	1	Møller & Ibañez-Alamo (2012)
Song post height	Sexual selection	*F*	172.27	<0.0001	0.316	0.327	1554	1	Møller (2011)
Difference in song post height and time since urbanization	Sexual selection	*F*	11.98	0.0012	0.436	0.467	55	1	Maller (2011)
Sexual dimorphism	Sexual selection	χ^2	0.03	<1	−0.027	−0.027	40	0	Kark et al. (2007)
Seasonal dimorphism	Sexual selection	χ^2	0.42	<1	0.102	0.103	40	0	Kark et al. (2007)
Nest building by both sexes	Sexual selection	χ^2	2.04	<0.20	−0.226	−0.230	40	0	Kark et al. (2007)
Flocking	Sociality	χ^2	4.94	<0.05	0.351	0.367	40	0	Kark et al. (2007)
Colonial nesting	Sociality	χ^2	6.11	<0.03	0.391	0.413	40	0	Kark et al. (2007)
Territorial	Sociality	χ^2	0.22	<1	−0.074	−0.074	40	0	Kark et al. (2007)

CHAPTER 6

Acoustic communication in the urban environment: patterns, mechanisms, and potential consequences of avian song adjustments

Diego Gil and Henrik Brumm

6.1 The function of bird songs and calls

Birds are vocal beings. We recognize bird species by their songs, we find them in the forest when they call, we marvel at the beauty and musical quality of their notes and trills. But these vocalizations are not directed to us. Vocal communication is of vital importance for birds, as it is helps them passing and acquiring crucial information. Birds use acoustic signals in various ways: between mates, between parents and offspring, between rivals, among conspecifics and even between different species. The two main functions of bird song are territory defence and mate attraction (Catchpole & Slater, 2008). In addition to their often very elaborate and versatile songs, birds also utter a number of shorter and simpler vocalizations. These calls are used to keep in touch with conspecifics, to warn of danger, to call for food, to signal aggression, etc. (Marler, 2004).

Vocal communication has several advantages over other kinds of communication: it works at a distance, without the need of eye contact, it can be used at night, in dense habitats or in atmospheric conditions when other types of messages could not be transmitted, it covers long ranges and can be used in subtle battles of many individuals at the same time. Moreover, unlike visual signals, sounds also go round corners, making them well suited to dense habitats, such as forests (Catchpole & Slater, 2008). Despite any advantages of a given signal modality, all communication processes are constrained by a crucial factor that decreases their effectiveness. This factor is called noise. Signal detection theory defines noise as anything that increases the likelihood of performance errors in the receiver, or in other words, anything that influences a receiver's receptors other than a signal of interest (Wiley, 2013). Although noise may take many forms, such as neuronal receptor noise, semantic or psychological noise, in the case of acoustic communication, the term refers to its main common meaning: an extraneous sound that interferes with the communication process.

6.2 Urbanization and anthropogenic noise

The urban environment is rich in unwanted sounds, and it can almost be defined by it. Humans have become very efficient at generating noise by many means, and we tend to accumulate a diverse number of noise sources in our habitats, such as cars, trains, factories, machines, turbines, but also loudspeaker announcements, music, and so on. Most cities around the world are heavily polluted by noise emissions and a good number of laws are passed every year trying to limit and punish their excesses. Anyone who studies animal sounds and tries to get good clean recordings in urban areas is well aware of this problem.

Avian Urban Ecology. Edited by Diego Gil and Henrik Brumm

Noise can be detrimental to the health and fitness of humans and other animals in many direct and indirect ways, which are not necessarily connected to communication. We will not discuss these additional effects in detail, but just briefly mention some that are considered crucial for animal conservation such as physiological stress, hearing damage or exclusion from important habitats (McGregor et al., 2014).

One of the main noise sources in urban areas is vehicular traffic. Noise produced by cars is typically low pitched, with most of its energy below 1 kHz (Can et al., 2010). Nevertheless, emission levels in urban areas can be very high, which means that the high-frequency components of traffic noise can still reduce the active space of bird songs. In extreme examples, traffic noise may even extend into the ultrasonic range. Although most of our chapter will deal with vocal changes in response to noise, it is important to consider that interfering sound is not the only important characteristic of the acoustic environment of cities. There are several other ways in which urban environmental acoustics can also affect vocal communication independent of noise pollution (Warren et al., 2006). One important acoustic feature of urban areas is the presence of many sound-reflective surfaces, such as buildings, streets, and paved areas. Mockford and co-authors (2011) measured the sound transmission properties of woodland and city environments and found that, in addition to the background noise, the characteristics of the acoustic environment in cities plays an important role in leading to the observed changes in urban song.

There are also other, non-acoustic, characteristics of urban habitats that may affect acoustic communication of birds. For instance, most bird species living in urban environments are confined to tiny islands of vegetation, where dispersal is compromised. This process of fragmentation can have large repercussions for song learning and sharing that could be decisive in determining song traditions (Laiolo & Tella, 2007). Additionally, other alternative explanations, none tested so far, have been put forward to explain the peculiarities of urban song (Section 6.4). For instance, differences in densities could affect singing (Hamao et al., 2011), and indeed some species occur at higher densities in cities (Francis & Chadwick, 2012). Also, different hormonal profiles in the response to stress or in testosterone levels could affect song (Nemeth & Brumm, 2009). However, although some studies show differences in corticosteroid levels in response to stress between urban and rural habitats (Partecke et al., 2006), a recent review suggests that these differences are not widespread (Bonier, 2012). In the case of testosterone, a most likely candidate to influence singing behaviour, most studies show no difference between habitats or, if anything, decreased levels of testosterone in urban areas (Bonier, 2012).

However, future work should carefully analyse possible confounding variables and not assume, somewhat hastily, that any urban song trait is a response to noise.

6.3 Consequences of signal masking

Anthropogenic noise in urban areas may lead to the masking of bird vocalizations. Acoustic masking refers to the increase in thresholds for detection or discrimination of sounds caused by the presence of noise. Whether or not masking occurs depends on the level and spectrum of the signal and the noise as well as the perceptual abilities of the receiver (Dooling & Blumenrath, 2014). Based on these, information models have been published that predict the effect of song masking by noise in different species (Lohr et al., 2003; Nemeth & Brumm, 2010). Because bird calls are often involved with immediate issues of life and death, and songs are linked to reproduction, the masking of bird vocalizations can have severe consequences, not only for the vocalizing bird but potentially also for population dynamics.

6.3.1 Impairment of communication

Studies reporting lower reproductive success in bird populations exposed to high noise levels have suggested that this is caused by the interference of vocal communication between mating partners (Habib et al., 2007; Halfwerk et al., 2011a). Detailed models have shown a large reduction of transmission distances for bird songs subjected to urban noise levels (Nemeth & Brumm, 2010). Such a reduction in active space may lead to smaller territories, higher densities than optimal, and higher levels of territorial intrusions and fights between males. Moreover, close-range communication, such as that used between pair members to coordinate activities, or between offspring and parents, may also be compromised by masking, leading to non-optimal parental behaviour. In the tree swallow (*Tachycineta bicolor*) it has been shown that an increase in noise leads to a modification of begging calls (Leonard & Horn, 2005), and to a mismatch between begging and provisioning (Leonard & Horn, 2012). Similarly, female house sparrows (*Passer domesticus*) have been found to reduce nestling provisioning in a noisy area compared to quiet sites (Schroeder et al., 2012). However, it is very difficult to know whether the observed effects are actually the outcome of signal masking or, for example, the result of noise-induced stress, and much more work is needed to disentangle the underlying causes.

Experiments with bats have shown that traffic noise can also impair the detection of the acoustic cues that these animals use to find their prey (Siemers & Schaub, 2011). Likewise, birds that use sound to localize their food might be similarly affected. American robins (*Turdus migratorius*) for instance, use acoustic cues to detect buried prey, and noise may limit their ability to exploit this resource (Montgomerie & Weatherhead, 1997). Another example would be the potential problems for raptors and owls that rely on sounds to find their prey animals (Knudsen & Konishi, 1979; Rice, 1982). On the other hand, potential prey also may suffer from signal masking through anthropogenic noise, as many bird species use alarm calls to warn other conspecifics about the proximity of a predator (Marler, 2004).

6.3.2 Role for vocalizations in the urban filtering of species?

It has been proposed that differences in species composition between natural and urban habitats, characterized by a decrease in species richness and an increase in homogenization among sites (Francis & Chadwick, 2012) may be partially due to some bird species not being able to make their vocalizations heard in noise (Rheindt, 2003). Several studies have indeed found that species differ greatly in their geographical distribution with respect to noise levels (Goodwin & Shriver, 2011; Paton et al., 2012) suggesting the potential role of noise tolerance as a species-specific trait.

But actually how strong is the evidence for a relationship of song with species filtering (i.e. the selective disappearance of certain bird species in noise polluted areas)? Several lines of evidence point in this direction. The origin of this idea can be traced to a classic study in Germany in which it was shown that species in areas close to a noisy motorway were more likely to have higher-pitched songs than those further away from the road (Rheindt, 2003). As we will see later on (Section 6.4.2), high-pitched songs can be detected more easily in urban noise conditions. In addition, a comparative study has suggested that birds that colonise urban habitats have higher dominant song frequencies than those that do not (Hu & Cardoso, 2009). It is tempting to assume that this is due to a pre-adaptation of some species to colonize urban habitats where higher frequencies may be favoured, but the possible confounding factors, such as nesting ecology, phylogeny, and body size need to be taken into account before firm conclusions can be drawn.

Perhaps the most convincing study so far about an effect of noise on species composition is the experiment in which Francis and co-authors were able to assess the effects of noise levels in several forest plots by using an existing industrial natural gas extraction setup (Francis et al., 2009). The results of the study showed a reduction of species richness in noisy plots in comparison with control plots. However, this effect was not random with respect to trophic categories: seed dispersers and predators decreased whereas pollinators increased in noise-polluted areas (Francis et al., 2012). This modification of the balance of ecosystem services provided by birds suggests that noise pollution can indeed alter the functioning of the ecosystem by non-random filtering of species with different capacities to adapt to noise conditions. Furthermore, a relationship between species resilience and song characteristics was found in species inhabiting the study site, suggesting that the capacity of birds to modify their songs in response to noise may be linked to their capacity to withstand human perturbations (Francis et al., 2011a).

6.4 Changes in urban bird vocalizations

A large and increasing body of literature has explored differences in bird song characteristics between urban habitats and natural environments. Many of those differences that constitute the 'acoustic phenotype' of urban birds (Slabbekoorn, 2013), are thought to maintain communication efficiency (see Wallschläger (1985) for perhaps the earliest reference to this hypothesis), although experimental work addressing these advantages is still largely lacking. Furthermore, most studies blame masking by noise as the causal factor behind these urban–forest differences and, although this is perhaps the likeliest hypothesis, it is by no means the only plausible one. We will consider the evidence for each of these ideas in turn.

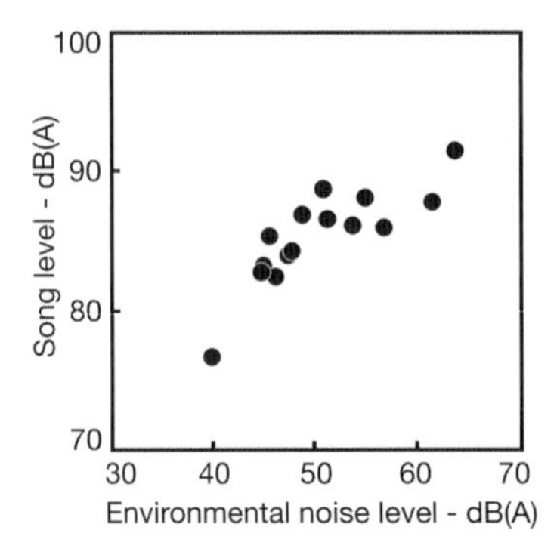

Figure 6.1 The Lombard effect in urban songbirds. Nightingales regulate the sound level of their songs depending on the noise level of the environment. Each data point represents the median value of one male nightingale recorded in the city of Berlin (Germany), (N = 15 males, dB ref. 20 μPa; dB(A), decibels measured with the A-filter). The variation in environmental noise was mainly due to differences in the amount of car traffic (from Brumm, 2004, used with permission).

6.4.1 Song amplitude

Birds and mammals, including humans, increase the vocal amplitude of their voices when the background noise level raises, a phenomenon known as the Lombard effect (Brumm & Zollinger, 2011). Experimental work in birds has shown that the Lombard effect is very robust and elicited most strongly by noise that overlaps with the frequency band of the signal (Brumm & Todt, 2002; Manabe et al., 1998). Birds use the Lombard effect to make themselves heard over the din in urban areas; nightingales (*Luscinia megarhynchos*) exposed to high levels of traffic noise sang songs that were up to 14 dB higher in amplitude than the songs of birds from more quiet sites (Figure 6.1),. Given the logarithmic scaling of level measurements, this range in vocal amplitude corresponds to a more than five-fold increase in vocal sound pressure.

The Lombard effect is not only a very robust and powerful means to deal with masking noise but it is also a very basal mechanism of vocal plasticity. So far it has been found in songbirds, parrots, Galliformes, and Tinamiformes, suggesting that it is probably a common trait of all living bird species (Schuster et al., 2012). It was present even in nestling birds immediately after they hatched, when their begging calls were overlapped by experimental noise (Leonard & Horn, 2005).

6.4.2 Frequency shifts

In 2003, Slabbekoorn and Peet reported that great tits (*Parus major*) in the city of Leiden (The Netherlands) sang songs with higher minimum frequencies in territories in which noise levels were higher (Slabbekoorn & Peet, 2003). They suggested that the increased minimum frequency facilitates communication between birds through release from masking by the low-frequency traffic noise. Since then, this pattern of song divergence has been found repeatedly in many field studies. For instance a similar frequency shift was reported in other European populations of great tits (Figure 6.2) (Salaberria & Gil, 2010; Slabbekoorn & den Boer-Visser, 2006), song sparrows (*Melospiza melodia*) (Wood & Yezerinac, 2006), house finches (*Carpodacus mexicanus*) (Bermudez-Cuamatzin et al., 2009; Fernández-Juricic et al., 2005), or common blackbirds (Figure 6.3a). Interestingly, most studies have found that it is the minimum frequency and not the peak or the maximum frequency, that increases with greater low-pitched noise levels. This may be accounted for by the fact that lower frequencies are more strongly masked by low-frequency traffic noise. On the other hand, some of the studies measured song frequencies by eye in spectrograms, a method which is prone to measuring artefacts and can yield spurious results (Zollinger et al., 2012). Indeed a recent study

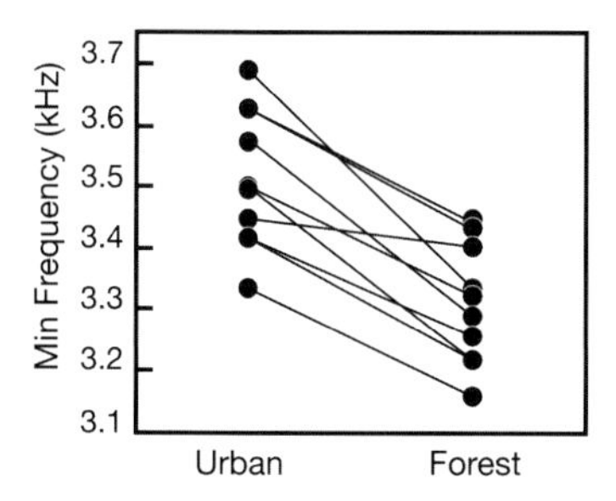

Figure 6.2 Population means of minimum song frequencies of 10 European city-forest pairs of great tits. On average, the birds at the urban site had a higher minimum frequency compared to the forest companion site in each country (from Slabbekoorn and den Boer-Visser, 2006, used with permission). Similar increases in urban song frequencies have been found in the meantime in many different bird species in Europe, Asia, the Americas, and Australia.

found differences in the song frequency—noise levels relationship depending on whether frequency analysis was done by eye or by analysis of power spectra (Rios-Chelen et al., 2013). Additionally, the biological significance of unmasking minimum frequencies remains to be shown, since most of the energy of the song is contained in the peak frequency, which is often much higher than the song minimum frequency.

Evidence supporting the notion of frequency adjustments in response to noise come from a completely different system; a frequency shift in the opposite direction (i.e. towards lower frequencies), has been found in green hylias (*Hylia prasina*), a warbler that occurs in areas with high insect activity in African forests (Kirschel et al., 2009). In this case, the main source of noise is the intense high-pitched buzzing of cicadas, and the melodic whistle of green hylias was found to be transposed to a lower key in areas of great insect activity.

Between-species differences in the magnitude of pitch changes are likely because species differ in song frequency and thus in the extent by which their songs would be affected by the low-frequency pattern of urban noise. Thus, it is to be expected that species with low-pitched songs should display stronger upward shifts than those that sing higher-pitched song. However, two studies to date have not found a linear relationship between these variables. A study with an Australian bird data set (Hu & Cardoso, 2010) found a quadratic relationship relating minimum frequency to song frequency shift in urban habitats, whereas a study conducted in the Neotropics found no relationship between the magnitude of change and species-specific minimum frequency (Rios-Chelen et al., 2012).

Experimental manipulation of noise levels has shown that song frequency may vary with noise in several cases. For instance, in chiffchaffs (*Phylloscopus collybita*), increases in noise were followed by increases in song frequency (Verzijden et al., 2010). In the red-winged blackbird (*Agelaius phoeniceus*) a similar manipulation lead birds to an increase in tonality, with a narrower frequency band (Hanna et al., 2011). Finally, a playback study in black-capped chickadees (Poecile atricapillus) showed that birds were more likely to shift their song upwards when their song frequency was overlapped by noise (Goodwin & Podos, 2013). However, none of these studies measured vocal amplitude and it is currently debated whether the observed increases in vocal pitch in urban birds are simply a side effect of the increased amplitude of songs in noise (Brumm & Naguib, 2009; Nemeth & Brumm, 2010; Nemeth et al., 2012; Slabbekoorn et al., 2012). Moreover, a recent study on common blackbirds, a successful urban colonizer, showed that amplitude and frequency are coupled during song production, and that city birds preferentially sang higher-frequency elements that can be produced at higher sound intensities (Nemeth et al., 2013; Figure 6.3). Thus, by choosing higher elements, urban birds may increase their capacity to sing at high amplitudes to mitigate acoustic masking by noise.

However, increased song frequencies have been shown to yield some release from masking. In great tits, for instance, perceptual experiments have shown that extremely high-pitched songs can be better detected in urban noise (Pohl et al., 2012). Furthermore, this study found that the discrimination of high-frequency differences was easier than that of low frequencies. A sound transmission study found that urban great tit song transmits better in urban environments than in the countryside (Mockford et al., 2011). However, increases in frequency alone were not responsible for this and the authors conclude that other song modifications account for this difference.

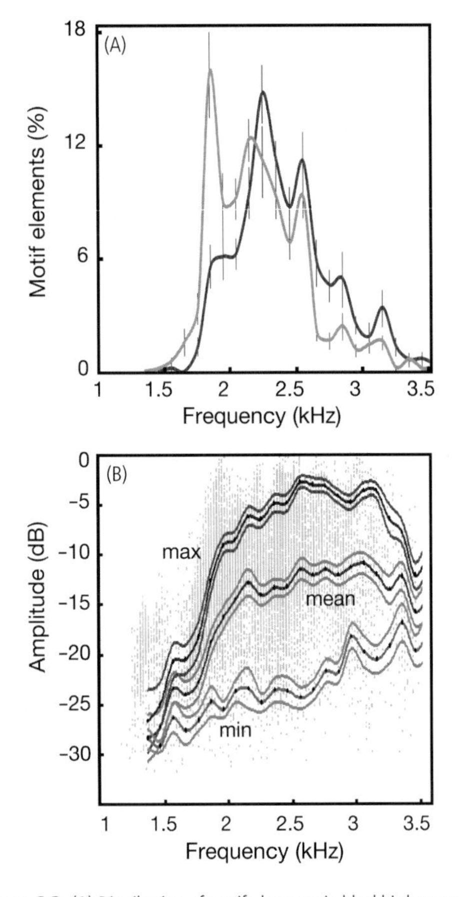

Figure 6.3 (A) Distribution of motif elements in blackbird songs. The curves give average percentages of peak frequency values in 100 Hz intervals for 16 urban (dark gray) and 17 forest birds (light gray), the error bars show standard errors of these values (adapted from Nemeth et al., 2013; used with permission). (B) Relationship between peak song frequency and amplitude in common blackbirds (N = 12 males recorded in sound-shielded chambers). Amplitude and peak frequency are coupled in the critical frequency band that is masked most heavily by traffic noise (below 2.2 kHz). In this band, higher-pitched song elements can be produced at higher amplitudes. Minimum peak frequency (min); maximum peak frequency (max) and mean peak frequency curves are based on the weighted amplitudes averages of all males, measured in 100 Hz intervals (black dots). Upper and lower lines denote standard errors above and below these averages. Individual peak frequencies of all measured motif elements of all males (i.e. 13,298 elements in 20 Hz intervals) are shown in grey.

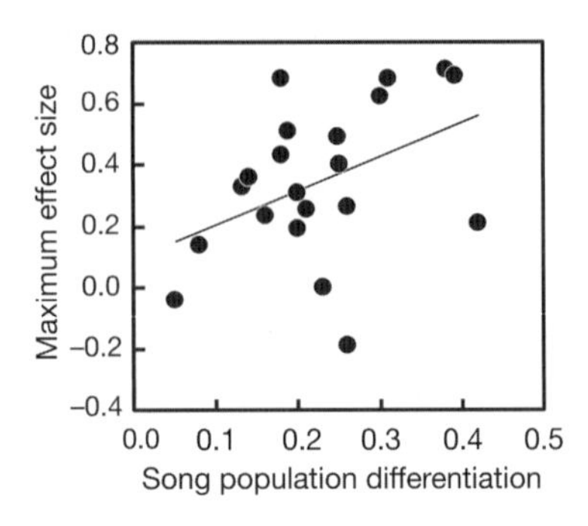

Figure 6.4 Relationship between population differentiation in song frequency within 14 Neotropical oscine and suboscine species and effect size (Pearson's correlation coefficient between minimum song frequency and noise levels). The degree of between-population song differentiation in song frequency was related to effect size: species with greater differences between populations in song frequency had larger increases in minimum song frequency when noise levels increased (from Rios-Chelen et al., 2012, used with permission).

Not all species show frequency modifications, or not to the same extent. In two different suboscine species, the same experimental noise exposure led to a higher frequency in one species (*Myiarchus cinerascens*), but no changes in another (*Empidonax wrightii*) (Francis et al., 2011b). A more systematic approach to this theme was conducted in a comparative study of oscine and suboscine species, in which it was found that oscines showed a stronger frequency effect than suboscines, suggesting that learning and or vocal versatility are important correlates of this capacity (Rios-Chelen et al., 2012). Furthermore, it was found that species in which song was more variable between populations also showed stronger frequency modifications in noise conditions (Figure 6.4). These results strongly suggest that patterns of change in response to noise perturbation differ between species, and thus that the capacity of resilience to changes may be determined by differences in learning ability or song consistency (Rios-Chelen et al., 2012).

6.4.3 Song duration and timing

In addition to changes in song amplitude and frequency, birds can also modify when and for how long they sing in order to increase the effectiveness of the signal. For instance, delivery of shorter songs could be advantageous when noisy conditions are not continuous but intermittent, since this would lead to a greater probability of the song being perceived during a window of silence. Indeed, shorter songs have been found in noisy conditions in blackbirds (Nemeth & Brumm, 2009), plumbeous vireos

(*Vireo plumbeus*) (Francis et al., 2011c), chiffchaffs (Verzijden et al., 2010), and great tits (Slabbekoorn & den Boer-Visser, 2006). However, for these shorter songs to be beneficial, they need to be sung at the right time in the fluctuating noise (i.e. when noise levels are low). Birds do this to avoid overlap with the songs of other species (Brumm, 2006b) but it is not known whether they use this capacity to reduce overlap with fluctuating urban noise. If a singing bird does not wait for a silent time-window to produce its song, it would actually be advantageous to produce longer songs, or longer bouts of song, because this would increase the probability of hitting such a window in fluctuating noise. Furthermore, even in constant noise, a receiver may extract increasing amounts of information from longer signals, an ability that has evolved alongside a number of other noise-reducing perceptual mechanisms that help birds to detect and recognize signals in noise (Dooling & Blumenrath, 2014). Indeed, grey vireos (*Vireo vicinior*) sang longer songs in noisy conditions (Francis et al., 2011c). Likewise, a study on serins (*Serinus serinus*) showed that birds sing longer bouts on weekdays than during weekends (Diaz et al., 2011). Thus, the current evidence yields conflicting patterns with some species producing shorter songs when exposed to intense anthropogenic noise and others producing longer songs. Perhaps different species use different strategies to make themselves heard in noise, but which strategy is adopted under which circumstance and how this relates to patterns of urbanization is not known.

In addition to short-term adjustments of song timing birds may also shift their general singing activity over the course of the day to avoid noise peaks in urban habitats. A study from the city of Sheffield (UK) found, after controlling for differences in nocturnal lighting, that European robins (*Erithacus rubecula*) sang for longer during the night in locations with high levels of daytime noise (Fuller et al., 2007). This suggests that these birds may partially avoid noise by shifting their singing routines earlier, when noise levels are reduced. Similarly, a study analysing song routines in a major park within the German city of Dortmund found that great tits, blue tits, and common blackbirds sang earlier in the city than in the forest nearby (Bergen & Abs, 1997). However, in this case it is not possible to tell whether differences in nocturnal lighting or noise were responsible for these differences.

Further work is necessary to grasp the ultimate consequences of this change in behaviour, since it could have a large influence on fitness if time budgets allocated for singing and feeding need to be adjusted on account of the noise. Warren and colleagues have predicted that if the dawn chorus is shifted as a result of noise produced at rush hour, a relevant test should be the comparison of cities at different latitudes in which the extent of the overlap between dawn and rush hour differs greatly (Warren et al., 2006). As far as we know, such a test has not yet been done.

6.4.4 Song contents

Few studies have examined whether song repertoires in urban birds differ from countryside populations in richness or composition. Song learning is often from neighbours and may involve local song traditions (Catchpole & Slater, 2008), so that a reduction in habitat patch size, or a large degree of isolation as can happen in urban islands, may lead to a situation of cultural bottleneck. For instance, in the Dupont's lark (*Chersophilus duponti*), habitat fragmentation results in the impoverishment of individual and population-wide song repertoires (Laiolo & Tella, 2005; Laiolo & Tella, 2007). A study in the rufous-collared sparrow (*Zonotrichia capensis*) has shown a reduction in between-individual variability in an Argentinian suburban population compared to a forest population (Laiolo, 2011). We expect these effects to be highly species and population-specific, because differences in inter-territory distances, dispersal patterns and population sizes should be highly decisive in the establishment of song cultures (Lachlan & Slater, 1999).

Information theory predicts that song repetition rate or redundancy within song bouts should increase with increasing noise levels, a pattern that has been found before in non-urban populations (Brumm & Slater, 2006; Diaz et al., 2011; Lengagne et al., 1999). Similarly, in European robins, a reduction in song complexity under experimental noisy treatment has been found (McLaughlin & Kunc, 2013).

The fact that urban habitats select particular types of syllables may lead to urban-specific dialects, and thus to convergent repertoires across large area. A continent-wide study conducted across cities in Australia has found that songs of urban silvereyes show a moderate degree of convergence in note selection (Potvin et al., 2013; see also Chapter 13).

6.5 Underlying mechanisms

One of the most fascinating aspects of the urban song phenotype is its origin and maintenance, since in this analysis we need to consider and tease apart explanations derived from physiological, ontogenetic and evolutionary processes: the Gordian knot of ethology (Tinbergen, 1963). Many of the song modifications that we have described above (amplitude, frequency and duration) could be due to either immediate vocal plasticity or else brought about by longer-term mechanisms, including differential syllable copying, cultural transmission, or genotypic selection. We will examine here the evidence for the potential mechanisms that underlie the observed patterns of urban bird song and consider possible evolutionary aspects in Section 6.6.

6.5.1 Phenotypic plasticity

Several studies using experimental manipulation of noise levels or comparing quiet and noisy days at the same location have found modifications of frequency and amplitude, suggesting that a high percentage of change in song in many species is due to immediate changes (e.g. Bermudez-Cuamatzin et al., 2011; Brumm & Todt, 2002). For instance, a study with reed warblers buntings (Emberiza schoeniclus) at two locations with different noise regimes showed that differences in the minimum frequency of song between sites almost disappeared on quiet days (Gross et al., 2010). In the case of amplitude modulations, several studies have shown immediate modifications of song amplitude when noise is broadcast to the birds (reviewed by Brumm and Zollinger (2011)). In the case of frequency modifications, two possibilities for this change have been examined. First, birds might choose songs with high frequencies among those available in their repertoire. This mechanism is expected to happen in birds with rather rigid song types, such as great tits. An experiment in which noise was played back to singing great tits showed that although birds did not choose the 'best' song type from their repertoire for a given noise regime, some individuals sang higher pitched songs during or after they were exposed to low frequency noise (Halfwerk & Slabbekoorn, 2009). Second, birds may actively modify the pitch of a given syllable. This pattern is to be expected in birds with less stereotyped song elements that allow various modifications. For instance, in the house finch, correlative and experimental evidence of this type of plasticity was found in two different studies (Bermudez-Cuamatzin et al., 2009; Bermudez-Cuamatzin et al., 2011). However, whether these forms of phenotypic plasticity are a direct response to low-frequency noise is not clear. As stated above (Section 6.4.2) changes in song pitch could be a side-effect of the increase in of song amplitude in noise (Lombard effect) and the usage of higher pitched syllables from the repertoire may help to increase amplitude when amplitude and pitch are coupled in phonation. Both effects would lead to higher-frequency songs that happen to be less masked in low-frequency traffic noise.

6.5.2 Learning and ontogeny

Oscine songbirds need to learn their songs from conspecifics. This process introduces a component of cultural evolution with its own peculiarities, including meme mutations and different transmission properties of song variants (Lachlan & Slater, 1999; Lynch & Baker, 1993). In 1979, Poul Hansen proposed that song learning could explain why birds use songs that match local habitat acoustics (Hansen, 1979). He suggested that the probability that a learner would learn a given song should increase with the capacity of transmission of this song. This kind of natural song selection has recently been shown empirically in the swamp sparrow (*Melospiza georgiana*): a tutees preferentially copied these songs that were less degraded, demonstrating a role for cultural selection in acoustic adaptation of learned signals (Peters et al., 2012). In the case of urban songs, we would expect this type of cultural transmission to lead to differential copying of those songs that are best heard in that environment, and thus variants with higher song frequency would be copied preferentially (Brumm, 2006a;

Patricelli & Blickley, 2006; Wood & Yezerinac, 2006). A relevant comparison here would be to compare modifications of song frequency in species that learn and those that do not learn. Two large groups of songbirds, the oscines and the suboscines differ in several song attributes: oscines typically learn songs and require exposure to conspecific songs to produce good copies of species-specific song, whereas suboscines can produce perfect copies of their song without learning or exposure to conspecifics (Kroodsma & Konishi, 1991). However, recent evidence suggests that this pattern may have several exceptions (Leitner et al., 2002; Saranathan et al., 2007). In a study comparing these two groups of birds in several cities in Brazil and Mexico, Rios Chelen and colleagues found that oscines showed stronger correlations between minimum frequency and noise levels than suboscines (Rios-Chelen et al., 2012). This could be taken to suggest that learning capacity explains a large part of the difference in the urban song phenotype. However, many suboscine birds have also less complex syringes than oscine birds (Ames, 1971), which might allow less variability in sound production, and thus the results are not conclusive on this respect. A comparison of song syllables between urban and rural populations of dark-eyed juncos (*Junco hyemalis*) revealed that about half of the differences between urban and rural songs were due to cultural mutations (i.e. modifications of the same syllable), whereas the other half consisted of syllable choice and creation of new syllables (Cardoso & Atwell, 2011). Taken together, the current evidence suggests that both differential learning and vocal plasticity are basic mechanisms behind the observed changes in urban bird songs, but detailed experimental work comparing species with different song learning abilities is needed to tease apart the different contributions of each mechanism to each specific situation.

6.6 Evolutionary implications

6.6.1 Sexual selection

Modifications of song brought about by the urban environment may lead urban song characteristics to deviate from female preference peaks (Patricelli & Blickley, 2006). It is difficult to predict how severe this effect might be, since the advantages of increased transmission of urban song may compensate for that deviation through increases in acoustic range and detectability that should be advantageous in sexual selection. Although many studies have addressed the relationship between song characteristics and male body condition, immunocompetence, and other resource-holding attributes (Gil & Gahr, 2002), most of them have analysed complex song characters, such as repertoire size or vocal performance. We know of no study to have addressed the relationship between the ability of individual birds to modify their songs in noisy environments and condition-dependence or resource-holding estimates.

Song frequency

It has been proposed that differences in song frequency may convey condition-dependent information about the singer (Gil & Gahr, 2002). For instance, low frequencies in songs may be preferred by females if large males are preferable and body-size is a limiting factor for frequency (Cardoso, 2012; Hall et al., 2013; Ryan & Brenowitz, 1985). However, song frequency could also be related to other indicators, hormonal for instance, that have not been investigated so far in connection with urbanization.

Mockford and Marshall (2009) investigated the responses of male great tits in 20 different cities across Britain and found that males from sites with high levels of noise (city centre) exhibited a significantly stronger response when hearing high-pitched songs from another territory holder from a site with high background noise than low-pitched songs from those with low background noise. Vice versa, individuals at sites with low noise levels (further away from the city centre) showed the opposite pattern. This behavioural difference hints at a possible difference in the strength of intra-sexual selection of urban bird songs through male–male territorial competition.

Halfwerk and colleagues suggested that anthropogenic noise can alter the inter-sexual selection of great tit song based on the observation that males that were not cuckolded by their females produced lower song frequencies during the peak of female fertility, and that females emerged faster from their nest box when high-pitched songs were played back in noisy conditions (Halfwerk et al., 2011b). However, as Eens et al. pointed out (Eens et al., 2012), this conclusion might not be fully warranted, since the control playback of the study clearly showed that female great tits did not respond more strongly to low-frequency than to high-frequency songs. Whereas low-frequency songs are indeed masked by low-frequency anthropogenic noise (Pohl et al, 2012), a recent review indicated that female birds may generally respond more strongly to high- than to low-frequency songs (Cardoso, 2012). Moreover, extra-pair fathers were not identified in the Halfwerk et al. study, and therefore the realized reproductive success could not be determined, making conclusions about whether low-frequency singing is sexually selected premature (Eens et al., 2012).

Song amplitude

Only a couple of studies have examined female preference for song amplitude, but evidence so far suggests directional preferences for higher amplitudes: the louder the better, at least for zebra finches (Ritschard et al., 2010) and red-winged blackbirds (Searcy, 1996). It would make sense thus to assume that increases in song amplitude in urban environments would be advantageous for males in terms of female attraction. However, although no relationship between body size and song amplitude has been found (Brumm, 2009), increases in song amplitude are very likely to be constrained by social aggression (Brumm & Ritschard, 2011; Ritschard et al., 2012), or increased predation (Gil & Gahr, 2002). It has also been suggested that males that have to increase their song amplitude for a long period would suffer higher energetic costs than those singing at a more moderate volume (Patricelli & Blickley, 2006) but measurements of metabolic costs of bird song in relation to song amplitude showed that the energetic costs of singing at high amplitudes are negligibly small (Zollinger et al., 2011).

Song contents

Although evidence so far is limited to a few species (Laiolo & Tella, 2007), we believe that it is important to underline the potential effects on sexual selection that an impoverishment of song repertoires in fragmented landscapes such as some urban habitats may bring about. Female preferences for song repertoires and certain song traditions (Baker, 1983) may set such a standard for song that males from suboptimal areas may simply not pass female thresholds. A link between sexual selection and extinction likelihood is exemplified by a study of New Zealand bird introductions, showing that species with strong sexual selection are less likely to be successful at colonizing new areas (Sorci et al., 1998). This process constitutes an Allee effect that could also potentially limit the survival of urban populations. Alternatively, if dispersal is limited and female preferences are directed to local song traditions, as it is the case for some species, male characters and female preference may evolve hand in hand.

6.6.2 Implications for song evolution and speciation

Despite habitat isolation and fragmentation, urban species are typically common and have very large distribution ranges (Francis & Chadwick, 2012). This should in principle preclude potential urban speciation, although it is possible that song drift may lead to a certain degree of population genetic differentiation (Potvin & Parris, 2012). However, learning about song in cities may help us better understand the evolution of song in nature.

Comparative analyses across species show that species-specific song frequency patterns are adapted to the local acoustics of the areas inhabited by birds. Thus, species occupying open habitats produce songs with a higher pitch, and with a higher number of frequency modulations than those species which are typical of forests and close environments (Boncoraglio & Saino, 2007; Ryan & Brenowitz, 1985). Experimental work shows that these differences can be explained by the differential capacity of different frequencies to travel in the environments, suggesting positive selection for memes that are well adapted to a particular habitat (Brumm & Naguib, 2009).

Although conventional genetic selection may account for these differences, several lines of evidence suggest that cultural processes may be at the origin of song divergences. Cultural transmission of preferences might result in cultural assimilation, akin to genetic assimilation (sensu Waddington, 1953), that would foster later selective processes leading to prezygotic isolation and eventual speciation (Slabbekoorn & Smith, 2002). As we mentioned in the previous section, differential copying and transmission of the most audible and less degraded songs would induce cultural evolution of acoustically adapted variants (Hansen, 1979).

It is to be expected that the characteristics of noise-adapted song that we have described above could be subject in nature to a similar evolutionary process. Two compelling examples from noisy natural habitats suggest that this may indeed be the case. The first example is that of the rufous-faced warbler, *Phylloscopus magnirostris*, which lives in areas that surround Himalayan streams and torrents, a habitat characterized by a loud and continuous rumbling of white-waters, and whose song has been proposed to constitute an adaptation to these acoustic conditions (Dubois & Martens, 1984; Mahler & Gil, 2009). In this case, it is not a frequency shift, but rather the abolition of those frequency modulations typical of *Phylloscopus* warblers that has been proposed to serve communication in this habitat. A more recent example is that of a Neotropical wren, the grey-breasted wood-wren (*Henicorhina leucophrys*), the two subspecies of which inhabit different slopes of the Andes, which are characterized by large differences in noise levels. Dingle and coauthors found that songs of the subspecies that is found in environments with typical high-pitched white water noise produces songs that lack high-pitched notes, as expected if songs had been selected not to overlap with noise frequency spectra (Dingle et al., 2008). In these two cases, species-specific or subspecific differences go in the predicted direction, suggesting that selection may have favoured songs with a lower overlap with noise.

In the only longitudinal study of urban song changes to date, Luther compared the current song minimum frequency of the white-crowned sparrow (*Zonotrichia leucophrys*), with recordings collected by Luis Baptista 30 years earlier (Luther & Baptista, 2010). Although information on noise levels was not available, evidence suggests that noise levels have increased in this area of California over the last 30 years. As expected, today's song types were sung with a higher minimum song frequency than thirty years before. Also, new dialects that had appeared had a higher pitch too, suggesting selection for birds to increase their song frequency (or selection for high-amplitude songs that lead indirectly to higher frequencies, see Section 6.4.2).

6.7 Conclusions

One consequence of urbanization that has received comparatively little attention yet is the massive level of anthropogenic noise that animals have to cope with. Because birds use acoustic signals to communicate vital information, the masking of their vocalizations has potentially major effects on their fitness. Research of the last decade has shown that birds can adjust the properties of their songs and calls to mitigate masking from noise. This allows researchers to use cities as outdoor laboratories to study adaptation and the phenotypic plasticity of animal behaviour in fluctuating environments. However, we are only beginning to understand the proximate causes that underlie the observed changes in urban bird vocalizations. Likewise, we need more studies that pinpoint the evolutionary consequences of communication impairments in urban areas. Considered from a more applied point of view, anthropogenic noise has been suspected to cause a variety of adverse effects on birds. Among other things, such as stress or hearing damage, these effects are related to masking of acoustic signals and other important biological sounds. A precise understanding of these effects should be of interest to many groups including biologists, environmentalists, regulators, as well as roadway and construction engineers.

Acknowledgements

We thank Peter Slater for his very helpful comments on a previous version of this chapter and Sue Anne Zollinger for editorial help. H.B.'s research programme is funded by the German Research Foundation (awards BR 2309/6-1, BR 2309/8-1). D.G.'s research has been funded by the Spanish Ministry of Competitiveness and Economy and the BBVA Foundation.

References

Ames, P. L. (1971). *The Morphology of the Syrinx in Passerine Birds*. Yale University Press, Yale, Peabody Museum of Natural History.

Baker, M. C. (1983). The behavioural response of female Nuttall's white-crowned sparrows to male song of natal and alien dialects. *Behavioral Ecology and Sociobiology*, **12**, 309–315.

Bergen, F. and Abs, M. (1997). Verhaltensökologische Studie zur Gesangsaktivität von Blaumeise (*Parus caeruleus*), Kohlmeise (*Parus major*) und Buchfink (*Fringilla coelebs*) in einer Großtadt. *Journal für Ornithologie*, **138**, 451–467.

Bermudez-Cuamatzin, E., Rios-Chelen, A. A., Gil, D. and Garcia, C. M. (2009). Strategies of song adaptation to urban noise in the house finch: syllable pitch plasticity or differential syllable use? *Behaviour*, **146**, 1269–1286.

Bermudez-Cuamatzin, E., Rios-Chelen, A. A., Gil, D. and Garcia, C. M. (2011). Experimental evidence for real-time song frequency shift in response to urban noise in a passerine bird. *Biology Letters*, **7**, 36–38.

Boncoraglio, G. and Saino, N. (2007). Habitat structure and the evolution of bird song: a meta-analysis of the evidence for the acoustic adaptation hypothesis. *Functional Ecology*, **21**, 134–142.

Bonier, F. (2012). Hormones in the city: endocrine ecology of urban birds. *Hormones and Behavior*, **61**, 763–772.

Brumm, H. (2004). The impact of environmental noise on song amplitude in a territorial bird. *Journal of Animal Ecology*, **73**, 434–440.

Brumm, H. (2006a). Animal communication: city birds have changed their tune. *Current Biology*, **16**, R1003–R1004.

Brumm, H. (2006b). Signalling through acoustic windows: nightingales avoid interspecific competition by short-term adjustment of song timing. *Journal of Comparative Physiology a-Neuroethology Sensory Neural and Behavioral Physiology*, **192**, 1279–1285.

Brumm, H. (2009). Song amplitude and body size in birds. *Behavioral Ecology and Sociobiology*, **63**, 1157–1165.

Brumm, H. and Naguib, M. (2009). Environmental acoustics and the evolution of bird song. *Advances in the Study of Behavior*, **35**, 151–209.

Brumm, H. and Ritschard, M. (2011). Song amplitude affects territorial aggression of male receivers in chaffinches. *Behavioral Ecology*, **22**, 310–316.

Brumm, H. and Slater, P. J. B. (2006). Ambient noise, motor fatigue, and serial redundancy in chaffinch song. *Behavioral Ecology and Sociobiology*, **60**, 475–481.

Brumm, H. and Todt, D. (2002). Noise-dependent song amplitude regulation in a territorial songbird. *Animal Behaviour*, **63**, 891–897.

Brumm, H. and Zollinger, S. A. (2011). The evolution of the Lombard effect: 100 years of psychoacoustic research. *Behaviour*, **148**, 1173–1198.

Can, A., Leclercq, L., Lelong, J. and Botteldooren, D. (2010). Traffic noise spectrum analysis: Dynamic modeling vs. experimental observations. *Applied Acoustics*, **71**, 764–770.

Cardoso, G. C. (2012). Paradoxical calls: the opposite signaling role of sound frequency across bird species. *Behavioral Ecology*, **23**, 237–241.

Cardoso, G. C. and Atwell, J. W. (2011). Directional cultural change by modification and replacement of memes. *Evolution*, **65**, 295–300.

Catchpole, C. K. and Slater, P. J. B. (2008). *Bird Song: Biological Themes and Variations*. Cambridge University Press, Cambridge, UK.

Diaz, M., Parra, A. and Gallardo, C. (2011). Serins respond to anthropogenic noise by increasing vocal activity. *Behavioral Ecology*, **22**, 332–336.

Dingle, C., Halfwerk, W. and Slabbekoorn, H. (2008). Habitat-dependent song divergence at subspecies level in the grey-breasted wood-wren. *Journal of Evolutionary Biology*, **21**, 1079–1089.

Dooling, R. J. and Blumenrath, S. H. (2014). Avian sound perception in noise. In H. Brumm, ed., *Acoustic Communication and Noise*. Springer, Heidelberg, in press.

Dubois, A. and Martens, J. (1984). A case of possible vocal convergence between frogs and a bird in Himalayan torrents. *Journal für Ornithologie*, **125**, 455–463.

Eens, M., Rivera-Gutierrez, H. F. and Pinxten, R. (2012). Are low-frequency songs sexually selected, and do they lose their potency in male–female interactions under noisy conditions? *Proceedings of the National Academy of Sciences of the United States of America*, **109**, E208–E208.

Fernández-Juricic, E., Poston, R., De Collibus, K., Morgan, T. J., Bastain, B., Martin, C. A., Jones, K. and Treminio, R. (2005). Microhabitat selection and singing behavior patterns of male house finches (*Carpodacus mexicanus*) in urban parks in a heavily urbanized landscape in the Western US. *Urban Habitats*, **3**, 49–69.

Francis, C. D., Kleist, N. J., Ortega, C. P. and Cruz, A. (2012). Noise pollution alters ecological services: enhanced pollination and disrupted seed dispersal. *Proceedings of the Royal Society B-Biological Sciences*, **279**, 2727–2735.

Francis, C. D., Ortega, C. P. and Cruz, A. (2009). Noise pollution changes avian communities and species interactions. *Current Biology*, **19**, 1415–1419.

Francis, C. D., Ortega, C. P. and Cruz, A. (2011c). Different behavioural responses to anthropogenic noise by two closely related passerine birds. *Biology Letters*, **7**, 850–852.

Francis, C. D., Ortega, C. P. and Cruz, A. (2011a). Noise pollution filters bird communities based on vocal frequency. *PLoS ONE*, **6**, e27052.

Francis, C. D., Ortega, C. P. and Cruz, A. (2011b). Vocal frequency change reflects different responses to anthropogenic noise in two suboscine tyrant flycatchers. *Proceedings of the Royal Society B-Biological Sciences*, **278**, 2025–2031.

Francis, R. A. and Chadwick, M. A. (2012). What makes a species synurbic? *Applied Geography*, **32**, 514–521.

Fuller, R. A., Warren, P. H. and Gaston, K. J. (2007). Daytime noise predicts nocturnal singing in urban robins. *Biology Letters*, **3**, 368–370.

Gil, D. and Gahr, M. (2002). The honesty of bird song: multiple constraints for multiple traits. *Trends in Ecology and Evolution*, **17**, 133–141.

Goodwin, S. E. and Podos, J. (2013). Shift of song frequencies in response to masking tones. *Animal Behaviour*, **85**, 435–440.

Goodwin, S. E. and Shriver, W. G. (2011). Effects of traffic noise on occupancy patterns of forest birds. *Conservation Biology*, **25**, 406–411.

Gross, K., Pasinelli, G. and Kunc, H. P. (2010). Behavioral plasticity allows short-term adjustment to a novel environment. *American Naturalist*, **176**, 456–464.

Habib, L., Bayne, E. M. and Boutin, S. (2007). Chronic industrial noise affects pairing success and age structure of ovenbirds *Seiurus aurocapilla*. *Journal of Applied Ecology*, **44**, 176–184.

Halfwerk, W., Bot, S., Buikx, J., Van Der Velde, M., Komdeur, J., Ten Cate, C. and Slabbekoorn, H. (2011b). Low-frequency songs lose their potency in noisy urban conditions. *Proceedings of the National Academy of Sciences of the United States of America*, **108**, 14549–14554.

Halfwerk, W., Holleman, L. J. M., Lessells, C. M. and Slabbekoorn, H. (2011a). Negative impact of traffic noise on avian reproductive success. *Journal of Applied Ecology*, **48**, 210–219.

Halfwerk, W. and Slabbekoorn, H. (2009). A behavioural mechanism explaining noise-dependent frequency use in urban birdsong. *Animal Behaviour*, **78**, 1301–1307.

Hall, M. L., Kingma, S. A. and Peters, A. (2013). Male songbird indicates body size with low-pitched advertising songs. *PLoS ONE*, **8**, e56717.

Hamao, S., Watanabe, M. and Mori, Y. (2011). Urban noise and male density affect songs in the great tit Parus major. *Ethology Ecology and Evolution*, **23**, 111–119.

Hanna, D., Blouin-Demers, G., Wilson, D. R. and Mennill, D. J. (2011). Anthropogenic noise affects song structure in red-winged blackbirds (*Agelaius phoeniceus*). *Journal of Experimental Biology*, **214**, 3549–3556.

Hansen, P. (1979). Vocal learning: its role in adapting sound structures to long-distance propagation, and a hypothesis on its evolution. *Animal Behaviour*, **27**, 1270–1271.

Hu, Y. and Cardoso, G. C. (2009). Are bird species that vocalize at higher frequencies preadapted to inhabit noisy urban areas? *Behavioral Ecology*, **20**, 1268–1273.

Hu, Y. and Cardoso, G. C. (2010). Which birds adjust the frequency of vocalizations in urban noise? *Animal Behaviour*, **79**, 863–867.

Kirschel, A. N. G., Blumstein, D. T., Cohen, R. E., Buermann, W., Smith, T. B. and Slabbekoorn, H. (2009). Birdsong tuned to the environment: green hylia song varies with elevation, tree cover, and noise. *Behavioral Ecology*, **20**, 1089–1095.

Knudsen, E. I. and Konishi, M. (1979). Mechanisms of sound localization in the barn owl (*Tyto alba*). *Journal of Comparative Physiology*, **133**, 13–21.

Kroodsma, D. E. and Konishi, M. (1991). A suboscine bird (Eastern Phoebe, *Sayornis phoebe*) develops normal song without auditory-feedback. *Animal Behaviour*, **42**, 477–487.

Lachlan, R. F. and Slater, P. J. B. (1999). The maintenance of vocal learning by gene-culture interaction: the cultural trap hypothesis. *Proceedings of the Royal Society B-Biological Sciences*, **266**, 701–706.

Laiolo, P. (2011). Homogenisation of birdsong along a natural-urban gradient in Argentina. *Ethology Ecology and Evolution*, **23**, 274–287.

Laiolo, P. and Tella, J. L. (2005). Habitat fragmentation affects culture transmission: patterns of song matching in Dupont's Lark. *Journal of Applied Ecology*, **42**, 1183–1193.

Laiolo, P. and Tella, J. L. (2007). Erosion of animal cultures in fragmented landscapes. *Frontiers in Ecology and the Environment*, **5**, 68–72.

Leitner, S., Nicholson, J., Leisler, B., Devoogd, T. J. and Catchpole, C. K. (2002). Song and the song control pathway in the brain can develop independently of exposure to song in the sedge warbler. *Proceedings of the Royal Society B-Biological Sciences*, **269**, 2519–2524.

Lengagne, T., Aubin, T., Lauga, J. and Jouventin, P. (1999). How do king penguins (*Aptenodytes patagonicus*) apply the mathematical theory of information to communicate in windy conditions? *Proceedings of the Royal Society B-Biological Sciences*, **266**, 1623–1628.

Leonard, M. L. and Horn, A. G. (2005). Ambient noise and the design of begging signals. *Proceedings of the Royal Society B-Biological Sciences*, **272**, 651–656.

Leonard, M. L. and Horn, A. G. (2012). Ambient noise increases missed detections in nestling birds. *Biology Letters*, **8**, 530–532.

Lohr, B., Wright, T. F. and Dooling, R. J. (2003). Detection and discrimination of natural calls in masking noise by birds: estimating the active space of a signal. *Animal Behaviour*, **65**, 763–777.

Luther, D. and Baptista, L. (2010). Urban noise and the cultural evolution of bird songs. *Proceedings of the Royal Society B-Biological Sciences*, **277**, 469–473.

Lynch, A. and Baker, A. J. (1993). A population memetics approach to cultural evolution in chaffinch song: meme diversity within populations. *American Naturalist*, **141**, 597–620.

Mahler, B. and Gil, D. (2009). The evolution of song in the *Phylloscopus* leaf warblers (Aves: Sylviidae): a tale of sexual selection, habitat adaptation, and morphological constraints. *Advances in the Study of Behavior*, **40**, 35–66.

Manabe, K., Sadr, E. I. and Dooling, R. J. (1998). Control of vocal intensity in budgerigars (*Melopsittacus undulatus*): differential reinforcement of vocal intensity and the Lombard effect. *Journal of the Acoustical Society of America*, **103**, 1190–1198.

Marler, P. (2004). Bird calls: a cornucopia for communication. In P. Marler and H. Slabbekoorn, eds., *Nature's Music: The Science of Birdsong*. Academic Press/Elsevier, San Diego.

McGregor, P. K., Horn, A. G., Leonard, M. and Thomsen, F. (2014). Anthropogenic noise and conservation. In H. Brumm, ed., *Acoustic Comunication and Noise*. Springer, Heidelberg, in press.

Mclaughlin, K. E. and Kunc, H. P. (2013). Experimentally increased noise levels change spatial and singing behaviour. *Biology Letters*, **9**, 20120771.

Mockford, E. J. and Marshall, R. C. (2009). Effects of urban noise on song and response behaviour in great tits. *Proceedings of the Royal Society B-Biological Sciences*, **276**, 2979–2985.

Mockford, E. J., Marshall, R. C. and Dabelsteen, T. (2011). Degradation of rural and urban great tit song: testing transmission efficiency. *PLoS ONE*, **6**, e28242.

Montgomerie, R. and Weatherhead, P. J. (1997). How robins find worms. *Animal Behaviour*, **54**, 143–151.

Nemeth, E. and Brumm, H. (2009). Blackbirds sing higher-pitched songs in cities: adaptation to habitat acoustics or side-effect of urbanization? *Animal Behaviour*, **78**, 637–641.

Nemeth, E. and Brumm, H. (2010). Birds and anthropogenic noise: are urban songs adaptive? *American Naturalist*, **176**, 465–475.

Nemeth, E., Pieretti, N., Zollinger, S. A., Geberzahn, N., Partecke, J., Miranda, A. C. and Brumm, H. (2013). Bird song and anthropogenic noise: vocal constraints may explain why birds sing higher-frequency songs in cities. *Proceedings of the Royal Society B-Biological Sciences*, **280**, 20122798.

Nemeth, E., Zollinger, S. A. and Brumm, H. (2012). Effect sizes and the integrative understanding of urban bird song (a reply to Slabbekoorn et al.). *American Naturalist*, **180**, 146–152.

Partecke, J., Schwabl, I. and Gwinner, E. (2006). Stress and the city: Urbanization and its effects on the stress physiology in European Blackbirds. *Ecology*, **87**, 1945–1952.

Paton, D., Romero, F., Cuenca, J. and Escudero, J. C. (2012). Tolerance to noise in 91 bird species from 27 urban gardens of Iberian Peninsula. *Landscape and Urban Planning*, **104**, 1–8.

Patricelli, G. L. and Blickley, J. L. (2006). Avian communication in urban noise: Causes and consequences of vocal adjustment. *Auk*, **123**, 639–649.

Peters, S., Derryberry, E. P. and Nowicki, S. (2012). Songbirds learn songs least degraded by environmental transmission. *Biology Letters*, **8**, 736–739.

Pohl, N. U., Leadbeater, E., Slabbekoorn, H., Klump, G. M. and Langemann, U. (2012). Great tits in urban noise benefit from high frequencies in song detection and discrimination. *Animal Behaviour*, **83**, 711–721.

Potvin, D. A. and Parris, K. M. (2012). Song convergence in multiple urban populations of silvereyes (*Zosterops lateralis*). *Ecology and Evolution*, **2**, 1977–1984.

Potvin, D. A., Parris, K. M. and Mulder, R. A. (2013). Limited genetic differentiation between acoustically divergent populations of urban and rural silvereyes (*Zosterops lateralis*). *Evolutionary Ecology*, **27**, 381–391.

Rheindt, F. E. (2003). The impact of roads on birds: Does song frequency play a role in determining susceptibility to noise pollution? *Journal für Ornithologie*, **144**, 295–306.

Rice, W. R. (1982). Acoustical location of prey by the Marsh Hawk: adaptation to concealed prey. *Auk*, **99**, 403–413.

Rios-Chelen, A. A., Quiros-Guerrero, E., Gil, D. and Garcia, C. M. (2013). Dealing with urban noise: vermilion flycatchers sing longer songs in noisier territories. *Behavioral Ecology and Sociobiology*, **67**, 145–152.

Rios-Chelen, A. A., Salaberria, C., Barbosa, I., Garcia, C. M. and Gil, D. (2012). The learning advantage: bird species that learn their song show a tighter adjustment of song to noisy environments than those that do not learn. *Journal of Evolutionary Biology*, **25**, 2171–2180.

Ritschard, M., Riebel, K. and Brumm, H. (2010). Female zebra finches prefer high-amplitude song. *Animal Behaviour*, **79**, 877–883.

Ritschard, M., Van Oers, K., Naguib, M. and Brumm, H. (2012). Song amplitude of rival males modulates the territorial behaviour of great Tits during the fertile period of their mates. *Ethology*, **118**, 197–202.

Ryan, M. J. and Brenowitz, E. A. (1985). The role of body size, phylogeny, and ambient noise in the evolution of bird song. *American Naturalist*, **126**, 87–100.

Salaberria, C. and Gil, D. (2010). Increase in song frequency in response to urban noise in the great tit *Parus major* as shown by data from the Madrid (Spain) city noise map. *Ardeola*, **57**, 3–11.

Saranathan, V., Hamilton, D., Powell, G. V. N., Kroodsma, D. E. and Prum, R. O. (2007). Genetic evidence supports song learning in the three-wattled bellbird *Procnias tricarunculata* (Cotingidae). *Molecular Ecology*, **16**, 3689–3702.

Schroeder, J., Nakagawa, S., Cleasby, I. R. and Burke, T. (2012). Passerine birds breeding under chronic noise experience reduced fitness. *PLoS ONE*, **7**, e39200.

Schuster, S., Zollinger, S. A., Lesku, J. A. and Brumm, H. (2012). On the evolution of noise-dependent vocal plasticity in birds. *Biology Letters*, **8**, 913–916.

Searcy, W. A. (1996). Sound-pressure levels and song preferences in female red-winged blackbirds (*Agelaius phoeniceus*) (Aves, Emberizidae). *Ethology*, **102**, 187–196.

Siemers, B. M. and Schaub, A. (2011). Hunting at the highway: traffic noise reduces foraging efficiency in acoustic predators. *Proceedings of the Royal Society B-Biological Sciences*, **278**, 1646–1652.

Slabbekoorn, H. (2013). Songs of the city: noise-dependent spectral plasticity in the acoustic phenotype of urban birds. *Animal Behaviour*, **85**, 1089–1099.

Slabbekoorn, H. and Den Boer-Visser, A. (2006). Cities change the songs of birds. *Current Biology*, **16**, 2326–2331.

Slabbekoorn, H. and Peet, M. (2003). Birds sing at a higher pitch in urban noise. *Nature*, **424**, 267–267.

Slabbekoorn, H. and Smith, T. B. (2002). Bird song, ecology and speciation. *Philosophical Transactions of the Royal Society of London, Series B-Biological Sciences*, **357**, 493–503.

Slabbekoorn, H., Yang, X. J. and Halfwerk, W. (2012). Birds and anthropogenic noise: singing higher may matter (a comment on Nemeth and Brumm, 'Birds and anthropogenic noise: are urban songs adaptive?'). *American Naturalist*, **180**, 142–145.

Sorci, G., Moller, A. P. and Clobert, J. (1998). Plumage dichromatism of birds predicts introduction success in New Zealand. *Journal of Animal Ecology*, **67**, 263–269.

Tinbergen, N. (1963). On aims and methods of ethology. *Zeitschrift für Tierpsychologie*, **20**, 410–433.

Verzijden, M. N., Ripmeester, E. A. P., Ohms, V. R., Snelderwaard, P. and Slabbekoorn, H. (2010). Immediate spectral flexibility in singing chiffchaffs during experimental exposure to highway noise. *Journal of Experimental Biology*, **213**, 2575–2581.

Waddington, C. H. (1953). Genetic assimilation of an acquired character. *Evolution*, **7**, 118–126.

Wallschläger, D. (1985). Der Einfluβ struktureller und abiotischer ökologischer Faktoren auf den Reviergesang von Passeriformes. *Mitteilungen aus dem Zoologischen Museum in Berlin.*, **61**, Suppl. Ann. Orn. 9 39–69.

Warren, P. S., Katti, M., Ermann, M. and Brazel, A. (2006). Urban bioacoustics: it's not just noise. *Animal Behaviour*, **71**, 491–502.

Wiley, R. H. (2013). Signal detection, noise and the evolution of communication. In H. Brumm, ed., *Acoustic Comunication and Noise*. Springer, Heidelberg, in press.

Wood, W. E. and Yezerinac, S. M. (2006). Song sparrow (*Melospiza melodia*) song varies with urban noise. *Auk*, **123**, 650–659.

Zollinger, S. A., Podos, J., Nemeth, E., Goller, F. and Brumm, H. (2012). On the relationship between, and measurement of, amplitude and frequency in birdsong. *Animal Behaviour*, **84**, E1–E9.

CHAPTER 7

The impact of anthropogenic noise on avian communication and fitness

Wouter Halfwerk and Hans Slabbekoorn

7.1 Introduction

7.1.1 Rising noise levels and acoustic signals

Worldwide noise levels are rising due to the ongoing process of urbanization (Barber et al., 2009). The urban acoustic environment is typically characterized by high levels of low-frequency noise produced by traffic and heavy machinery (Slabbekoorn & Ripmeester, 2008). Furthermore, the expanding transport network of highways, train tracks, shipping lanes and flight ways connecting urban areas is increasingly invading natural habitats, exposing large areas to anthropogenic noise (Barber et al., 2009; Forman, 2000; Hildebrand, 2009; Reijnen & Foppen, 2006).

High noise levels can disturb animals and can interfere with detection, discrimination and recognition of species-specific vocalizations (Brumm & Slabbekoorn, 2005; Klump, 1996). Noise at extreme levels and extended periods of exposure can lead to physical damage, physiological stress and ultimately death (Slabbekoorn et al. 2010; Kight & Swaddle, 2011). However, most animals living in urban areas or in the vicinity of transport lines will be exposed to moderate levels of anthropogenic noise and, therefore, the biggest environmental impact in these areas is expected to come from the masking interference of noise on communication or more subtle noise effects on species distribution or individual fitness (Slabbekoorn & Ripmeester, 2008; Slabbekoorn et al. 2010).

Many animals, including birds, rely heavily on sounds for their survival and reproduction (Bradbury & Vehrencamp, 1998). Birds may pick up on acoustic cues to assess predation risk or to locate prey, but they also produce many different sounds themselves to communicate with one another through begging or alarm calls or advertisement song (Catchpole & Slater, 2008; Collins, 2004). Song may be related to male attractiveness for female partners and may have associated consequences for reproductive success (see e.g. Gil & Gahr, 2002; Hasselquist et al., 1996; Rivera-Gutierrez et al., 2010). Birds also use song to keep neighbouring rivals out of their territories (Krebs et al., 1978), or to attract neighbouring females to gain extra-pair copulations (Poesel et al., 2006). We may therefore expect that masking of important song components by noise will have direct fitness consequences for singing individuals (Slabbekoorn & Ripmeester, 2008).

7.1.2 Noise impact studies on birds

Roadside and urban studies have reported reduced density and diversity in breeding bird communities (Forman, 2000; Reijnen & Foppen, 2006; Tratalos et al., 2007). This impact on bird breeding behaviour has been linked to high traffic noise levels, although most of these studies have not been able to exclude confounding variables associated with roads and cities such as edge effects in vegetation type or variation in traffic collision rate (Halfwerk et al., 2011b; Warren et al., 2006). One way to avoid such problems is to investigate similar environments that vary only in noise conditions and investigate correlations between the spatial distribution

Avian Urban Ecology. Edited by Diego Gil and Henrik Brumm

of anthropogenic noise and bird breeding activity (c.f. Bayne et al., 2008; Francis et al., 2009, 2011a). Another way to gain insight is to experimentally test processes that can reveal detrimental effects of artificially noisy conditions on signals or breeding more directly (e.g. Blickley et al., 2012; Halfwerk et al., 2011a).

In this chapter, we first review the current insights into the impact of anthropogenic noise on species distributions and individual reproductive success. Subsequently, we specifically address the evidence for an impact of anthropogenic noise through interference of communication by spectral and temporal overlap with acoustic signals. We will illustrate aspects on the sender and receiver side in our model species the great tit (*Parus major*), before we address species comparisons and potential for generalization and extrapolation. We will close with some notes on interesting avenues for future studies.

7.2 Noise impact on distribution and reproduction

The effect of anthropogenic noise on the occurrence of bird breeding activity is best studied without other confounding parameters that are typically associated with cities or highways. Several recent studies have investigated gas compressor stations of the energy industry comparing species distributions at quiet and noisy stations that were otherwise identical (Bayne et al., 2008; Francis et al., 2009; Habib et al., 2007). These studies confirmed that anthropogenic noise can reduce the number of breeding individuals of specific species as well as the number of different species (Bayne et al., 2008; Francis et al., 2009).

The compressor noise studies suggested that masking of signals and cues were likely to be an important determinant for the observed patterns of noise-related decline (Francis et al., 2011a). Interestingly, these studies also revealed effects on ecological interactions among species, which sometimes yield unexpected shifts such as some species becoming more abundant under noisy conditions (Francis et al., 2009; Slabbekoorn & Halfwerk 2009). Another approach was taken in an experimental exposure study with drilling and road noise impact on greater sage grouse (*Centrocercus urophasianus*). The long-term playback of continuous and intermittent noise during the breeding season in this study revealed detrimental effects on bird attendance at leks (Blickley et al., 2012).

Absence of birds may indicate a negative effect, but mere presence is no evidence yet for successful breeding. The individual birds or species that persist in noisy areas may not always represent healthy populations or communities. Habib et al. (2007) showed that male ovenbirds (*Seiurus aurocapilla)* at noisy compressor stations were of relatively low quality, as the more experienced, older males occupied the quiet stations. Furthermore, taking both individual and territory quality into account, singing male ovenbirds at noisy sites were significantly less successful in attracting a partner than those at quiet sites. Presumably, either females rated noisy sites as lower-quality territories, or the songs of males at noisy sites attracted fewer females due to masking. Similar findings were reported for reed bunting males (*Emberiza schoeniclus*), which remained more often unpaired at noisy than at more quiet lake shores in Switzerland (Gross et al., 2010).

Evidence for other post-settlement effects of anthropogenic noise on reproductive success through an impact during the laying or nestling stage is still limited. Reduced growth and fledgling success have been suggested as possible outcome of a direct physiological impact on chick development (Kight & Swaddle, 2011), or to an indirect impact through interference with parental behaviour and feeding effort (Halfwerk et al., 2011b; Leonard & Horn 2012). A recent case-study by Schroeder et al. (2012) on house sparrows (*Passer domesticus*) also showed a negative correlation between clutch size, body weight, and recruitment rate and exposure to generator noise close to the nestboxes in which the sparrows were breeding. These results suggest a direct effect of noise on female feeding rate and chick development through masking of parent–offspring communication. More studies are needed with respect to the direct and indirect effects on chicks to understand whether or not masking of begging calls or other aspects of parent-offspring communication are affected critically by noise exposure. More is already known about the masking of other more prominent acoustic signals, which we will address below.

7.3 Noise impact on communication

7.3.1 Overlap between anthropogenic noise and birdsong

The impact of anthropogenic noise on communication and breeding behaviour depends on the overlap in time and space between high noise levels and the species-specific advertisement sounds that are important for territorial defence, mate attraction, and pair bonding (Halfwerk et al., 2011b). Many (sub)urban areas are for instance characterized by a pronounced rise in noise levels caused by a morning rush hour (e.g. Arroyo et al., 2013). Most bird species are also acoustically active in the morning, a pattern often referred to as the dawn chorus (Mace, 1987).

Another important factor is the overlap in frequency range between background noise and acoustic signals (Brumm & Slabbekoorn, 2005). The more overlap, the more interference from masking is expected. Anthropogenic noise is typically biased in spectral energy towards the lower frequencies (Pohl et al., 2012) and several roadside studies, as well as gas compressor studies have shown that species vocalizing at lower frequencies are more affected by noisy areas (Francis et al., 2011a; Goodwin & Shriver, 2011; Proppe et al., 2013; Rheindt, 2003). However, birds have also been shown to sing louder, longer, or higher in noisy areas (Brumm & Slater, 2006; Brumm, 2004; Slabbekoorn & Peet, 2003), presumably to decrease the masking impact of noise (see Chapter 6 for more details). These studies provide additional evidence that noise affects communication of birds, although it remains unclear how signal adjustment is related to reproductive success.

7.3.2 Costs and benefits of noise-dependent song adjustments

A singing bird produces a song that transmits through a particular environment before it is detected by a receiver bird (Bradbury & Vehrencamp, 1998). The production, transmission as well as perception of such acoustic signals all have their own specific limitations that warrant attention when trying to understand the consequences of noise-dependent signal change in terms of cost and benefits.

The production of acoustic characteristics by senders can be constrained by morphological structures or physiological demands (Gil & Gahr, 2002; Podos, 2001). Very low- or very high-frequency songs, for example, can be difficult to produce (Cardoso, 2011; Lambrechts, 1997) and loud songs may require more energy than soft songs (but see Zollinger et al., 2011). The transmission properties of the habitat influence the distance over which individuals can communicate and may affect signal detection (Morton, 1975; Slabbekoorn, 2004; Wiley & Richards, 1982). In most habitats high-frequency sounds are more attenuated compared to low-frequency sounds, favouring selection of low-frequency long-range signals. Finally, receivers are typically perceptually tuned to specific acoustic features that match their signal characteristics while sounds outside the perceptual tuning range are less well perceived (Dooling et al., 2000).

The increased noise levels associated with the urban habitat can mask particular signals at the side of the receiver, consequently forcing senders to vocalize louder or at higher frequencies (Patricelli & Blickley, 2006). However, the benefits of such signal adjustment may come at the costs of, for example, reduced transmission or production efficacy, or negative fitness consequences, such as increased predation risk or reduced male attractiveness to females. Therefore, understanding the impact of communication in noise requires an integrative approach, looking at processes found at the level of senders, receivers and transmission properties of the habitat, and both on a short- and a long-term.

7.4 Insights from great tits

7.4.1 Model species from communication to fitness

The great tit (*Parus major*) is an ideal species for an integrated approach on questions related to anthropogenic noise, acoustic communication, and reproductive success. Great tits can be found in high numbers in relatively quiet forests as well as in urban, noise-impacted areas, such as cities and along highways (Reijnen et al., 1995; Tratalos et al., 2007). Great tits also function as key species in many studies regarding behaviour, including breeding ecology and performance (Kluyver, 1951). Their preference for artificial nest boxes over natural cavities, makes them easier to monitor and measure compared to many other species. Song of the great tit has been related to the acoustic properties of the habitat (Hunter & Krebs, 1979; Slabbekoorn & Peet, 2003) as well as to reproductive success (McGregor et al., 1981; and see Halfwerk et al., 2011b), which allows us to evaluate the impact of noise on communication and the potentially negative consequences for fitness.

The singing behaviour of great tits has been extensively studied and we know that males use their songs during communication with males and females. Great tits have a small repertoire of different song types (typically less than 10), which are delivered with so-called eventual variety: the same song type is sung for several minutes before an individual switches to another song type (see Figure 7.1). Playback experiments have revealed that male singing increases territory tenure (Krebs et al., 1978) and mediates aggressive interactions (Peake et al., 2001). In particular, repertoire use and counter-singing contests can be an important source of information about an individual's condition or status for neighbouring males and have been shown to affect female choice in extra-pair mating (Halfwerk et al., 2011a; Otter et al., 1999; Peake et al., 2001) Also, birds that sing relatively inconsistent or that slow down at the end of a song type bout generally have lower reproductive success (Lambrechts & Dhondt, 1986; Rivera-Gutierrez et al., 2010; 2011).

7.4.2 Male quality rated by female ears

Females typically roost inside the nest box at the start of the breeding season and listen to singing males outside (Halfwerk et al., 2012; Mace, 1987). Females under these conditions have highly accurate perceptual abilities—it has been shown that they can discriminate between shared song types of high similarity from their own mate and those from neighbours (Blumenrath et al., 2007). They may use the variation in acoustic performance of both to make reproductive decisions (Otter et al., 1999). Males typically reside very close to their mates nest cavity prior to sunrise to sing for prolonged periods, during which they are occasionally answered by the female. This, so called dawn chorus ritual, typically continues until the female emerges, after which the pair often copulates (Gorissen & Eens, 2004).

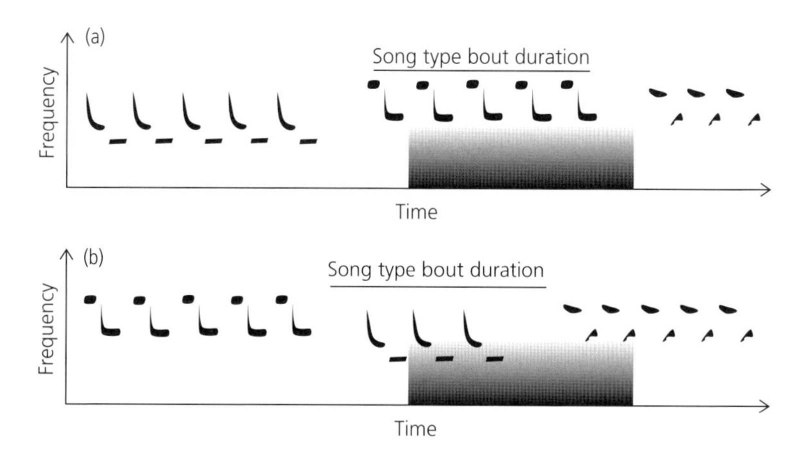

Figure 7.1 Great tit song type switching is masking-dependent. Shown are schematic representations of three different song types from a male's repertoire during noise exposure experiments. Experimental noise exposure affected the duration a male would use a particular song type from its repertoire and this effect depended on the amount of overlap between song and urban noise. (a) Males singing a high-frequency song type prior to noise exposure used this song type for long durations during noise (grey shaded area). (b) Males singing a low-frequency song type prior to exposure used it for much shorter durations and consequently switched to another, on average higher-frequency song type, during the exposure. These data suggest that great tits adjust the duration they use a particular song type, depending on the masking impact of urban noise and can explain why males in noisy areas sing on average higher frequency songs.

The male-female vocal interactions during the dawn chorus ritual peak with female fertility and likely contribute to intra-pair synchronization of breeding efforts. The interactions require the presence of both male and female at a time during which effective extra-pair copulations could take place and may thereby also serve at least to some extent as mutual mate guarding. However, some females escape this mate guarding effect by emerging very early from the nest box under poor ambient light conditions. Although it is not clear when and how intra-pair interactions are resumed, such females were reported to engage in extra-pair copulations more often than later-emerging females as determined by variation in extra-pair paternity rates (Halfwerk et al., 2011a).

The fact that females listen carefully to what their own males sing and possibly also what their neighbours sing increases the potential for acoustically based mate choice decisions. Interestingly, we found out that the males sang more of their lowest song types just a few days prior to the start of egg-laying. Males that used their low-frequency song types less often, and consequently had a relatively high weighted frequency, had a higher chance of ending up with extra-pair chicks in their own nest (Halfwerk et al., 2011a). These are correlative data and we should therefore be cautious about causation, but they suggest that singing low may have signal value as it is associated with male fitness benefits through increased female fidelity.

7.4.3 Noise-dependent singing by male great tits

Great tit song variation has been related to urban noise in several observational studies. Great tits have been found to produce higher songs in cities compared to nearby forests (Mockford & Marshall, 2009; Slabbekoorn & den Boer-Visser, 2006) and within cities they use higher frequency song types in noisy territories compared to quiet ones (Hamao et al., 2011; Slabbekoorn & Peet, 2003). These patterns suggest that great tits possess a signalling strategy that allows them to reduce the masking effect of urban noise. The causal relationship between song frequency use and low-frequency urban noise was investigated experimentally to reveal the underlying behavioural mechanism (Halfwerk & Slabbekoorn, 2009). Males singing a low-frequency song type from their repertoire switch sooner to another song type (by chance higher in frequency) when exposed to low-frequency noise (Figure 7.1). As a consequence, males will sing high-frequency song types for a larger proportion of time (Halfwerk & Slabbekoorn, 2009). Interestingly, a similar response was found when reversing the noise: males tend to switch sooner to low-frequency songs (by chance) when exposed to high-frequency noise when they initially sang a high-frequency song type.

7.4.4 Perceptual benefits for high-frequencies in urban noise

Communication requires a receiver to extract the relevant signal of a sender from the more or less complex background of ambient noise (Brumm & Slabbekoorn, 2005; Dooling et al., 2000). This ability depends heavily on the signal-to-noise ratio at the position of the receiver and its *masked* auditory threshold. These masked thresholds can be calculated using filter bands of particular frequency ranges (known as critical bands), which are species-specific and which have been estimated for the great tit to be around 300 Hz across the entire audible frequency range (Langemann et al., 1998). A recent psycho-acoustic study in the laboratory confirmed that critical bands can be used to estimate the masked threshold, which is around signal-to-noise ratios of 0 dB and showed that high-frequency song types are better detected compared to low-frequency song types under urban noise conditions (Pohl et al., 2012).

The laboratory study showed that the use of high-frequency song types aid signal perception under noisy urban conditions. However, under field conditions a number of additional challenges need to be taken into account such as the impact of sound transmission on song amplitude and signal degradation (Nemeth & Brumm, 2010; Slabbekoorn et al., 2012; Wiley & Richards, 1982). Males communicate with their neighbours over large distances and high-frequency song types may attenuate faster compared to low-frequency song types, negating the masking avoidance. Furthermore, males also communicate with females who are inside nest boxes, and although males can sing in close proximity of females, the complex acoustics of the nest box can potentially have a big impact on sound transmission.

We recorded the male dawn song at song posts used by neighbouring males in a natural woodland area to account for natural variation in song amplitude. At the same time, we played artificially created urban noise from the ground to the recording microphone. We selected several recorded repetitions of the highest and the lowest song type from a focal male's repertoire, both under experimental noise conditions as well as natural background noise control conditions. Using great tit critical bands we could calculate the signal-to-noise ratios of both song types under both noise conditions and found that low-frequency song types have lower signal-to-noise-ratios at the position of rivalling receivers compared to high-frequency song types in noise (Figure 7.2). Differences were in the order of ~5 dB but, as distances between song posts used were relatively short (~70 m), recordings of both song types were still well above the threshold for detection.

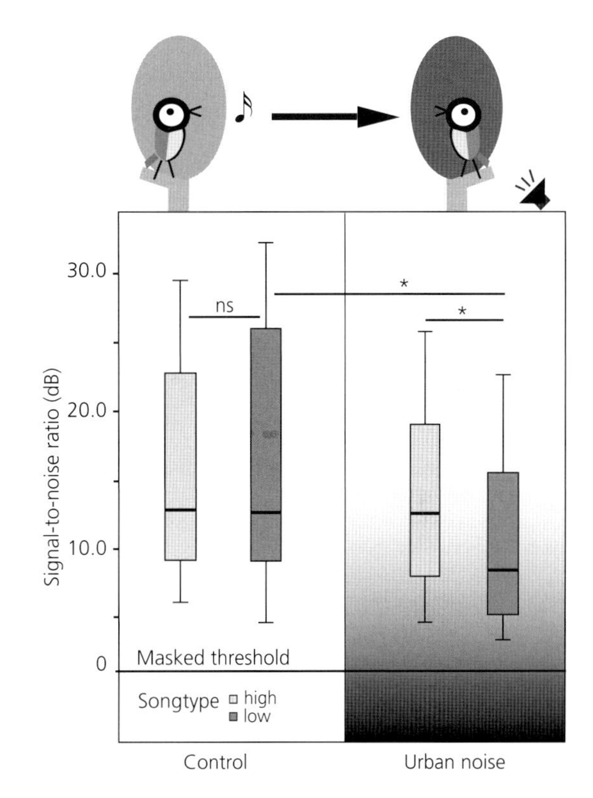

Figure 7.2 Male-male interactions benefit from high song types in urban noise conditions. We recorded songs from a focal male at a known song post of a neighbouring male and at the same time we played urban noise towards the microphone (~65 dB; A-weighted) from a speaker on the ground. We selected several repeats of high- and low-frequency song types from a male's repertoire, during experimental noise exposure (urban noise), as well as during natural background noise (control). We used critical bands of great tits (after Langemann et al. 1998) centred on the peak frequency of the lowest note of the song types to calculate signal-to-noise ratios. Ratios below 0 dB are presumed undetectable by great tits (Pohl et al. 2012), which is indicated by the *masked threshold* line. Boxplots show median signal-to-noise ratios and standard deviations. Urban noise exposure only affects the low-frequency song types (Generalized linear mixed model [GLMM], song type X noise treatment; $N_{individuals} = 6$, $N_{song\ types} = 58$, $\chi^2 = 14.0$, $P = 0.03$; pair-wise comparison $*P < 0.05$), but detection values are still well above the masked threshold line.

7.4.5 Noise-dependent responses by female great tits

We also recorded male song types at the female position, inside the nest box. These recordings were part of a larger study on the impact of noise on females, for which we designed an experimental nest box to control noise level and profile. This set-up allowed us to expose only females to urban noise levels, leaving the singing males outside unexposed (see Figure 7.3). We played the highest and lowest song type from a male's repertoire towards the nest boxes, at amplitude levels mimicking a male singing from 8–16 m distance. Song types were recorded at the position of the female under experimental, as well as natural daytime noise conditions. The experimental noise conditions mimicked inside sound levels of a nest box situated ~50 m from a busy highway (Halfwerk et al., 2011a). We found the signal-to-noise-ratios of the highest frequency song types to be higher than the lowest frequency song types. These differences were mainly caused by frequency-dependent attenuation into the nest box.

Additionally, our recordings from within the nest-box showed that the lowest frequency song types of natural song bouts of males outside the nest-box were dropping to the masked detection threshold of 0 dB under the experimentally noisy urban conditions (Figure 7.4). We also tested the impact of these noise-dependent detection thresholds on female

Figure 7.3 Experimental setup for noise exposure inside nest boxes. (a) artificial nest box, made of wood and a metal plate around the entrance to keep out woodpeckers. (b) experimental nest box to play noise and to record song inside. We extended the normal nest boxes in our study area with a wooden box, made from the same material. Inside the extended box we placed a speaker for noise playback, at ~15 cm away from the position of the female when roosting. The speaker was connected to a mp3-player and battery-pack, hidden under the leave litter. Noise was played at 65–68 dB (A-weighted) at the position of the female, which is similar to levels recorded inside nest boxes situated ~50 m from a busy highway. An automatic recorder was attached to the tree, opposite to the nest box. One microphone recorded sounds outside and another inside the nest box.

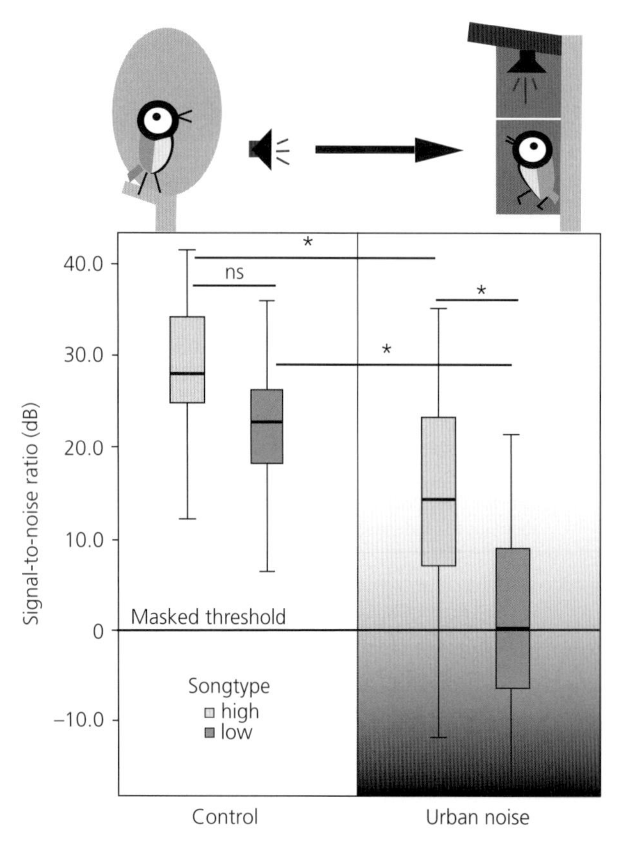

Figure 7.4 Male-female communication benefits from high song types in urban noise conditions. We played high- and low-frequency song types towards a nest box, mimicking a male singing from ~16 m distance, the average distance between song post and nest box in our study area. At the same time, we played noise from a speaker placed in an extended nest box (see Figure 7.3). We analysed signal-to-noise ratios of both song types under experimental (Urban noise) as well as natural background noise conditions (Control). The high song types from a male's repertoire were still detectable, whereas signal-to-noise ratios of the low song types dropped below the critical threshold for detection (GLMM, song type X noise treatment; $N_{\text{individuals}} = 16$, $N_{\text{song types}} = 58$, $\chi^2 = 5.4$, $P = 0.02$; pair-wise comparison $^*P < 0.001$). Overall differences between high-and low song types (GLMM; $\chi^2 = 10.1$, $P = 0.001$) can be ascribed to the complex transmission properties of nest cavities, which for our nest boxes favour high song types over low song types.

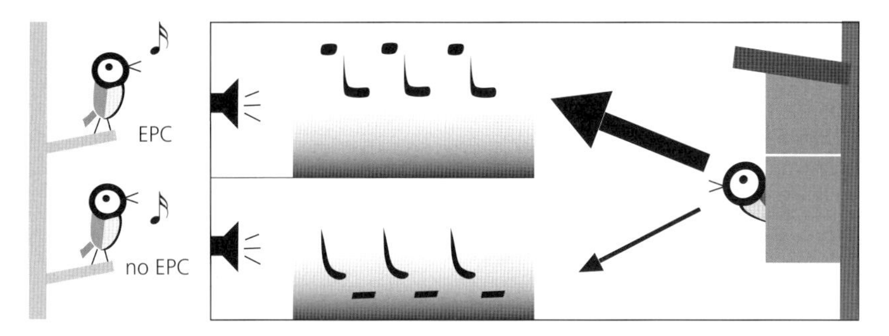

Figure 7.5 Loss of potential attractiveness of low-frequency songs in noise. Song type use during the dawn chorus is related to paternity loss in great tits. Males that are cuckolded by another male sing the high-frequency song types from their repertoire more often during male–female interactions, whereas males that secured within-pair paternity use the low-frequency song types from their repertoire more often. Females experimentally exposed to noise inside their nest box still responded to high-frequency song types, but hardly to low-frequency song types. This may have important fitness consequences, as males living in urban areas may have to choose between detection and attraction.

response in a playback experiment (Figure 7.5). Experiments were carried out during the day and in the incubation phase to avoid aggressive responses of the male to the playbacks (Halfwerk et al., 2011a). We did not find a significant noise-dependent decline in response for the high-frequency song types while the female response to low-frequency song types dropped almost to zero.

7.4.6 Low to be loved or high to be heard

A generally positive picture seems to emerge for the use of high-frequency song types by great tits in noisy urban environments. Higher song frequencies are associated with higher noise levels, under which conditions higher frequencies are also perceptually advantageous. Consequently, noise-dependent production and perception of songs appear to only benefit communication among great tits. However, we still lack measurements of fitness benefits due to singing high under noisy urban conditions. Currently, we only have correlative evidence for a negative impact of traffic noise on reproductive success, suggestive data for an indirect effect of increased predation risk, and a speculative indication for a noise-induced compromise

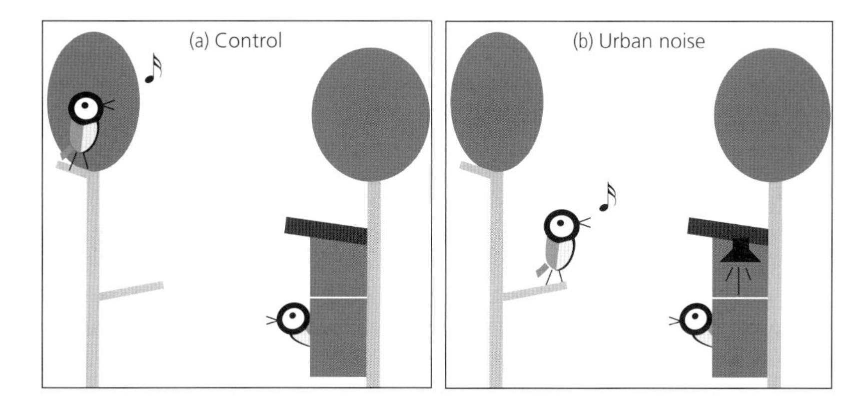

Figure 7.6 Communication trade-offs under noisy conditions. Here we exclusively exposed females to noise during dawn chorus interactions with their mates. Males did not receive direct exposure but responded by singing from closer perches, which restored signal-to-noise ratios inside. However, this benefit came at a cost through singing further away from territory boundaries and from lower perches, which increased transmission distance and signal attenuation to neighbouring males and females. This may be especially important for the low-frequency song types and their potential to attract extra-pair copulations. Furthermore, lower perches are often below the canopy and the spatial adjustment may have lead to increased predation risk.

on signal efficiency, which may restrict male sexual opportunities.

We conducted a correlative study on the impact of highway traffic noise in a nest box population of great tits for which there were long-term data on reproductive success (over a decade; Halfwerk et al., 2011b). Noise levels were negatively correlated to the distance from the cars typically zooming by at high speed, but we also found spatial variation in noise levels independent of the proximity to the highway. This allowed us to demonstrate a negative relationship between noise levels and reproductive success. We analysed laying date, clutch size, number of hatchlings, number of fledglings, and fledging mass (average weight of chicks at the age of 15 days). Traffic noise had a negative effect on reproductive success, with females laying smaller clutches and pairs fledging fewer young in noisier areas. The effect on clutch size was best explained by traffic noise in the frequency band that overlapped most with the lower frequency part of great tit song (Halfwerk et al., 2011b).

As a second link to fitness, anthropogenic noise may affect predation risk and vigilance behaviour of birds (Chan et al., 2010; Quinn, 2006). During experiments with long-term noise exposure inside the nest box, we found a negative effect of noise on female response behaviour to singing males on the first day of full exposure. Nevertheless, these differences with females in control nest boxes disappeared rapidly in the following days because males with females inside the noise-treated nest box were found to sing in closer proximity (Halfwerk et al., 2012). The male approach within those first few days led to higher song amplitudes at the female and compensated for the negative effect of noise exposure on signal detection. In order to achieve this, however, singing males had to get closer and had to move out of the protective canopy layer, making themselves more vulnerable to aerial predators (Figure 7.6).

The third negative link between noise and fitness may be related to the upward shift in frequency, which may induce a compromise with respect to signal value for female choice (Gross et al., 2010; Slabbekoorn & Ripmeester, 2008). Urban males may be faced with a trade-off between singing songs that favour female attraction and mate guarding or singing songs that favour signal detection. As mentioned earlier, and although we do not fully understand why, singing low is associated with fitness benefits for male great tits through female fidelity (Halfwerk et al., 2011a). Therefore, singing high may benefit singers through masking avoidance, but at the same time the songs of these singers may suffer reduced signal value through reduced attractiveness to females. Although we still need to determine what females breeding under noisy conditions prefer, at present it seems that male great tits may have to choose between singing low to be loved or high to be heard.

7.5 Generalization and extrapolation

7.5.1 Translation to other species

The great tit is one of the most abundant species throughout Europe and has been shown to breed readily in cities as well as in the proximity of busy highways (Halfwerk et al, 2011b; Junker-Bornholdt et al., 1998). Clearly, although they may suffer from some decline in reproductive success, great tit populations are not under threat of extinction through a negative impact of noise. However, several issues addressed in this chapter may be valuable to predict a critical impact of noise in other species, whether or not they are locally or globally threatened at present. Consequently, data on great tits provide insights that could be useful to conservation efforts and in taking mitigating measures.

The work on great tits and other species has shown that there are general and more specific acoustic strategies possible in response to noisy conditions. General features are for example amplitude adjustment (Brumm & Zollinger, 2011; Brumm, 2004) and spatial adjustment (Halfwerk et al., 2012; McLaughlin & Kunc, 2013), which seem independent of specific singing styles or song features and are likely to be widespread among taxa. Specific features relate to, for example, the species-specific frequency ranges and species-typical times of singing during the day. These features determine the spectral and temporal overlap with the anthropogenic noise spectra and diurnal cycling in noise amplitude levels respectively (Halfwerk et al., 2011b). The spectral overlap and spectral plasticity data on great tits may yield opportunities for translating to other species to predict species-specific noise impact.

The great tit exposure study with low-frequency city noise and high-frequency inverse city noise (Halfwerk & Slabbekoorn, 2009) provide experimental support for the critical role of spectral overlap in the effect of anthropogenic noise on birds. The detrimental perceptual consequences for spectral overlap were tested and confirmed in the laboratory (Pohl et al, 2012) and in the field (Halfwerk et al., 2011a). These effects present a proof of principle and extrapolation to other species seems reasonable, since comparative studies have shown that species with relatively high-frequency songs demonstrate higher resistance to noisy conditions (Francis et al., 2011a; Goodwin & Shriver, 2011; Hu & Cardoso, 2009; Rheindt, 2003). Comparative data on effects of variation in temporal overlap are much scarcer, but may be of equal importance, in particular for species that depend on a short timeframe to attract mates, such as many migratory songbirds.

7.5.2 Acoustic flexibility may matter

Related to the critical feature of species-specific overlap with noise is the ability to modify song behaviour with fluctuations in noise level. We would expect that those species who are able to adjust their songs in ways that will increase signal detection are more likely to survive in noisy urban environments (Brumm & Slabbekoorn, 2005; Slabbekoorn & Peet, 2003). Our great tit work suggests that the possession of a repertoire with considerable frequency differences among song types can be an important prerequisite to escape masking of noise (Halfwerk & Slabbekoorn, 2009). However, there are now several experimental noise exposure studies on other species that persist in noisy environments (Bermudez-Cuamatzin et al., 2010; Gross et al., 2010; Hanna et al., 2011; Verzijden et al., 2010). These studies show that behavioural flexibility of spectral modification occurs in species with quite a variety of singing styles in which the noise-dependent song type switching as found for great tits is not necessarily the underlying mechanism (reviewed in Slabbekoorn, 2013).

The only direct support of fitness benefits associated with acoustic flexibility comes from a small-scale comparative study, in which two different species of tyrant flycatchers from North-America were shown to differ in their response flexibility to noise, with the least flexible species being more affected in terms of breeding numbers at noisy sites (Francis et al., 2011b). In a similar study on two Australian bird species that were both common along noisy roads, only the one with relatively low-frequency songs exhibited noise-dependent frequency shifts, while the phenomenon was absent and apparently irrelevant for the one with relatively high-frequency songs (Parris & Schneider, 2009). Species may also differ in their flexibility of other acoustic adjustments to noise, such as timing or amplitude. Although most birds are likely to raise song amplitude with overlapping levels of noise, species may differ in magnitude or the energetic costs underlying such increase.

Finally, song features are likely to covary as a result of the shared production mechanism and associated motor limitations (see Montague et al., 2012; Nemeth et al., 2013; Verzijden et al., 2010). Song amplitude and frequency can be coupled, such that songs of particular frequencies can be easier produced at high amplitude (Nemeth et al., 2013). Likewise, song speed and song frequency can correlate as a result of limitations in the filter mechanism (Podos, 1997). Birds may therefore respond to rising noise levels by changing one acoustic parameter, but at the same time alter other parameters in the process. Whether acoustic flexibility ultimately benefits senders therefore depends on the combined effects of all acoustic parameters changed in response to noise.

7.5.3 Future directions and conclusion

Further experimental exposure studies on a greater variety of species should provide a better picture on taxonomic distribution and whether there are specific behavioural features, such as singing style, associated with the capacity of adjustment to noise level (Halfwerk, 2012). Additionally, we need more empirical data that provide a better insight into the magnitude of benefits as well as costs of singing birds that do or do not adjust acoustically to noise (Nemeth et al., 2012; Slabbekoorn et al., 2012). This will require integrating both short-term behavioural assessments of acoustic flexibility and communicative consequences with long-term assessments of associated consequences for reproductive success and overall fitness.

It seems also critical to test birds from urban and non-urban areas to get a better picture of whether the extent of acoustic flexibility varies geographically and whether adjustments can emerge over immediate, ontogenetic, or evolutionary time scales (Patricelli & Blickley, 2006; Ripmeester et al., 2010). Variation among individuals in the amount of juvenile and adult exposure to anthropogenic noise may affect the capacity and tendency to adjust spectrally through selective use of their song type repertoire (Halfwerk & Slabbekoorn, 2009; Slabbekoorn 2013). Urban nest-box populations would also have high potential to get more data on noise-dependent fitness consequences, although long-term noise exposure experiments are required to disentangle the impact of noise from other confounding factors.

In conclusion, we have made a start in getting insight into the impact of anthropogenic noise on acoustic communication and fitness in birds, but many interesting and important questions remain. It is also unlikely that the world will get quieter soon as urbanization is expected to continue in the coming decades. More and more traffic will fuel a continuation of the global rise and spread in anthropogenic noise levels. The need to understand the nature and magnitude of potential problems for birds as well as other animals will therefore only grow in the near future. Against this background of human-altered soundscapes, conservation value and potential for practical applications, we hope that more studies will follow that address fundamental questions. We believe their importance goes beyond the value for scientific progress as they also increase awareness, and we hope that more insight may make that people enjoy and respect natural ambient noise conditions more often.

References

Arroyo-Solís, A., Castillo, J. M., Figueroa, E., López-Sánchez, J. L., and Slabbekoorn, H. (2013). Experimental evidence for an impact of anthropogenic noise on dawn chorus timing in urban birds. *Journal of Avian Biology*, **44**, 288–296.

Barber, J. R., Crooks, K. R., and Fristrup, K. M. (2009). The costs of chronic noise exposure for terrestrial organisms. *Trends in Ecology & Evolution*, **25**, 180–189.

Bayne, E. M., Habib, L., and Boutin, S. (2008). Impacts of chronic anthropogenic noise from energy-sector activity on abundance of songbirds in the boreal forest. *Conservation Biology*, **22**, 1186–1193.

Bermudez-Cuamatzin, E., Rios-Chelen, A.A., Gil, D., and Garcia, C.M. (2010). Experimental evidence for real-time song frequency shift in response to urban noise in a passerine bird. *Biology Letters*, **7**, 36–38.

Blickley, J. L., Blackwood, D. and Patricelli, G. L. (2012). Experimental evidence for the effects of chronic anthropogenic noise on abundance of greater sage-grouse at leks. *Conservation Biology*, 26, 461e471.

Blumenrath, S. H., Dabelsteen, T., and Pedersen, S. B. (2007). Vocal neighbour-mate discrimination in female great tits despite high song similarity. *Animal Behaviour*, **73**, 789–796.

Bradbury, J. W. and Vehrencamp, S. L. (1998). *Principles of Animal Communication*. Sinauer Associates, Sunderland, MA.

Brumm, H. (2004). The impact of environmental noise on song amplitude in a territorial bird. *Journal of Animal Ecology*, **73**, 434–440.

Brumm, H. and Slabbekoorn, H. (2005). Acoustic communication in noise. *Advances in the Study of Behavior*, **35**, 151–209.

Brumm, H. and Slater, P. J. B. (2006). Ambient noise, motor fatigue, and serial redundancy in chaffinch song. *Behavioral Ecology and Sociobiology*, **60**, 475–481.

Brumm, H. and Zollinger, S. A. (2011). The evolution of the Lombard effect: 100 years of psychoacoustic research. *Behaviour*, **148**, 1173–1198.

Cardoso, G.C. (2011). Paradoxical calls: the opposite signaling role of sound frequency across bird species. *Behavioural Ecology*, **23**, 237–241.

Catchpole, C. K. and Slater, P. J. B. (2008). *Bird Song: Biological themes and Variations*. Cambridge University Press, Cambridge.

Chan, A. A. Y. H., Giraldo-Perez, P., Smith, S., and Blumstein, D. T. (2010). Anthropogenic noise affects risk assessment and attention: the distracted prey hypothesis. *Biology Letters*, **6**, 458–461.

Collins, S. (2004). Vocal fighting and flirting: the functions of birdsong. In P. Marler and H. Slabbekoorn, eds., *Nature's Music: the Science of Birdsong*, pp. 39–79. Elsevier Academic Press, San Diego.

Dooling, R. J., Lohr, B., and Dent, M. L. (2000). Hearing in birds and reptiles. In R. J. Dooling, R. R. Fay, and A. N. Popper, eds, *Comparative Hearing in Birds and Reptiles*. Springer, Berlin.

Forman, R. T. T. (2000). Estimate of the area affected ecologically by the road system in the United States. *Conservation Biology*, **14**, 31–35.

Francis, C. D., Ortega, C. P., and Cruz, A. (2009). Cumulative consequences of noise pollution: noise changes avian communities and species interactions. *Current Biology*, **19**, 1415–1419.

Francis, C. D., Ortega, C. P., and Cruz, A. (2011a). Noise pollution filters bird communities based on vocal frequency. *PLoS ONE*, **6**, e27052.

Francis, C. D., Ortega, C. P., and Cruz, A. (2011b) Vocal frequency change reflects different responses to anthropogenic noise in two suboscine tyrant flycatchers. *Proceedings of the Royal Society B-Biological Sciences*, **278**, 2025–2031.

Gil, D. and Gahr, M. (2002), The honesty of bird song: multiple constraints for multiple traits. *Trends in Ecology and Evolution*, **17**, 133–141.

Goodwin, S. E. and Shriver, W. G. (2011). Effects of traffic noise on occupancy patterns of forest birds. *Conservation Biology*, **25**, 406–411.

Gorissen, L. and Eens, M. (2004). Interactive communication between male and female Great Tits (*Parus major*) during the dawn chorus. *Auk*, **121**, 184–191.

Gross, K., Pasinelli, G., and Kunc, H.P. (2010). Behavioral plasticity allows short-term adjustment to a novel environment. *American Naturalist*, **176**, 456–464.

Habib, L., Bayne, E. M., and Boutin, S. (2007). Chronic industrial noise affects pairing success and age structure of ovenbirds *Seiurus aurocapilla*. *Journal of Applied Ecology*, **44**, 176–184.

Halfwerk, W. (2012). Tango to Traffic: A field study into consequences of noisy urban conditions for acoustic courtship interactions in birds. PhD thesis, Leiden University, The Netherlands.

Halfwerk, W., Bot, S., and Slabbekoorn, H. (2012). Male great tit song perch selection in response to noise-dependent female feedback. *Functional Ecology*, **26**, 1339–1347.

Halfwerk, W., Bot, S., Buikx, J., van der Velde, M., Komdeur, J., ten Cate, C., and Slabbekoorn, H. (2011a). Low songs lose potency in urban noise conditions. *Proceedings of the National Academy of Sciences of the United States of America*, **108**, 14549–14554.

Halfwerk, W., Holleman, L. J. M., Lessells, C. M., and Slabbekoorn, H. (2011b). Negative impact of traffic noise on avian reproductive success. *Journal of Applied Ecology*, **48**, 210–219.

Halfwerk, W. and Slabbekoorn, H. (2009). A behavioural mechanism explaining noise-dependent frequency use in urban birdsong. *Animal Behaviour*, **78**, 1301–1307.

Hamao, S., Watanabe, M., and Mori, Y. (2011). Urban noise and male density affect songs in the great tit *Parus major*. *Ethology Ecology & Evolution*, **23**, 111–119.

Hanna, D., Blouin-Demers, G., Wilson, D. R., and Mennill, D. J. (2011). Anthropogenic noise affects song structure in red-winged blackbirds (*Agelaius phoeniceus*). *Journal of Experimental Biology*, **214**, 3549e3556.

Hasselquist, D., Bensch, S., and vonSchantz, T. (1996). Correlation between male song repertoire, extra-pair paternity and offspring survival in the great reed warbler. *Nature*, **381**, 229–232.

Hildebrand, J. A. (2009). Anthropogenic and natural sources of ambient noise in the ocean. *Marine Ecology-Progress Series*, **395**, 5–20.

Hu, Y. and Cardoso, G. C. (2009). Which birds adjust the frequency of vocalizations in urban noise? *Animal Behaviour*, **79**, 863–867.

Hunter, M. L. and Krebs, J. R. (1979). Geographic variation in the song of the great tit (*Parus major*) in relation to ecological factors. *Journal of Animal Ecology*, **48**, 759–785.

Junker-Bornholdt, R., Wagner, M., and Zimmerman M., et al. (1998). The impact of a motorway in construction and after opening to traffic on the breeding biology of great tit (*Parus major*) and blue tit (*P. caeruleus*). *Journal of Ornithology*, **139**, 131–139.

Kight, C. R. and Swaddle, J. P. (2011). How and why environmental noise impacts animals: an integrative, mechanistic review. *Ecology Letters* **14**, 1052–1061.

Klump, G. M. (1996). Bird communication in the noisy world. In D. E. Kroodsma and E. H. Miller, eds., *Ecology and Evolution of Acoustic Communication in Birds*, pp. 321–338. Cornell University Press, New York.

Kluyver, H. N. (1951). The population ecology of the great tit, *Parus major* L. *Ardea*, **39**, 1–135.

Krebs, J., Ashcroft, R., and Webber, M. (1978). Song repertories and territory defence in great tit. *Nature*, **271**, 539–542.

Lambrechts, M. and Dhondt, A. A. (1986) Male quality, reproduction, and survival in the great tit (*Parus major*). *Behavioral Ecology and Sociobiology*, **19**, 57–63.

Lambrechts, M. M. (1997) Song frequency plasticity and composition of phrase versions in Great Tits *Parus major*. *Ardea*, **85**, 99–109.

Langemann, U., Gauger, B., and Klump, G. M. (1998). Auditory sensitivity in the great tit: perception of signals in the presence and absence of noise. *Animal Behaviour*, **56**, 763–769.

Leonard, M. L. and A. G. Horn. (2012). Ambient noise increases missed detections in nestling birds. *Biology Letters*, **8**, 530–532.

Mace, R. (1987). The dawn chorus in the great tit *Parus major* is directly related to female fertility. *Nature*, **330**, 745–746.

McGregor, P. K., Krebs, J. R., and Perrins, C. M. (1981). song repertoires and lifetime reproductive success in the great tit (*Parus major*). *American Naturalist*, **118**, 149–159.

McLaughlin, K. and Kunc, H. P. (2013). Experimentally increased noise levels change spatial and singing behaviour. *Biology Letters*, e20120771.

Mockford, E. J. and Marshall, R. C. (2009). Effects of urban noise on song and response behaviour in great tits.

Proceedings of the Royal Society B-Biological Sciences, **276**, 2979–2985.

Montague, M. J., Danek-Gontard, M., and Kunc, H. P. (2012). Phenotypic plasticity affects the response of a sexually selected trait to anthropogenic noise. *Behavioral Ecology*, doi.org/10/1093/beheco/ars169.

Morton, E. S. (1975). Ecological sources of selection on avian sounds. *American Naturalist*, **109**, 17–34.

Nemeth, E. and Brumm, H. (2010). Birds and anthropogenic noise: Are urban songs adaptive? *American Naturalist*, **176**, 465–475.

Nemeth, E., Pieretti, N., Zollinger, S. A., Geberzahn, N., Partecke, J., Miranda, A. C., and Brumm, H. (2013). Bird song and anthropogenic noise: vocal constraints may explain why birds sing higher-frequency songs in cities. *Proceedings of the Royal Society B Biological Sciences*, **280**, e20122798.

Nemeth, E., Zollinger, S. A., and Brumm, H. (2012). Effect sizes and the integrative understanding of urban bird song. *American Naturalist*, **180**, 146–152.

Otter, K., McGregor, P. K., Terry, A. M. R., Burford, F. R. L., Peake, T. M., and Dabelsteen, T. (1999). Do female great tits (*Parus major*) assess males by eavesdropping? A field study using interactive song playback. *Proceedings of the Royal Society of London Series B-Biological Sciences*, **266**, 1305–1309.

Parris, K. M. and Schneider, A. (2009). Impacts of traffic noise and traffic volume on birds of roadside habitats. *Ecology and Society*, **14**, e29.

Patricelli, G. L. and Blickley, J. L. (2006). Avian communication in urban noise: Causes and consequences of vocal adjustment. *Auk*, **123**, 639–649.

Peake, T. M., Terry, A. M. R., McGregor, P. K., and Dabelsteen, T. (2001). Male great tits eavesdrop on simulated male-to-male vocal interactions. *Proceedings of the Royal Society of London Series B-Biological Sciences*, **268**, 1183–1187.

Podos, J. (1997). A performance constraint on the evolution of trilled vocalizations in a songbird family (Passeriformes:Emberizidae). *Evolution*, **51**, 537–551.

Podos, J. (2001). Correlated evolution of morphology and vocal signal structure in Darwin's finches. *Nature*, **409**, 185–188.

Poesel, A., Kunc, H. P., Foerster, K., Johnsen, A., and Kempenaers, B. (2006). Early birds are sexy: male age, dawn song and extrapair paternity in blue tits, *Cyanistes* (formerly *Parus*) *caeruleus*. *Animal Behaviour*, **72**, 531–538.

Pohl, N. U., Leadbeater, E., Slabbekoorn, H., Klump, G. M., and Langemann, U. (2012). Great tits in urban noise benefit from high frequencies in song detection and discrimination. *Animal Behaviour*, **83**, 711–721.

Proppe, D. S., Sturdy, C. B., and Cassady StClair, C. (2013). Anthropogenic noise decreases urban songbird diversity and may contribute to homogenization. *Global Change Biology*, **19**, 1075–1084.

Quinn, J. L., Whittingham, M. J., Butler, S. J., and Cresswell, W. (2006). Noise, predation risk compensation and vigilance in the chaffinch *Fringilla coelebs*. *Journal of Avian Biology*, **37**, 601–608.

Reijnen, R. and Foppen, R. (2006). Impact of road traffic on breeding bird populations. In J. Davenport and J. L. Davenport, eds, *The Ecology of Transportation: Managing Mobility for the Environment*, pp. 255–274. Springer-Verlag, Heidelberg.

Reijnen, R., Foppen, R., Ter Braak, C., and Thissen, J. (1995). The effects of car traffic on breeding bird populations in woodland. III. Reduction of density in relation to the proximity of main roads. *Journal of Applied Ecology*, **32**, 187–202.

Rheindt, F. E. (2003). The impact of roads on birds: Does song frequency play a role in determining susceptibility to noise pollution? *Journal Fur Ornithologie*, **144**, 295–306.

Ripmeester, E. A. P., Kok, J., van Rijssel, J., and Slabbekoorn, H. (2010). Habitat-related birdsong divergence: a multi-level study on the influence of territory density and ambient noise. *Behavioral Ecology and Sociobiology*, **64**, 409–418.

Rivera-Gutierrez, H. F., Pinxten, R., and Eens, M. (2010). Multiple signals for multiple messages: great tit, *Parus major*, song signals age and survival. *Animal Behaviour*, **80**, 451–459.

Rivera-Gutierrez, H. F., Pinxten, R., and Eens, M. (2011). Songs differing in consistency elicit differential aggressive response in territorial birds. *Biology Letters*, **7**, 339–342.

Schroeder, J., Nakagawa, S., Cleasby, I. R., and Burke, T. (2012). Passerine birds reeding under chronic noise experience reduced fitness. *PLoS One*, **7**, e39200.

Slabbekoorn, H. (2004). Singing in the wild: the ecology of birdsong. In P. Marler and H. Slabbekoorn, eds, *Nature's Music: the science of birdsong*. pp. 178–205. Elsevier-Academic Press, San Diego, CA.

Slabbekoorn, H. and den Boer-Visser, A. (2006). Cities change the songs of birds. *Current Biology*, **16**, 2326–2331.

Slabbekoorn, H. and Halfwerk, W. (2009). Behavioural ecology: noise annoys at community level. *Current Biology*, **19**, R693–R695.

Slabbekoorn, H. and Peet, M. (2003). Ecology: birds sing at a higher pitch in urban noise. *Nature*, **424**, 267–267.

Slabbekoorn, H. and Ripmeester, E. A. P. (2008). Birdsong and anthropogenic noise: implications and applications for conservation. *Molecular Ecology*, **17**, 72–83.

Slabbekoorn, H., Bouton, N., van Opzeeland, I., Coers, A., ten Cate, C., and Popper, A. N. (2010). A noisy spring:

the impact of globally rising underwater sound levels on fish. *Trends in Ecology & Evolution*, **25**, 419–427.

Slabbekoorn, H., Yang, X.-J., and Halfwerk, W. (2012). Birds and anthropogenic noise; Singing higher may matter. *American Naturalist*, **180**, 142–145.

Slabbekoorn, H. 2013. Songs of the city: noise-dependent spectral plasticity in the acoustic phenotype of urban birds. *Animal Behaviour*, **85**, 1089–1099.

Tratalos, J., Fuller, R. A., Evans, K. L., Davies, R. G., Newson, S. E., Greenwood, J. J. D., and Gaston, K. J. (2007). Bird densities are associated with household densities. *Global Change Biology*, **13**, 1685–1695.

Verzijden, M. N., Ripmeester, E. A. P., Ohms, V. R., Snelderwaard, P., and Slabbekoorn, H. (2010). Immediate spectral flexibility in singing chiffchaffs during experimental exposure to highway noise. *Journal of Experimental Biology*, **213**, 2575–2581.

Warren, P. S., Katti, M., Ermann, M., and Brazel, A. (2006). Urban bioacoustics: it's not just noise. *Animal Behaviour*, **71**, 491–502.

Wiley, R. H., and Richards, D.G. (1982). Adaptations for acoustic communication in birds: sound transmission and signal detection. In D. E. Kroodsma and E. H. Miller, eds, *Acoustic Communication in Birds*, Vol. I, pp. 131–181. Academic Press, New York.

Zollinger, S. A., Goller, F., and Brumm, H. (2011). Metabolic and respiratory costs of increasing song amplitude in zebra finches. *PLoS ONE* **6**, e23198.

CHAPTER 8

Reproductive phenology of urban birds: environmental cues and mechanisms

Pierre Deviche and Scott Davies

8.1 Introduction

The Earth's land surfaces are being converted to urban spaces at an ever increasing rate, irreversibly altering habitats, local climates, and biodiversity (Seto et al., 2011). One of the clearest and most consistent patterns emerging from studies of the effects of urbanization on birds is that it is associated with advancement in the timing of seasonal reproduction (reproductive phenology). Indeed, a recent meta-analysis showed that many urban birds lay significantly earlier than non-urban birds (Chamberlain et al., 2009). To further demonstrate this conclusion, we performed a literature search using the methods of Chamberlain et al., (2009) and, in addition to including investigations of lay date, expanded the search to encompass studies of reproductive physiology and morphology (Table 8.1). We identified 27 studies that included a paired, intraspecific comparison of bird species inhabiting urban and non-urban areas. The analysis of these studies revealed that the pattern of advanced lay date in urban birds is mirrored in the phenology of reproductive morphology. For instance, reproductive phenology was advanced in 20 of 26 comparisons and in 16 of 21 studies using a significance test (Figure 8.1). Furthermore, advanced reproductive phenology was reported in 13 of 18 species; three species did not differ and only two species, Acadian flycatcher, *Empidonax virescens*, and European starling, *Sturnus vulgaris*, showed differences in the opposite direction. It is notable that these studies come from just three ecoregions: temperate broadleaf and mixed forest, temperate conifer forest, and desert and xeric shrubland. Moreover, most (23 of 27) studies were performed in cities located in the first of these ecoregions, with only two studies each coming from the temperate coniferous forest and from desert and xeric shrubland ecoregions. Whether findings from these regions can be generalized to other regions remains, therefore, unknown.

A number of environmental factors including artificial lighting, ambient temperature, and food abundance and/or type are hypothesized to cause the above disparity in reproductive phenology. However, experimental tests of these hypotheses are, for the most part, lacking and the underlying cause of the observed trend is poorly understood. In this chapter, we aim to (a) describe how urbanization modifies three environmental factors (light, temperature, and food availability), (b) analyse the role of these modifications in alterations of avian reproductive phenology, and (c) discuss mechanisms that underlie these modifications.

8.2 Control and timing of reproduction in birds

One of the most critical decisions for seasonally breeding birds is when to initiate reproduction in a given year. Timing seasonal reproduction to occur when conditions are optimal is crucial to maximize reproductive success (Both et al., 2006). Most

Avian Urban Ecology. Edited by Diego Gil and Henrik Brumm

Table 8.1 Studies comparing reproductive phenology of birds inhabiting urban and non-urban environments. Studies are sorted by the parameter measured, so a given study may appear in more than one category.

	Species	City	Ecoregion	Parameter	Reproductive phenology (urban relative to non-urban)	Sex
Lay date	Black-billed Magpie (*Pica pica*)[1]	Sheffield, UK	Temperate broadleaf and mixed forest	Lay date	Advanced	M & F
	Black-billed Magpie (*Pica pica*)[2]	Sofia, Bulgaria	Temperate broadleaf and mixed forest	Lay date	Advanced	M & F
	Blue tit (*Cyanistes caeruleus*)[3]	Oxford, UK	Temperate broadleaf and mixed forest	Lay date	Advanced	M & F
	Blue tit (*Cyanistes caeruleus*)[4]	Ghent, Belgium	Temperate broadleaf and mixed forest	Lay date	Lay date earliest in suburban habitat, latest in one rural location. Urban and remaining rural habitats intermediate.	M & F
	Eurasian blackbird (*Turdus merula*)[5]		Temperate broadleaf and mixed forest	Lay date	Advanced	
	Eurasian starling (*Sturnus vulgaris*)[6]	UK (national network of nest record cards)	Temperate broadleaf and mixed forest	Lay date	Advanced	M & F
	Great tit (*Parus major*)[3]	Oxford, UK	Temperate broadleaf and mixed forest	Lay date	Advanced	M & F
	Great tit (*Parus major*)[7]	Frankfurt, Germany	Temperate broadleaf and mixed forest	Lay date	Advanced	M & F
	Great tit (*Parus major*)[4]	Ghent, Belgium	Temperate broadleaf and mixed forest	Lay date	Advanced (gradient of lay date from urban to suburban to rural)	M & F
	Great tit (*Parus major*)[8]	Cardiff, UK (Urban); Wytham Woods, Oxfordshire, UK (rural)	Temperate broadleaf and mixed forest	Lay date	Advanced	M & F
	House sparrow (*Passer domesticus*)[6]	UK (national network of nest record cards)	Temperate broadleaf and mixed forest	Lay date	Advanced	M & F
	Florida scrub-jay (*Aphelocoma coerulescens*)[9,10]	Lake Placid, Florida USA	Temperate coniferous forest	Lay date	Lay date advanced in suburban population.	F
	Blue tit (*Cyanistes caeruleus*)[8]	Cardiff, UK (Urban); Wytham Woods, Oxfordshire, UK (rural)	Temperate broadleaf and mixed forest	Lay date	No difference	M & F
	House wren (*Troglodytes aedon*)[11]	Washington, DC–Baltimore, Maryland metropolitan area, USA	Temperate broadleaf and mixed forest	Lay date	No difference	M & F
	Acadian flycatcher (*Empidonax virescens*)[12,13]	Columbus, Ohio USA	Temperate broadleaf and mixed forest	Lay date; last nest initiation of breeding season	Lay date delayed; termination of breeding activity advanced	M & F
	Eurasian starling (*Sturnus vulgaris*)[14]	Rennes, France	Temperate broadleaf and mixed forest	Lay date	Delayed	M & F
	House sparrow (*Passer domesticus*)[15]		Temperate broadleaf and mixed forest	Lay date	Delayed	M & F

continued

Table 8.1 *Continued*

	Species	City	Ecoregion	Parameter	Reproductive phenology (urban relative to non-urban)	Sex
Reproductive endocrinology	Eurasian blackbird (*Turdus merula*)[16]	Munich, Germany	Temperate broadleaf and mixed forest	Plasma LH, T, and E2	*Males*: lower plasma LH during recrudescence and regression; lower plasma T during testicular development, similar plasma T during regression. *Females*: similar plasma LH during follicular recrudescence and regression; plasma E2 similar during follicular recrudescence and regression.	M & F
	Florida scrub-jay (*Aphelocoma coerulescens*)[9,10]	Lake Placid, Florida USA	Temperate coniferous forest	Plasma LH and estradiol	No difference in plasma LH or estradiol when controlling for days before laying.	F
Gonadal development	Abert's towhee (*Melozone aberti*)[17]	Phoenix, Arizona USA	Desert and xeric shrublands	Testicular development	Advanced	M
	Curve-billed thrasher (*Toxostoma curvirostre*)[18]	Phoenix, Arizona USA	Desert and xeric shrublands	Testicular development	Advanced	M
	Eurasian blackbird (*Turdus merula*)[16]	Munich, Germany	Temperate broadleaf and mixed forest	Follicular and testicular development	Advanced follicular and testicular development; similar follicular and testicular regression.	M & F
Nest building, clutch incubation, and hatching	Australian magpie (*Gymnorhina tibicen*)[19]	Brisbane, Australia	Temperate broadleaf and mixed forest	Initiation of nest building and clutch incubation	Advanced	M & F
	Carrion crow (*Corvus corone*)[20]	Lausanne, Switzerland	Temperate broadleaf and mixed forest	Hatching date	No difference	M & F
	Eurasian blackbird (*Turdus merula*)[21]	Czech Republic and Slovakia (national nestling ringing datasets)	Temperate broadleaf and mixed forest	Breeding date	Advanced	M & F
	Northern mockingbird (*Mimus polyglottos*)[22]	Gainesville, Florida USA	Temperate coniferous forest	Start of nesting behaviour and first clutch completion date	Advanced	M & F

continued

Table 8.1 *Continued*

Species	City	Ecoregion	Parameter	Reproductive phenology (urban relative to non-urban)	Sex
Song thrush (*Turdus philomelos*) [21]	Czech Republic and Slovakia (national nestling ringing datasets)	Temperate broadleaf and mixed forest	Breeding date	Advanced	M & F
White-winged chough (*Corcorax melanorhamphos*)[23]	Canberra, Australia	Temperate broadleaf and mixed forest	First nest attempt	Advanced	M & F

[1]Eden, S.F. 1985. *J. Zool.* **205**, 305–334; [2]Antonov, A. and Atanasova, D. 2003. *Ornis Fenn.* **80**, 21–30; [3]Perrins, C. M. 1965. *J. Anim. Ecol.* **34**, 601–647; [4]Dhondt, A.A. et al. 1984. *Ibis* **126**, 388–397; [5]Lack, D. 1968. *Population Studies of Birds.* Clarendon Press; [6]Crick, H.Q.P. et al. 2002. *Investigation into the Causes of the Decline of Starlings and House Sparrows in Great Britain.* British Trust for Ornithology; [7]Schmidt, K.-H. and Steinbach, J. 1983. *J. Orn.* **124**, 81–83; [8]Cowie, R. J. and Hinsley, S. A. 1987. *Ardea* **75**, 81–90; [9]Schoech, S.J. and Bowman, R. 2001. *Avian Ecology and Conservation in an Urbanizing World.* Kluwer, New York; [10]Schoech, S.J. and Bowman, R. 2003. *Auk* **120**, 1114–1127; [11]Newhouse, M.J. et al. 2008. *Wilson J. Ornithol.* **120**, 99–104; [12]Rodewald, A.D. and Shustack, D.P. 2008. *J. Anim. Ecol.* **77**, 83–91; [13]Shustack, D.P. and Rodewald, A.D. 2010. Auk. **127**, 421–429; [14]Mennechez, G. and Clergeau, P. 2006. *Acta Oecol.* **30**, 182–191; [15]Summers-Smith, J.D. 1963. *The House Sparrow.* Collins New Naturalist; [16]Partecke *et al.* 2005. *J. Avian Biol.* **36**, 295–305; [17]Davies and Deviche, unpublished data; [18]Deviche, P. et al. 2012 *Avian Testicular Structure, Function, and Regulation.* Academic Press; [19]Rollinson, D.J. and Jones, D.N. 2002. *Urban Ecosyst.* **6**, 257–269; [20]Richner, H. 1989. *J. Anim. Ecol.* **58**, 427–440; [21]Najmanova, L. and Adamik, P. 2009. *Bird Study* **56**, 349–356; [22]Stracey, C.M. and Robinson, S.K. 2012. *J. Avian Biol.* **43**, 50–60; [23]Beck, N.R. and Heinsohn, R. 2006. *Austral Ecol.* **31**, 588–596.

Ecoregions from Olson et al. (2001) Terrestrial ecoregions of the world: a new map of life on earth. *BioScience* 51, 933–938.

bird species undergo cycles of breeding activity and inactivity during which the reproductive system is regressed outside the breeding period and must develop before the start of the next breeding period. The physiological and morphological modifications associated with transition from reproductive inactivity to activity require considerable time, so birds must anticipate the start and end of optimal environmental conditions. To do this, they track environmental cues that predict conditions favourable for reproduction. As the annual change in day length is constant between years and reliably predicts favourable conditions, most birds use day length as the principal initiator of reproductive development. In most species, exposure to increasing day length in the spring stimulates the hypothalamic production and secretion of gonadotropin-releasing hormone (GnRH-I), the main neuropeptide that controls the secretion of anterior pituitary gland gonadotropins (luteinizing hormone, LH, and follicle-stimulating hormone, FSH; Dawson et al., 2001; Deviche et al., 2011). Increased secretion of these hormones into the general circulation stimulates gonadal development, resulting in gametogenesis and elevated circulating levels of gonadal hormones that, in turn, increase the probability of birds expressing reproductive behaviours such as singing, courtship, and nesting. The rate of gonadal development in response to long days increases, up to a maximum, as a function of the photoperiod duration (Follett & Maung, 1978). In many species, prolonged exposure to long days causes inactivation of the hypothalamo-pituitary-gonadal (HPG) axis, resulting in gonadal regression and loss of reproductive behaviour (Section 8.3.2) as normally observed at the end of a breeding cycle.

Naturally increasing vernal day length creates a temporal window of opportunity for seasonal reproduction. However, environmental conditions can vary markedly between years, so birds also use supplementary environmental cues, such as ambient temperature (Schaper et al., 2012), rainfall (Small et al., 2007), and food abundance and/or type (i.e. changes in availability of specific food sources; see later), to enhance or attenuate the effects of photoperiod and thereby fine-tune the precise timing of reproduction within this window. For example, clement conditions, such as favourable temperatures or high food abundance, can advance the initiation of reproduction and potentially delay the termination of reproductive activity (Nussey et al., 2005), thereby increasing the length of the

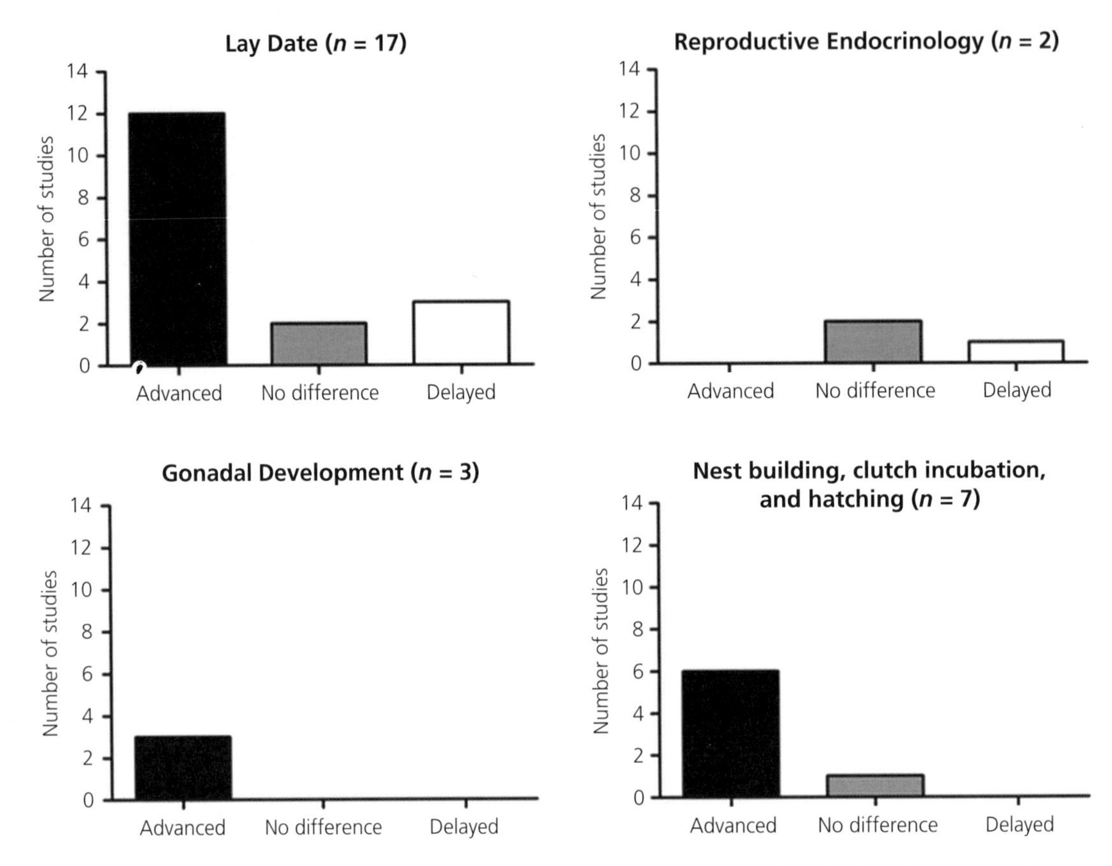

Figure 8.1 Summary of the number of paired, intraspecific studies comparing parameters associated with breeding phenology of bird populations inhabiting urban and non-urban areas. The difference between habitats is expressed as the change in urban population, relative to the corresponding non-urban population. All studies compare the population mean date of occurrence for each parameter. Lay date is the population mean date of laying the first egg. The panels on reproductive endocrinology and gonadal development encompass studies comparing the seasonal increase in reproductive hormones or gonad size, respectively. *n* indicates the number of studies quantifying each parameter. Studies that simultaneously quantified more than one parameter are included in multiple panels.

breeding period. Inversely, inclement conditions can delay the initiation of reproductive activity and potentially also curtail this activity, shortening the breeding period. The neuroendocrine pathways that mediate the reproductive effects of photoperiod have been studied extensively (Sharp, 2005), but we know much less about the mechanisms that underlie the effects of supplementary cues. Timing of reproduction is considered one of the major life history traits reflecting the adaptation of birds to local environmental characteristics. The importance of environmental cues to the timing of reproduction means that changes in these cues associated with urbanization have the potential to influence the HPG axis at any or all levels and, in turn, modify the length of breeding cycles in urban birds compared to their non-urban conspecifics. These changes may thus have fitness consequences for birds breeding in urban areas.

8.3 Light

8.3.1 Effects of urban lights

Urbanization is almost always associated with the use of artificial lights (Chapter 2). Effects of urban lights on the behaviour and aspects of reproduction of birds have been described in several species.

Miller (2006) measured singing by free-ranging American robins, *Turdus migratorius*, exposed to artificial nocturnal light of varying intensity and compared these birds with robins that presumably experienced little nocturnal light exposure. The beginning of the morning chorus in the latter males was generally synchronized with the civil twilight. By contrast and similar to the situation in other songbirds (Bergen & Abs, 1997; Kempenaers et al., 2010), robins exposed to artificial nocturnal light began to sing during the true night and well in advance of the civil twilight. In these birds, the beginning of the morning chorus correlated positively to the amount of nocturnal artificial lighting and, within an area, was further advanced on cloudy or misty days, which were associated with more light trapped close to the ground. These observations support the hypothesis in American robins that artificial light plays an important role to determine the daily onset of singing.

Artificial lights can also influence the phenology of seasonal reproduction events. For example, female blue tits, *Cyanistes caeruleus*, occupying territories exposed to street light begin to lay slightly earlier (average: 1.5 days) in the spring than females in territories without street light (Kempenaers et al., 2010). This difference may result from direct stimulatory effects of light on the reproductive system of these females (see below). However, as male blue tits occupying territories near street lights are more successful at obtaining extra-pair copulations than other males, changes in female breeding phenology may be indirect and consequent to changes in male behaviour. A study on black-tailed godwits, *Limosa l. limosa*, further illustrates the complexity of the effects of artificial lights on reproduction. In this species, the proximity to artificial lights has a weak but significant negative influence on nest density (de Molenaar et al., 2000). Furthermore, early-arriving birds select nest sites that are located further away from light sources than late-arriving birds.

Urbanization is generally associated with numerous alterations besides increased nocturnal illumination. As an example, it is frequently associated with increased ambient noise. Ambient anthropogenic noise has the potential to profoundly alter the composition of urban avian communities (review: Francis et al., 2009) through changes in reproductive behaviours such as the spectral characteristics of vocalizations (great tit, *Parus major*: Slabbekoorn & den Boer-Visser, 2006; Slabbekoorn & Peet, 2003). The potentially confounding influence of this factor is illustrated by studies on the European robin, *Erithacus rubecula* (Fuller et al., 2007) and other European songbirds (Bergen & Abs, 1997). This influence and the potential mechanisms involved (e.g. direct effects vs. effects mediated by noise-induced changes in female perception of male song characteristics, see Habib et al., 2007) are discussed elsewhere (Chapter 7). They will not be reviewed here but serve as a reminder that this type of influence must be considered in the interpretation of the role of artificial illumination on reproductive processes.

8.3.2 Mechanisms of action of urban light on reproductive functions

The above examples make it clear that artificial illumination can influence reproductive behaviours (e.g. singing) and physiology (e.g. lay dates), but what mechanisms mediate these effects? Little information is available on this subject, but it can be proposed that these effects are either direct (i.e. resulting from light-mediated changes in HPG axis activity), or indirect and resulting, for example, from light-mediated increases in daily foraging time and social interactions.

Direct effects

The effects of light on the avian HPG axis result primarily from direct activation of hypothalamic encephalic receptors (Sharp, 2005). The spectral sensitivity characteristics of these receptors have been determined in the Japanese quail, *Coturnix coturnix* (Foster et al., 1985) and chicken, *Gallus gallus domesticus* (Davies et al., 2012; Figure 8.2). In these species, photoreceptors are based on opsin photopigments with peak sensitivity in the blue (approx. 490 nm), which is close to the maximum wavelength of indirect sunlight illumination. By contrast, the emission of modern city lights (high-pressure sodium lamps and mercury vapour lamps) is low at wavelengths between 450 nm and 520 nm (Figure 8.2). The prominent contribution of high-pressure sodium lamp emission to the spectral properties of urban artificial

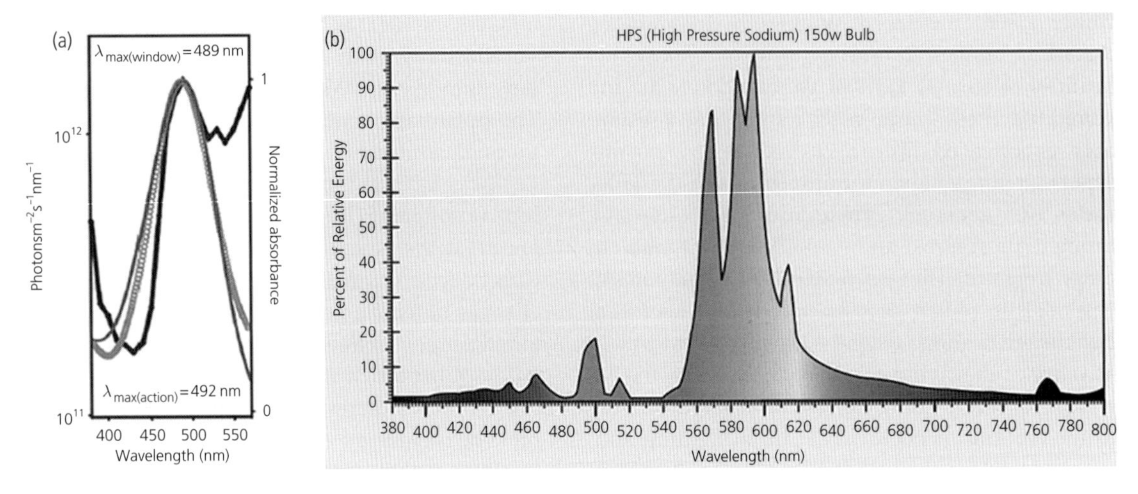

Figure 8.2 (a) Spectrum of light penetrance at the avian hypothalamus (black line; modified from Foster & Follett, 1985) showing peak penetrance at 489 nm. The profile of penetrance closely matches that of the avian photoperiodic response (dark grey line; modified after Foster et al., 1985). Reprinted from Davies et al. 2012, with permission of the Royal Society. (b) Emission spectrum of sodium, showing minimal emission at wavelengths shorter than 540 nm and maximal emission in the 550–620 nm range. Reprinted from Gotham Products (gothamhydroponics.com), with permission.

light is demonstrated by the striking similarity between the spectral distribution of incident nocturnal light on urban lakes and of these lamps (Moore & Kohler, 2002). Given the spectral distribution of nocturnal artificial light emissions and the absorption properties of avian brain photoreceptors, urban night lights may not be effective in stimulating these receptors and, therefore, the avian HPG axis.

In most avian species studied to date, exposure to long days exerts two temporally distinct effects on the HPG axis. The first effect of long days is to stimulate all aspects of this axis, leading to gonadal development and elevated secretion of gonadal steroids into circulation (Section 8.2). Prolonged exposure to long days, however, eventually inhibits the secretion and then also production of GnRH-I (Saldanha et al., 1994). Birds in this condition lose sensitivity to the stimulatory influence of long days and are said to be photorefractory (Dawson et al., 2001). With the onset of photorefractoriness, the HPG axis activity rapidly decreases, culminating in gonadal involution and curtailing of gonadal steroid secretion. If artificial illumination directly affects the HPG axis, we would predict urban birds to become photorefractory and undergo gonadal involution earlier in the year than non-urban birds. However, studies on this subject found no difference between urban and non-urban birds in the timing of seasonal testicular or ovarian regression, suggesting no difference in the onset of photorefractoriness (Partecke et al., 2006). This observation also argues against artificial lights being directly responsible for differences in reproductive phenology between urban and non-urban birds.

Indirect effects

Artificial illumination may, alternatively, influence reproductive functions indirectly. For example, urban birds, being exposed to artificially longer days than in natural conditions, may have increased time for foraging, resulting in enhanced reproductive system activity (see Section 8.5). Furthermore, artificial lights, by attracting phototaxic insects, may create enhanced foraging opportunities for urban birds. Through effects on behaviour (see above), artificial lights may also modify the amount of social interactions between mates (Kempenaers et al., 2010), which may in turn influence the activity of the reproductive system. Supporting this hypothesis, exposure of females to male songs has rapid endocrine (Maney et al., 2007) and behavioural (Maney et al., 2003) effects, and can induce ovarian development (Bentley et al., 2000). Reciprocally, males can respond to oestradiol-treated females by elevating their plasma testosterone (Wingfield & Monk, 1994). Importantly, the neural and endocrine effects of male song on conspecific females depend on the song quality (Marshall et al., 2005). In the Northern mockingbird, *Mimus polyglottos*, the structural characteristics of nocturnal and daytime songs differ in some respects (Derrickson, 1988). This observation suggests the intriguing possibility that nocturnal song associated with artificial lights may influence the female reproductive system differently than daytime song.

8.3.3 Fitness consequences of artificial lighting

Exposure of primarily diurnal organisms to night lights has the potential to disrupt circadian rhythms. By affecting endocrine and neurobiological parameters, this disruption can in turn interfere with biological processes related to immune responses, metabolism, social interactions, and susceptibility to diseases (Navara & Nelson, 2007). Little is known in birds about the effects of artificial lighting on these parameters or on fitness. Song is energetically expensive and calls that attract mates also attract predators (Krams, 2001). Even though birds at night are exposed to fewer diurnal predators than during the day, nocturnal singing may, therefore, incur a fitness cost. On the other hand and as already discussed, there is evidence in some situations that artificial illumination can promote changes in nest location and density (bar-tailed godwit: de Molenaar et al., 2000). No effect of these changes on fitness was detected in this species (de Molenaar et al., 2000), but research on this topic in other species is warranted. Furthermore, males of typically diurnal species that extend their activity into the night due to artificial lights may benefit from enhanced opportunities to forage (see above) and locate mates, and nocturnal singing may promote mate attraction and reproductive development (Kempenaers et al., 2010). Finally, in one study, predation of artificial nests decreased with urbanization, but this decrease was not associated with a difference in nesting success, possibly because urban birds have a reduced rate of nest abandonment than non-urban birds (Blair, 2004). Collectively, these studies demonstrate complex effects of artificial lights on reproduction, but further multidisciplinary research is needed to assess the fitness consequences of exposure to these lights.

8.4 Temperature

8.4.1 Temperature and the avian reproductive system

It has long been known in domesticated birds that temperature can influence the reproductive system. In maturing chickens, exposure to low ambient temperatures depresses testicular growth and delays spermatogenesis (Huston, 1975). In mature males of this species, fertility varies in a non-linear fashion as a function of ambient temperature, being higher at 19 °C than at 8 °C or 30 °C. In the male Japanese quail, which undergoes gonadal regression when transferred from long to short photoperiod, this regression requires exposure to low temperature, with long day-exposed birds maintaining developed gonads when transferred to short days while held at constant temperature (Wada et al., 1990).

Studies on free-ranging birds likewise indicate that ambient temperature influences reproduction. In the field, testes develop later in free-ranging song sparrows, *Melospiza melodia*, of a high altitude population, which are exposed to cold vernal temperatures, than in birds of a coastal population, which are exposed to milder temperatures (Perfito et al., 2005). That temperature partially accounts for this difference is given credibility by the observation that captive photostimulated sparrows from the high altitude population develop their testes later when exposed to cold than warm temperature. Similarly, testes recrudesce at a slower rate in long day-exposed captive willow tits, *Poecile montanus*, held at 4 °C or 10 °C than 20 °C (Silverin & Viebke, 1994). Ambient temperature affects not only the vernal development but also the regression of gonads that takes place when birds become photorefractory in response to prolonged long day exposure at the end of the breeding season (see earlier). In captive willow and great tits, exposure to low temperatures slows the rate of gonadal regression that occurs during photorefractoriness (Silverin & Viebke, 1994; Silverin et al., 2008), thereby extending the period of fertility relative to birds exposed to higher temperatures. Other studies show effects of temperature on lay dates and these effects are, therefore, not specific to males (Visser et al., 1998).

The effects of temperature on gonadal function can be population-specific. For example, temperature influences photoinduced gonadal development in high altitude but not coastal song sparrows (Perfito et al., 2005). Great tits from different populations also respond differently to temperature (Silverin et al., 2008), as do different subspecies of white-crowned sparrows, *Zonotrichia leucophrys* (Wingfield et al., 1997; 2003). This type of variation may reflect an adaptation to environmental predictability, with birds breeding in relatively predictable environments being less sensitive to temperature and other non-photic factors and, therefore, more dependent on photoperiod than birds breeding in relatively unpredictable environments (Wingfield et al., 1992).

8.4.2 Urbanization and ambient temperatures

Urbanization is often associated with changes in ambient temperature. In urban areas, increased heat production resulting from energy consumption, along with enhanced heat dissipation caused by changes in land use (increased pavement and buildings, decreased green areas), combine to elevate average air temperatures, thereby creating 'urban heat islands', relative to surrounding non-urban areas (Zhang et al., 2010; Figure 8.3). It is conceivable in some situations that a sustained increase in local temperatures resulting from urbanization in and by itself stimulates a shift in community composition, with warm climate species replacing colder climate species. This may, however, not be the case in other situations due to the influence of factors such as land use that also contribute to determining avian community compositions (Clavero et al., 2011). Illustrating the fact that elevated ambient temperatures affect avian community composition

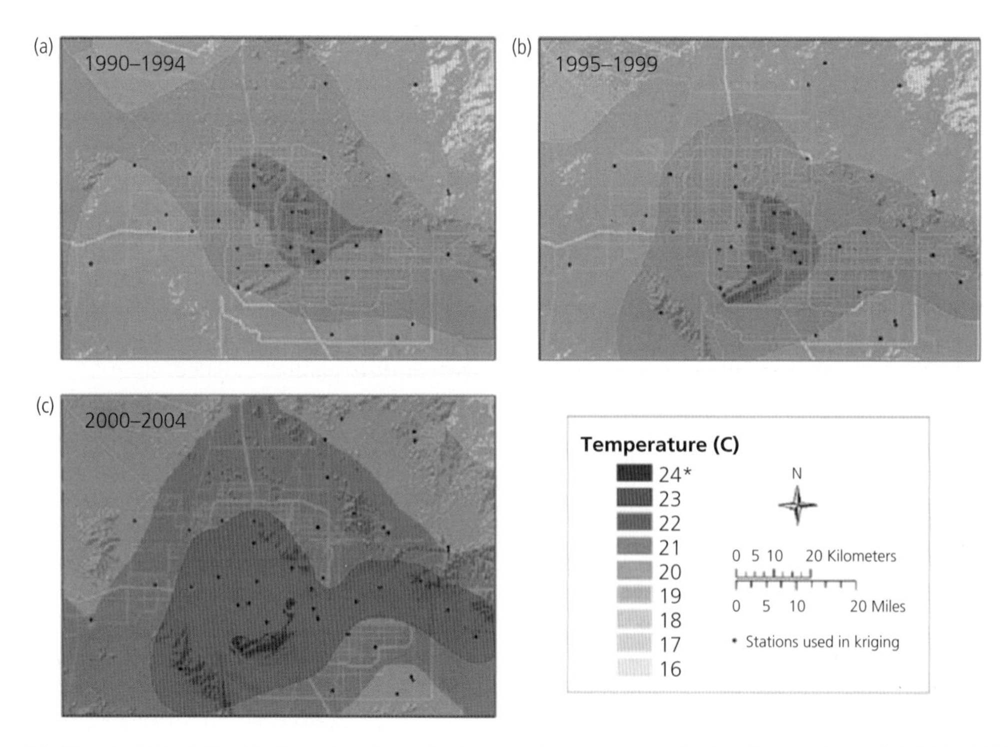

Figure 8.3 Urban heat island effect. The figure shows interpolated average June temperatures for three five-year time periods (1990–1994; 1995–1999; 2000–2004) in the Phoenix (Arizona, USA) metropolitan area. Note the increase over time of the extent and intensity of the urban heat island. Reproduced with permission from the American Meteorological Society.

is contingent upon land use dynamics, these authors propose that effects of increased ambient temperature on avian communities can be mitigated by forest expansion, the net outcome being an increase in avian species preferring cold temperature niches and having a northern distribution range. These findings exemplify the importance of incorporating interactions between land use dynamics and temperatures in the development of models aimed at predicting the effects of elevated temperatures, such as resulting from the urban heat island effects, on community compositions.

8.4.3 Effects of temperature on the reproductive system of urban birds

Direct effects

The early seasonal testicular recrudescence and early lay dates that often characterize urban compared to non-urban birds (Table 8.1) are consistent with the hypothesis that elevated urban temperatures influence the reproductive system. To our knowledge no incisive evidence is, however, available that relatively small (a few degrees) increases in ambient temperature, such as associated with urban heat islands, in and by themselves exert reproductive effects. Future research is needed to determine whether this is the case and whether temperature influences reproductive functions directly or indirectly. If ambient temperatures are found to directly affect these functions in urban birds, new research should also be directed at identifying the specific aspects of these functions that are influenced and unravelling the underlying neuroendocrine mechanisms.

Indirect effects

As the hypothesis that ambient temperature affects the avian reproductive system through direct neuroendocrine actions has not been tested experimentally, it is conjectural and the current data are open to alternative interpretations. It is, for example, important to consider that studies using captive birds obtained from wild populations reveal effects of temperature on the avian reproductive system, but these differences are usually found in response to exposure to a considerably wider range of temperatures than normally experienced by free-ranging urban vs. comparable non-urban birds. Additionally, temperature often influences the phenology of vegetation and food that birds depend on. As food is an important determinant of reproductive phenology (Section 8.5), temperature may affect the reproductive physiology of free-ranging birds by altering food availability. Also, it cannot be excluded that urbanization generates microhabitats that birds use to escape extreme temperatures.

The complexity of the mechanisms involved is illustrated by a recent study on captive great tits (Schaper et al., 2012). In this species, the average lay date of free-ranging females is earlier in warmer springs. In captivity, early lay can be induced by a gradual increase in temperature, but not by simply transferring females to constant higher temperatures. Furthermore, the advanced lay dates resulting from a gradual increase in temperature are not associated with an early spring increase in plasma LH or ovarian growth, suggesting that temperature differentially affects selective aspects of reproduction. Finally, studies on the reproductive effects of elevated temperature in the context of urbanization need to incorporate the fact that interactions between photoperiod and temperature can be population- and subspecies-specific (see above). As shown by previous studies, the timing of reproduction in mammals (Prendergast et al., 2004) and birds (Brommer et al., 2005) includes a heritable component. Similarly, a captive study on female great tits that compared siblings of known ancestry established that the response of these birds to ambient temperature, as measured by the timing of laying, is influenced by genetic differences (Schaper et al., 2012).

8.5 Food

Food supply is widely accepted as the primary ultimate determinant of reproductive phenology in birds. The breeding period of most birds is timed so that young are developing when food supply is at its seasonal peak. By synchronizing breeding with the seasonal peak in food supply, birds can maximize their reproductive success (Sheldon et al., 2007). In seasonal environments, photoperiod predicts when food will generally be most abundant. However, photoperiod does not reflect year to year variation in the phenology of food sources or predict the precise time of maximum food abundance. During the early stages of reproductive development birds, therefore, also use food supply as a proximate cue to fine-tune the timing of breeding activities. Studies on captive and free-ranging birds reveal that two aspects of the food supply play crucial roles in the timing of seasonal reproduction: the quantity and the quality of food.

8.5.1 Food quantity

Compelling evidence exists that the quantity of food available during the early stages of reproductive development modulates reproductive phenology. Greater food availability, be it due to natural spatial and temporal variation (Ligon, 1974) or to experimental food supplementation (Schoech, 2009), commonly advances the reproductive phenology of wild birds. Larger quantities of food available to birds in urban relative to neighbouring non-urban areas (Figure 8.4a) are hypothesized to play a role in the advanced reproductive phenology of urban birds (Chamberlain et al., 2009; Partecke et al., 2005; Schoech & Bowman, 2001). In North America and parts of Europe, the provisioning of food at private feeders is common. Indeed, bird feeders may be regularly used in up to 43% of US households and 75% of UK households (Cowie & Hinsley, 1988; Martinson & Flaspohler, 2003). Feeders potentially provide a predictable, energy-rich resource that may increase food intake. Perrins (1970) suggests that there is a minimum threshold of food intake necessary for birds to begin egg formation. Human-provided food may allow urban birds either to maintain this threshold year-round or to reach it earlier in the spring than their non-urban counterparts. However,

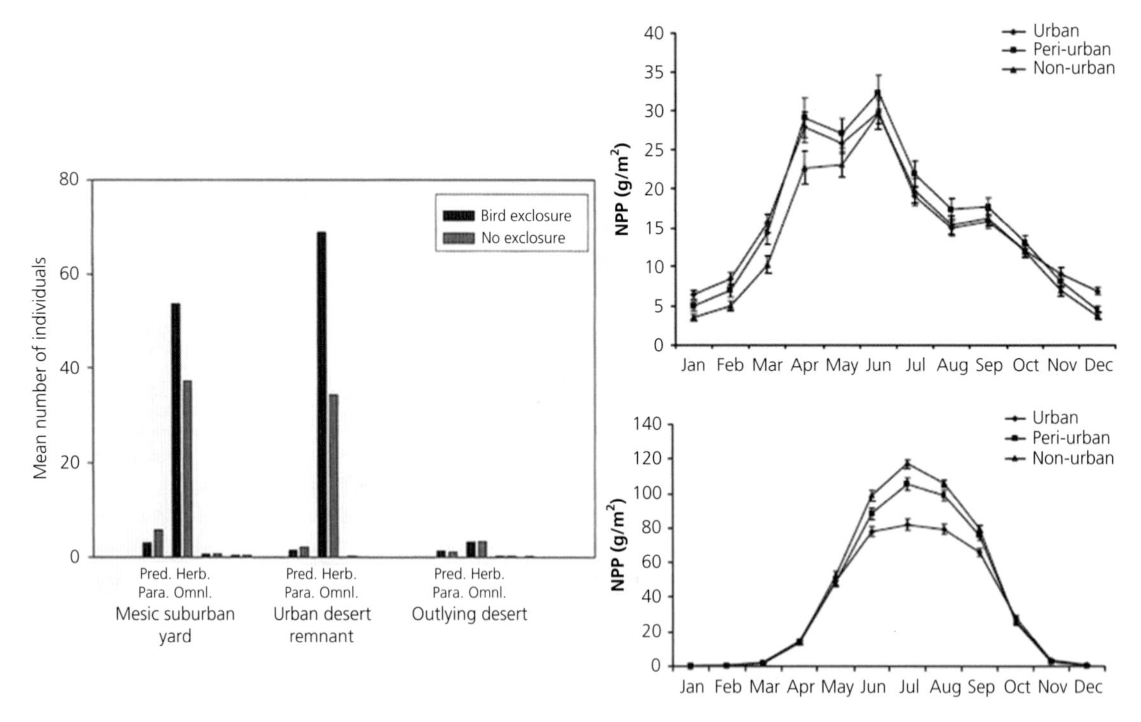

Figure 8.4 (a) Number of individual insect predators (Pred.), herbivores (Herb.), parasites (Para.), and omnivores (Omni.) in mesic suburban yards, urban desert remnants, and outlying Sonoran Desert in the Phoenix (Arizona, USA) metropolitan area. Note the large differences between urban/suburban areas and the surrounding native desert. These differences are observed whether birds are excluded (Bird exclosure) or not (No exclosure). Reprinted from Faeth et al. 2005, with permission of the American Institute of Biological Sciences. (b and c) Seasonal variation in net primary productivity (NPP) in urban, peri-urban, and non-urban areas of (b) the arid Southwest United States and (c) the temperate Northeast United States. Reproduced with permission from the American Institute of Biological Sciences.

to our knowledge, this hypothesis has yet to be tested experimentally. It is also unclear whether and how the effect of human-provided food on avian reproductive phenology interacts with changes in natural food abundance associated with urbanization.

An important determinant of the quantity of food available in an area is the net primary productivity (NPP, the rate at which plant biomass accumulates). Peaks in food availability are generally synchronized with peaks in NPP, which can therefore be used as a proxy for the quantity of locally available food. Considering the widespread advanced reproductive phenology of birds in urban areas (Table 8.1), if a general increase in the quantity of natural food causes this trend, we may expect to see increased NPP in urban areas. Evidence suggests, however, that the response of NPP to urbanization is variable and depends on the type of environment. For example, urban areas in arid, resource-limited environments have higher total annual NPP than do outlying non-urban areas (Imhoff et al., 2004; Figure 8.4b). Urban areas in high rainfall environments, on the other hand, have lower total annual NPP (Imhoff et al., 2004; Figure 8.4c). Therefore, changes in NPP associated with urbanization are probably not solely responsible for the widespread advanced reproductive phenology seen in urban birds. We must, however, point out that NPP does not incorporate human-provided food sources, such as from feeders and discarded food waste that potentially alter food availability independent of productivity. The separate and cumulative effects of these factors warrant further scrutiny.

For many bird species, arthropods are an important part of the diet and the availability of these organisms may be a more relevant measure of food availability than NPP. Therefore, the availability of arthropods potentially plays an important role in modulating reproductive phenology. However, like NPP, the response of arthropod abundance to urbanization varies widely among studies and is taxon-specific. For example, Faeth *et al.* (2005) identified six studies reporting an increase in the abundance of arthropods in urban areas, but five studies reported no effect and 15 studies reported a decrease. Once again, considering the widespread advanced reproductive phenology of urban birds, the evidence supporting the idea that this trend is solely due to increased abundance of arthropods in urban areas is weak.

Despite clear effects on reproductive phenology, the underlying physiological mechanism(s) by which food quantity influences the reproductive system remains obscure. In captivity, it is possible to precisely control the quantity of food available and the day length that a bird experiences. Transferring birds from short, winter-like day lengths to long, summer-like day lengths mimics the increasing day length during spring and, in most species, stimulates the reproductive system (Section 8.2). When male red crossbills, *Loxia curvirostra*, given unlimited food are transferred from short to long day lengths they show a surge in plasma LH after just one day whereas crossbills experiencing food restriction take 10 days to produce such a surge (Hahn, 1995). Interestingly, despite this effect, both experimental groups had similar testis sizes after 30 days of exposure to summer-like day lengths. In contrast to these findings, Perfito *et al.* (2008), using methods similar to those of Hahn (1995), found no effect of food restriction on plasma LH of male zebra finches, *Taeniopygia guttata*, but testis size increased in finches given unlimited food and not in those experiencing food restriction. These studies indicate that food quantity can influence the HPG axis, but the mechanism that mediates this influence remains equivocal.

Importantly, the studies showing that urbanization is associated with advanced reproductive phenology did not find a corresponding advancement in reproductive physiology. For example, in female Eurasian blackbirds, *Turdus merula*, and Florida scrub-jays, *Aphelocoma coerulescens*, urban and non-urban birds had similar plasma LH and oestradiol during reproductive development (Partecke et al., 2005; Schoech & Bowman, 2003). These species belong to different families (Turdidae and Corvidae, respectively) and use different native habitats. These observations suggest that the observed endocrine differences between urban and non-urban birds may be widespread across species. In male Eurasian blackbirds, plasma LH and testosterone during reproductive development were actually lower in urban birds than in their non-urban conspecifics (Partecke et al., 2005). Thus, the advanced reproductive phenology seen in urban birds is apparently not associated with an advancement of the seasonal rise in secretion of reproductive hormones. This could be explained if target tissues of urban birds, in particular the brain and gonads, were more sensitive (i.e. able to respond to lower circulating levels and, therefore, earlier in the spring) to a seasonal increase in reproductive hormones than is the case in non-urban conspecifics—a hypothesis that deserves scrutiny but has not yet been tested experimentally.

8.5.2 Food quality

In addition to seasonal changes in the quantity of food available in general, there are seasonal changes in specific types of food. For example, the growth of new green vegetation and the abundance of arthropods generally increase as spring progresses. It is therefore possible that the type of food plays a proximate role in fine-tuning reproductive phenology. Consistent with this proposition, the presence and/or the consumption of green vegetation or seeds can modulate reproductive phenology. For example, access to growing green vegetation in captivity advances gonadal development in male Atlantic canaries, *Serinus canaria* (Voigt et al., 2007), common redpolls, *Carduelis flammea* (Hahn et al., 2005), and female white-crowned sparrows, *Zonotrichia leucophrys* (Ettinger & King, 1981). Similarly, providing seeds to male red crossbills and pine siskins, *Spinus pinus*, during gonadal development increases their plasma LH and testis size (Hahn et al., 2005; Watts & Hahn, 2012). Furthermore, if pinyon jays, *Gymnorhinus cyanocephalus*, are given green pinyon cones during late summer they reverse the gonadal regression characteristic of the end of breeding (Ligon, 1974). Crucially, birds in all these studies received *ad libitum* maintenance food, strongly supporting the idea that food quality plays an important role in synchronizing reproductive phenology.

In urban areas, plant phenology is generally advanced and dormancy is delayed, extending the growing period (Fisher et al., 2006). This trend is mirrored in studies of the seasonality of NPP. Compared to non-urban areas, NPP of urban areas embedded in arid, resource-limited environments is dramatically higher during spring and autumn (Imhoff et al., 2004). Furthermore, the vernal rise in NPP occurs earlier in urban areas and the autumnal decline occurs later. During summer, however, NPP of urban areas located in arid, resource-limited environments declines to levels slightly lower than that of non-urban areas. A similar trend is also found in high rainfall, sylvan environments, although the NPP during spring and autumn is only slightly higher in urban areas and the levels of productivity in urban areas are much lower during summer (Imhoff et al., 2004). Overall, therefore, it appears that urbanization not only extends the length of the growing season, but also results in a seasonally earlier peak in plant growth.

Many songbirds modify their diet from consuming a high proportion of seeds in the winter to a high proportion of arthropods in the spring. Furthermore, young birds often have specific dietary requirements of small arthropods. Therefore, the availability of arthropods in particular may be a crucial factor that birds use to fine-tune reproductive phenology. One of the best demonstrations of the importance of arthropods to avian reproductive development is an experiment on the neotropical spotted antbird, *Hylophylax naevioides*. In this experiment, captive males received unlimited standard diet; however, only antbirds that additionally received live crickets increased gonadal development (Hau et al., 2000).

We are not aware of high temporal resolution studies examining the effect of urbanization on the seasonal pattern of arthropods. In a long-term study of ground arthropods in Phoenix, Arizona, USA, annual abundance in mesic yards, which have green lawns and are heavily watered, was considerably more consistent from year to year than at non-urban desert sites (Bang & Faeth, 2011). Interestingly, urban xeric gardens, which have arid-adapted plants, showed more year to year variation than non-urban desert sites, but abundance was generally higher in the xeric gardens throughout. Whether the seasonal pattern of arthropod abundance responds to urbanization in a similar way to plants is unclear and requires scrutiny, but a long-term study of caterpillar phenology found that abundance has effectively tracked changes in tree phenology caused by climate change (Visser et al., 2006). If, indeed, the seasonal pattern of arthropod abundance responds to urbanization in a similar way to plants, we would expect abundance to increase earlier in the year. As the availability of particular food types appears to influence reproductive development in most birds studied, the modified patterns of plant growth and arthropods abundance associated with urbanization may be responsible, at least in part, for the modified patterns of avian reproductive phenology.

8.6 Conclusions

Earth's natural ecosystems are being rapidly and, in many cases, irreversibly transformed by humans. These transformations are nowhere more evident than in cities, where humans alter virtually all the components of ecosystems including vegetation density and composition; food chain characteristics; and physical features such as illumination timing and level, temperature, noise; and water distribution, quality, and availability. These changes result in shifts in animal communities that often translate into decreased diversity (homogenization), although not necessarily total numbers of individuals. Mitigating and potentially remediating the effects of urbanization on organisms requires the identification of the specific environmental factors to which these organisms are sensitive and ultimately of their effects on fitness. Innumerable factors undoubtedly contribute to influencing urban organisms at the individual and community levels, but the influence of most of these factors generally remains poorly understood, as are their proximate mechanisms of action. The best studied environmental factors with the potential to modify fitness through reproductive actions are light, temperature, and food and this chapter, therefore, focuses on these factors. Detailed research using captive and free-ranging birds provides a reasonably good basis for discussing their actions (Figure 8.5). As the figure illustrates, a number of fundamental questions related the effects of light, but in particular temperature and food, on urban birds await further research. For example, little is known regarding how ambient temperature influences the avian HPG axis, and the respective roles of food quality *vs.* quantity on this axis remain poorly defined. Much remains to be gained from studies examining the influence of these factors in and by themselves, and how they potentially interact to regulate reproduction. For example, food and/or temperature may play more important or different roles in regions where

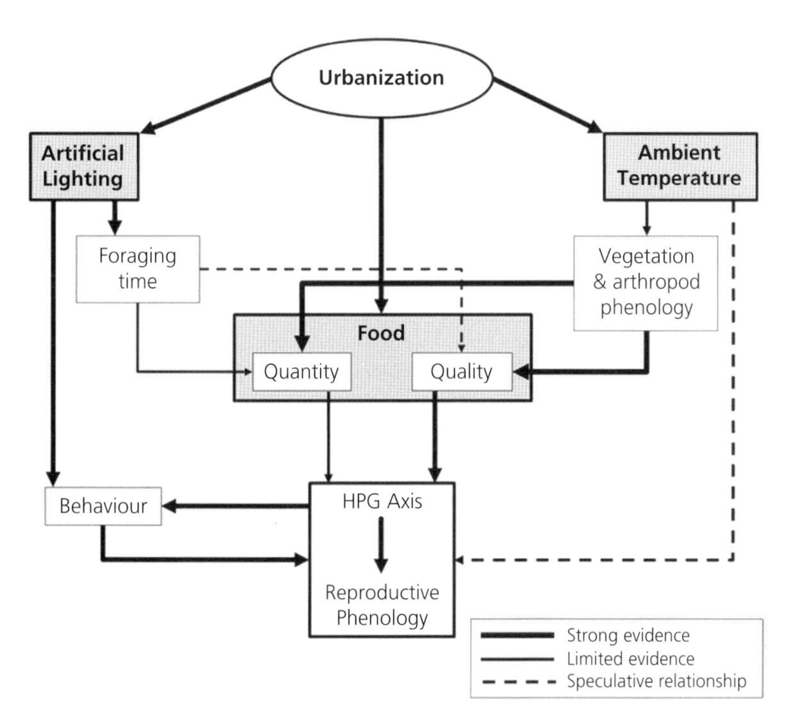

Figure 8.5 Conceptual framework visualizing potential mechanisms by which urbanization affects the hypothalamo-pituitary-gonadal (HPG) axis and, in turn, advances reproductive phenology of birds through modifications of artificial light, ambient temperature, and food availability. Thick lines represent interactions for which there is strong supporting evidence. Thin lines illustrate interactions where evidence is limited. Dashed lines represent speculative relationships.

Box 8.1 Future research directions

Promising future directions for research related to the topics discussed in this chapter include the following:

(a) Artificial lights, elevated urban temperatures, and urban food resources have the potential to profoundly affect reproductive functions. A challenge to researchers studying this question has to do with the fact that these factors may to some extent influence reproductive functions independently but also interactively (Figure 8.5). It follows that understanding the overall effects of urbanization on reproductive functions and disentangling the mechanisms involved requires an understanding of the role of each environmental parameter considered. Considerable progress along this line has already been made, but many questions are left to answer. Providing these answers will be facilitated by the judicious choice of study species, an integration of correlative and manipulative approaches, and comparative research aimed at identifying commonalities across urban gradients and across urban areas located in different ecoregions. Studies on this subject will continue to be most productive and informative if they draw from a wide range of disciplines, in particular ecology, population biology, physiology, animal behaviour, genetics, urban planning, and climatology.

(b) Many studies on urban avian populations have been correlative rather than manipulative and, therefore, not designed to isolate the selective contribution of specific environmental parameters. To address this issue, researchers are beginning to perform controlled field and laboratory studies aimed, for example, at determining the effects of experimental alterations of the timing or intensity of ambient lights or of food quality or quantity. Such studies hold great promise in terms of helping us unravel the contribution of these factors to the physiological and behavioural differences that have been reported between urban and non-urban avian populations.

(c) Urbanization often advances the breeding phenology of birds, but most information on this subject comes from studies in temperate regions. The effects of urbanization on ambient temperatures and urban food resources, in particular, are not uniform but vary as a function of the ecoregion where a city is located and even within an ecoregion, on the size, age, open water availability, amount of green space, etc., of a city. As the composition of bird communities also differs across ecoregions (with the exception of a few widely distributed generalists such as the house sparrow, *Passer domesticus*), generalizing conclusions from one situation to another proves quite difficult. Furthermore, we are aware of no studies focusing on cities in any tropical or arctic ecoregion. Cities in tropical ecoregions have special relevance because globally they possess high avian species density and are experiencing high rates of urbanization. Arctic ecoregions also have special relevance as they are the breeding ground for many bird species. Major progress in our overall understanding of the effects of urbanization will benefit from comparative investigations conducted in a diversity of urban settings, especially those located outside temperate regions.

(d) The physiological and behavioural responses of birds to environmental factors are often not linear. For example, blue tits lay earlier when exposed to gradually increasing ambient temperature at the appropriate stage of their breeding cycle, but not when experimentally transferred to and then held at a constantly higher temperature. New studies, probably best conducted in captive settings, are warranted to delineate the qualitative and quantitative determinants of such interactions and to identify the neuroendocrine processes that underlie their effects.

(e) Large inter-individual differences in the response to the same environmental factor (e.g. temperature) generally exist within a seemingly homogeneous population. These differences result from individual variation in the degree of phenotypic plasticity as well as from the individual genetic make-up. Further research on this subject to tease apart the respective role of phenotypic plasticity vs. genetic factors, and using various species as models, is critical and likely to be particularly enlightening. This research should help illuminate why some species adjust more or less well to urbanization. It should also assist us predict how populations will fare in response to factors such as the heat island effect and global climate change.

(f) It is increasingly apparent that a given environmental factor can influence reproductive processes (e.g. lay date) without necessarily also affecting upstream components of the HPG cascade (e.g. vernal increase in plasma LH). This difference may result from changes in sensitivity of target tissues (e.g. brain and gonads) to the relevant hormones, a mechanism that could explain the frequently observed difference (advancement) of breeding phenology in urban compared to non-urban species. It would be extremely fruitful and relatively easy to experimentally test this hypothesis.

photoperiod is relatively constant (i.e. close to the Equator) than at higher latitudes. Progress in our understanding of how urbanization affects avian populations at the individual and community levels should benefit greatly from new correlative and experimental research conducted in a diversity of ecosystems and integrating ecological, physiological, and behavioural perspectives (see Box 8.1).

8.7 Acknowledgements

P.D. and S.D. were supported by National Science Foundation grants DEB-0423704 and BCS-1026865 to Central Arizona—Phoenix Long-Term Ecological Research. The research was also supported by NSF award IOB 1026620 to PD.

References

Bang, C. and Faeth, S. H. (2011). Variation in arthropod communities in response to urbanization: Seven years of arthropod monitoring in a desert city. *Landscape and Urban Planning*, **103**, 383–399.

Bentley, G.E., Wingfield, J.C., Morton, M.L. and Ball, G.F. (2000). Stimulatory effects on the reproductive axis in female songbirds by conspecific and heterospecific male song. *Hormones and Behavior*, **37**, 179–189.

Bergen, F. and Abs, M. (1997). Verhaltensokologische Studie zur Gesangsaktivitat von Blaumeise (*Parus caeruleus*), Kohlmeise (*Parus major*) und Buchfink (*Fringilla coelebs*) in einer Grossstadt. *Journal of Ornithology*, **138**, 451–467.

Blair, R. (2004). The effects of urban sprawl on birds at multiple levels of biological organization. *Ecology and Society*, **9**, 2.

Both, C., Bouwhuis, S., Lessells, C. M. and Visser, M. E. (2006). Climate change and population declines in a long-distance migratory bird. *Nature*, **441**, 81–83.

Brommer, J.E., Merila, J., Sheldon, B.C., Gustafsson, L. (2005). Natural selection and genetic variation for reproductive reaction norms in a wild bird population. *Evolution*, **59**, 1362–1371.

Chamberlain, D. E., Cannon, A. R., Toms, M. P., Leech, D. I., Hatchwell, B. J. and Gaston, K. J. (2009). Avian productivity in urban landscapes: A review and meta-analysis. *Ibis*, **151**, 1–18.

Chow, W.T.L., Brennan, D., Brazel, A.J. (2012). Urban heat island research in Phoenix, Arizona. *Bulletin of the American Meteorological Society*, **93**, 517–530.

Clavero, M., Villero, D. and Brotons, L. (2011). Climate change or land use dynamics: do we know what climate change indicators indicate? *PLoS One*, **6**, e18581.

Cowie, R. J. and Hinsley, S. A. (1988). The provision of food and the use of bird feeders in suburban gardens. *Bird Study*, **35**, 163–168.

Davies, W.I., Turton, M., Peirson, S.N., Follett, B.K., Halford, S., Garcia-Fernandez, J.M., Sharp, P.J., Hankins, M.W. and Foster, R.G. (2012). Vertebrate ancient opsin photopigment spectra and the avian photoperiodic response. *Biology Letters*, **8**, 291–294.

Dawson, A., King, V.M., Bentley, G.E. and Ball, G.F. (2001). Photoperiodic control of seasonality in birds. *Journal of Biological Rhythms*, **16**, 365–380.

de Molenaar, J.G., Jonkers, D.A. and Sanders, M.E. (2000). Road illumination and nature. III Local influence of road lights on a black-tailed godwit (*Limosa l. limosa*) population. Green World Research, Alterra, The Netherlands.

Derrickson, K.C. (1988). Variation in repertoire presentation in northern mockingbirds. *Condor*, **90**, 592–606.

Deviche, P., Hurley, L.L. and Fokidis, B.H. (2011). Avian Testicular Structure, Function, and Regulation. In D. O. Norris and K. H. Lopez, eds, *Hormones and Reproduction of Vertebrates*, Vol. 4, pp. 27–69. Academic Press, San Diego, CA.

Ettinger, A. O. and King, J. R. (1981). Consumption of green wheat enhances photostimulated ovarian growth in white-crowned sparrows. *Auk*, **98**, 832–834.

Faeth, S. H., Warren, P. S., Shochat, E. and Marussich, W. A. (2005). Trophic dynamics in urban communities. *Bioscience*, **55**, 399–407.

Fisher, J. I., Mustard, J. F. and Vadeboncoeur, M. A. (2006). Green leaf phenology at landsat resolution: Scaling from the field to the satellite. *Remote Sensing of Environment*, **100**, 265–279.

Follett, B.K. and Maung, S.L. (1978). Rate of testicular maturation, in relation to gonadotrophin and testosterone levels, in quail exposed to various artificial photoperiods and to natural daylengths. *Journal of Endocrinology*, **78**, 267–280.

Foster, R.G., Follett, B.K. (1985). The involvement of a rhodopsin-like photopigment iin the photoperiodic response of the Japanese Quail. *Journal of Comparative Physiology [A]*, **157**, 519–528.

Foster, R.G., Follett, B.K. and Lythgoe, J.N. (1985). Rhodopsin-like sensitivity of extra-retinal photoreceptors mediating the photoperiodic response in quail. *Nature*, **313**, 50–52.

Francis, C.D., Ortega, C.P. and Cruz, A. (2009). Noise pollution changes avian communities and species interactions. *Current Biology*, **19**, 1415–1419.

Fuller, R.A., Warren, P.H. and Gaston, K.J. (2007). Daytime noise predicts nocturnal singing in urban robins. *Biology Letters*, **3**, 368–370.

Habib, L., Bayne, E.M. and Boutin, S. (2007). Chronic industrial noise affects pairing success and age structure of ovenbirds, *Seiurus aurocapilla*. *Journal of Applied Ecology*, **44**, 176–184.

Hahn, T. P. (1995). Integration of photoperiodic and food cues to time changes in reproductive physiology by an opportunistic breeder, the red crossbill, *Loxia curvirostra* (Aves: Carduelinae). *Journal of Experimental Zoology*, **272**, 213–226.

Hahn, T. P., Pereyra, M. E., Katti, M., Ward, G. M. and MacDougall-Shackleton, S. A. (2005). Effects of food availability on the reproductive system. In A. Dawson and P. J. Sharp, eds, *Functional Avian Endocrinology*, pp. 167–180. Narosa, New Delhi.

Hau, M., Wikelski, M. and Wingfield, J. C. (2000). Visual and nutritional food cues fine-tune timing of reproduction in a neotropical rainforest bird. *Journal of Experimental Zoology*, **286**, 494–504.

Huston, T.M. (1975). The effects of environmental temperature on fertility of the domestic fowl. *Poultry Sciences*, **54**, 1180–1184.

Imhoff, M. L., Bounoua, L., DeFries, R., Lawrence, W. T., Stutzer, D., Tucker, C. J. and Ricketts, T. (2004). The consequences of urban land transformation on net primary productivity in the United States. *Remote Sensing of Environment*, **89**, 434–443.

Kempenaers, B., Borgstrom, P., Loes, P., Schlicht, E. and Valcu, M. (2010). Artificial night lighting affects dawn song, extra-pair siring success, and lay date in songbirds. *Current Biology*, **20**, 1735–1739.

Krams, I. (2001). Communication in crested tits and the risk of predation. *Animal Behavior*, **61**, 1065–1068.

Ligon, J. D. (1974). Green cones of the pinon pine stimulate late summer breeding in the pinon jay. *Nature*, **250**, 80–82.

Maney, D.L., Goode, C.T., Lake, J.I., Lange, H.S. and O'Brien, S. (2007). Rapid neuroendocrine responses to auditory courtship signals. *Endocrinology*, **148**, 5614–5623.

Maney, D.L., MacDougall-Shackleton, E.A., MacDougall-Shackleton, S.A., Ball, G.F. and Hahn, T.P. (2003). Immediate early gene response to hearing song correlates with receptive behaviour and depends on dialect in a female songbird. *Journal of Comparative Physiology A. Neuroethology and Sensory Neural and Behavioral Physiology*, **189**, 667–674.

Marshall, R.C., Leisler, B., Catchpole, C.K. and Schwabl, H. (2005). Male song quality affects circulating but not yolk steroid concentrations in female canaries (*Serinus canaria*). *Journal of Experimental Biology*, **208**, 4593–4598.

Martinson, T. J. and Flaspohler, D. J. (2003). Winter bird feeding and localized predation on simulated bark-dwelling arthropods. *Wildlife Society Bulletin*, **31**, 510–516.

Miller, M.W. (2006). Apparent effects of light pollution on singing behaviour of American Robins. *Condor* **108**, 130–139.

Moore, M.V. and Kohler, S.J. (2002). Measuring light pollution in urban lakes and its effects on lake invertebrates. Abstract, conference on Ecological Consequences of Artificial Lighting, Los Angeles, California.

Navara, K.J. and Nelson, R.J. (2007). The dark side of light at night: physiological, epidemiological, and ecological consequences. *Journal of Pineal Research*, **43**, 215–224.

Nussey, D.H., Postma, E., Gienapp, P. and Visser, M.E. (2005) Selection on heritable phenotypic plasticity in a wild bird population. *Science*, **310**, 304–306.

Olson, D. M., Dinerstein, E., Wikramanayake, E. D., et al. (2001). Terrestrial ecoregions of the world: a new map of life on earth. *BioScience*, **51**, 933–938.

Partecke, J., Van't Hof, T. and Gwinner, E. (2005). Underlying physiological control of reproduction in urban and forest-dwelling European blackbirds *Turdus merula*. *Journal of Avian Biology*, **36**, 295–305.

Partecke, J., Schwabl, I. and Gwinner, E. (2006). Stress and the city: urbanization and its effects on the stress physiology in European blackbirds. *Ecology*, **87**, 1945–1952.

Perfito, N., Kwong, J. M. Y., Bentley, G. E. and Hau, M. (2008). Cue hierarchies and testicular development: Is food a more potent stimulus than day length in an opportunistic breeder (*Taeniopygia g. guttata*)? *Hormones and Behavior*, **53**, 567–572.

Perfito, N., Meddle, S.L., Tramontin, A.D., Sharp, P.J. and Wingfield, J.C. (2005). Seasonal gonadal recrudescence in song sparrows: response to temperature cues. *General and Comparative Endocrinology*, **143**, 121–128.

Perrins, C. (1970). The timing of birds' breeding seasons. *Ibis*, **112**, 242–255.

Prendergast, B.J., Renstrom, R.A., Nelson, R.J. (2004). Genetic analyses of a seasonal interval timer. *Journal of Biological Rhythms*, **19**, 298–311.

Saldanha, C.J., Deviche, P.J. and Silver, R. (1994). Increased VIP and decreased GnRH expression in photorefractory dark- eyed juncos (*Junco hyemalis*). *General and Comparative Endocrinology*, **93**, 128–136.

Schaper, S.V., Dawson, A., Sharp, P.J., Gienapp, P., Caro, S.P. and Visser, M.E. (2012). Increasing temperature, not mean temperature, is a cue for avian timing of reproduction. *American Naturalist*, **179**, E55–E69.

Schoech, S. J. (2009). Food supplementation experiments: A tool to reveal mechanisms that mediate timing of reproduction. *Integrative and Comparative Biology*, **49**, 480–492.

Schoech, S. J., and Bowman, R. (2001). Variation in the timing of breeding between suburban and wildland Florida scrub-jays: Do physiologic measures reflect different environments? In J. M. Marzluff, R. Bowman, and R. Donnelly, eds, *Avian Ecology and Conservation in an Urbanizing World*, pp. 289–306. Kluwer, New York.

Schoech, S. J., and Bowman, R. (2003). Does differential access to protein influence differences in timing of breeding of florida scrub-jays (*Aphelocoma coerulescens*) in suburban and wildland habitats? *Auk*, **120**, 1114–1127.

Seto, K.C., Fragkias, M., Guneralp, B. and Reilly, M.K. (2011). A meta-analysis of global urban land expansion. *PLoS One*, **6**, e23777.

Sharp, P.J. (2005). Photoperiodic regulation of seasonal breeding in birds. *Annals of the New York Academy of Sciences*, **1040**, 189–199.

Sheldon, B. C., Kruuk, L. E. B. and Merila, J. (2007). Natural selection and inheritance of breeding time and clutch size in the collared flycatcher. *Evolution*, **57**, 406–420.

Silverin, B. and Viebke, P.A. (1994). Low temperatures affect the photoperiodically induced LH and testicular cycles differently in closely related species of tits (*Parus* spp.). *Hormones and Behavior*, **28**, 199–206.

Silverin, B., Wingfield, J., Stokkan, K.A., Massa, R., Jarvinen, A., Andersson, N.A., Lambrechts, M., Sorace, A. and Blomqvist, D. (2008). Ambient temperature effects on photoinduced gonadal cycles and hormonal secretion patterns in Great Tits from three different breeding latitudes. *Hormones and Behavior*, **54**, 60–68.

Slabbekoorn, H. and den Boer-Visser, A. (2006). Cities change the songs of birds. *Current Biology*, **16**, 2326–2331.

Slabbekoorn, H. and Peet, M. (2003). Ecology: Birds sing at a higher pitch in urban noise. *Nature*, **424**, 267.

Small, T.W., Sharp, P.J. and Deviche, P. (2007). Environmental regulation of the reproductive system in a flexibly breeding Sonoran Desert bird, the Rufous-winged Sparrow, *Aimophila carpalis*. *Hormones and Behavior*, **51**, 483–495.

Visser, M. E., Holleman, L. J. M. and Gienapp, P. (2006). Shifts in caterpillar biomass phenology due to climate change and its impact on the breeding biology of an insectivorous bird. *Oecologia*, **147**, 164–172.

Visser, M.E., Van Noordwijk, A.J., Tinbergen, J.M. and Lessells, C.M. (1998). Warmer springs lead to mistimed reproduction in great tits (*Parus major*). *Proceedings of the Royal Society of London. B:Biological Sciences*, **265**, 1867–1870.

Voigt, C., Goymann, W. and Leitner, S. (2007). Green matters! growing vegetation stimulates breeding under short-day conditions in wild canaries (*Serinus canaria*). *Journal of Biological Rhythms*, **22**, 554–557.

Wada, M., Hatanaka, F., Tsuyoshi, H. and Sonoda, Y. (1990). Temperature modulation of photoperiodically induced LH secretion and its termination in Japanese quail (*Coturnix coturnix japonica*). *General and Comparative Endocrinology*, **80**, 465–472.

Watts, H. E. and Hahn, T. P. (2012). Non-photoperiodic regulation of reproductive physiology in the flexibly breeding pine siskin (*Spinus pinus*). *General and Comparative Endocrinology*, **178**, 259–264.

Wingfield, J.C. and Monk, D. (1994). Behavioral and hormonal responses of male song sparrows to estradiol-treated females during the non-breeding season. *Hormones and Behavior*, **28**, 146–154.

Wingfield, J.C., Hahn, T.P., Levin, R. and Honey, P. (1992). Environmental predictability and control of gonadal cycles in birds. *Journal of Experimental Zoology*, **261**, 214–231.

Wingfield, J.C., Hahn, T.P., Maney, D.L., Schoech, S.J., Wada, M. and Morton, M.L. (2003). Effects of temperature on photoperiodically induced reproductive development, circulating plasma luteinizing hormone and thyroid hormones, body mass, fat deposition and molt in mountain white-crowned sparrows, *Zonotrichia leucophrys oriantha*. *General and Comparative Endocrinology*, **131**, 143–158.

Wingfield, J.C., Hahn, T.P., Wada, M. and Schoech, S.J. (1997). Effects of day length and temperature on gonadal development, body mass, and fat depots in white-crowned sparrows, *Zonotrichia leucophrys pugetensis*. *General and Comparative Endocrinology*, **107**, 44–62.

Zhang, K., Wang, R., Shen, C. and Da, L. (2010). Temporal and spatial characteristics of the urban heat island during rapid urbanization in Shanghai, China. *Environmental Monitoring Assessment*, **169**, 101–112.

CHAPTER 9

The impacts of urbanization on avian disease transmission and emergence

Lynn B. Martin and Martyna Boruta

9.1 Introduction

Cities differ dramatically from natural environments, which may explain why they have higher prevalence of some diseases and are often the source of epizootics (Bradley & Altizer, 2007). For most of the year, urban ambient temperatures are higher and less variable than surrounding areas (Grimm et al., 2008). Precipitation in cities too differs from suburbs and wildlands, and the relative impermeability of ground surfaces can foster both parasite transmission and breeding of vectors (Githeko et al., 2000). Pollutants also tend to be more concentrated in cities than any other locale except agricultural lands (Ellis, 2006). Light and noise pollution are also common and can alter the time budgets and hormone regulation of city-dwellers (Longcore & Rich, 2004). Habitat quality too tends to be poorer in cities, especially for some species with certain diets (Longcore & Rich, 2004) or roost and/or nest preferences (Chace & Walsh, 2006). All of these factors may interact in complex ways to reduce biodiversity (McKinney, 2002), limit urban wildlife to dense populations of a few urban-adapted (and often non-native) species (Shochat et al., 2006), and release or encourage certain diseases.

Our goal in this chapter is to review how cities impact avian disease transmission and emergence. Our approach is to summarize the relevant literature then integrate it with ideas derived from other fields, namely disease ecology and ecological immunology (Martin et al., 2011). We favour this approach because it should provide conceptual unification and direction to an area that is simply too young to warrant synthesis. As such efforts involve dissimilar lexicons, we provide a glossary (Box 9.1), which we hope will enable scientists from various backgrounds to find value in this chapter.

9.2 Avian disease in cities

Only a few avian host–parasite systems have been studied with respect to urbanization (Delgado & French, 2012), so generalizations about avian disease risk in cities (i.e. city conditions elevate risk) would be premature. The need for urban (avian) disease ecology is obvious though, given that many avian diseases have recently emerged from cities (Bradley & Altizer, 2007). However, what requires explanation is why urban conditions seem to have positive impacts on some diseases and negative impacts on others. Table 9.1 summarizes what we know about parasites, part of what causes disease, in urban birds. Although other data likely exist, these studies represent all of the examples (we could find) comparing infection differences between urban and non-urban birds. In essence, Table 9.1 reveals that (a) we know very little about urbanization effects on most parasites and (b) for parasites on which we do have information, study design has limited inference (i.e. low replication of sites in many studies). Ideally going forward, studies will either study urbanization gradients or replicate urban-rural site pairs. Even in well-designed studies, data collection seems to have been motivated by a few factors: (i) methodological simplicity (e.g. ectoparasites), (ii) a pre-existing, robust literature

Avian Urban Ecology. Edited by Diego Gil and Henrik Brumm

Box 9.1 Glossary of terms

competency—the capacity for a host to transmit a parasite to another host (or vector) (Keesing et al., 2006)

dilution host—a host with a disproportionate ability to prevent infection of other hosts (Keesing et al., 2006)

disease dilution—a reduction in the risk of infection driven by changes in host community composition (Keesing et al., 2006)

exposure—an encounter between an uninfected host and a parasite/vector

parasite—any viral, bacterial, protist, helminth or comparable organism that has a net-negative effect on a host (Dobson & Hudson, 1986; Raffel et al., 2008)

parasite burden/load—the number of parasites of one type harbored by an individual host

parasite prevalence—the proportion of a population infected by one parasite type

parasite resistance—the capacity of a host to reduce parasite burden, including prevention of infection at the time of exposure

parasite tolerance—the relationship between parasite burden and host fitness, typically summarized as the slope coefficient between burden and fitness (Baucom & de Roode, 2011; Raberg et al., 2007)

superspreader—a host with a disproportionate capacity to transmit parasites (Lloyd-Smith et al., 2005)

synanthrope—an organism that thrives in human-modified habitats

synurbic—an organism occurring in higher density in urban than rural areas (Francis & Chadwick, 2012)

urban—more than 1000 humans km^{-2} with >50% land impervious to water (MacGregor-Fors, 2011)

urban-adapter—urban-dwelling organism that adjusts successfully (via plasticity or evolution) to an urban area (Croci et al., 2008)

urban-exploiter—urban-dwelling organism that is pre-adjusted to an urban area (Croci et al., 2008)

urban-avoider—urban-dwelling organism that, given time, is likely to be extirpated from an urban area (Croci et al., 2008)

on a disease or parasite (e.g. avian malaria, *Borrelia burgdorferi*), and/or (iii) a recent disease emergence (e.g., West Nile virus, WNV). Few of these studies seem to have taken a prospective approach to account systematically for mode of transmission (i.e. vector or contact), virulence, or other key characteristics of parasites, hosts, or environmental conditions in cities.

One would expect that parasite life history traits would influence disease patterns strongly, and several studies have invoked such traits as explanations for patterns in cities. For instance, parasites carried by vectors often mirror vector distribution patterns. Some but not all ticks, many of which vector other parasites, tend to be more common in natural than urban areas (Evans et al., 2009; Gregoire et al., 2002; Hamer et al., 2012). Such patterns are not altogether surprising, given that ticks (and other ectoparasites) often have strict microhabitat requirements (LoGiudice et al., 2003). As might be expected, Lyme disease, caused by the tick-borne (*Ixodes* sp.) pathogen, *Borrelia burdorferi*, is rarer in urban than intact woodland habitats in the northeastern US (Keesing et al., 2006).

For flying insect vectors, results are more mixed. West Nile prevalence (vectored by *Culex* sp. and other mosquitoes) is often higher in cities, although many results have been extrapolated from exposure data (WNV + seroprevalence), which is greater in some but not all urban birds (Bradley et al., 2008; Hamer et al., 2012; Kilpatrick, 2011). However, *Culex* densities (Loss et al., 2009) and WNV impacts on hosts (Koenig et al., 2010; LaDeau et al., 2007) tend to be higher in urban areas, suggesting that at least some vector-borne parasites are more common in cities. Others though, namely avian malarial parasites (e.g. *Plasmodium* sp. and *Haemoproteus* sp.), have been more variable (Geue & Partecke, 2008). In Britain, malaria prevalence was lower in 8 of 11 city/wildland pairs (Evans et al., 2009); in Brazil though, malaria prevalence (and diversity) was higher in an urban rather than a rural site (Belo et al., 2011),

Table 9.1 Summary of salient results from urban avian disease ecology studies.

Study type	Urbanization effect on disease prevalence	Parasite types			References
		Vector-borne	Contact		
Urbanization gradient	+	4	0	*Anaplasma phagocytophilum, Amblyomma aureolatum, Ixodes auritulus,* Flavivirus	Arzua et al., 2003, Bradley et al., 2008, Gibbs et al., 2006, Hamer et al., 2012
	–	1	0	*Philornis porteri*	LeGros et al., 2011
	no difference	0	3	*Borrelia burgdoferi, Escherichia coli, Salmonella* spp.	Hamer et al., 2012, Kullas et al., 2002
Replicate urban/rural populations	+	1	2	*Flavivirus, Trichomonas gallinae*	Boal & Mannan, 1999, Boal et al., 1998, Ringia et al., 2002
	–	4	0	*Ixodes auritulus, Haemoproteus* spp., *Plasmodium* spp., *Trypanosoma* spp., *Haemoproteus* spp., Microfilariae, Western Equine encephalitis	Evans et al., 2009, Fokidis et al., 2008, Gregoire et al., 2002, McLean et al., 1989
	no difference	3	2	*Trichomonas gallinae, Escherichia coli, Haemoproteus spp., Pasteurella multocida, Salmonella* spp., *Mycoplasma* spp., Flavivirus, New Castle disease virus, Avian influenza virus	Bonier et al., 2006, Loss et al., 2009, Morishita et al., 1999, Rosenfield et al., 2002, Sol et al., 2000
One urban-one rural site comparisons	+	0	0		
	–	2	0	*Haemoproteus* spp., *Plasmodium* spp., *Leucocytozoon* spp., *Trypanosoma* spp., Microfilariae	Bentz et al., 2006, Geue & Partecke, 2008
	no difference	2	1	*Trichomonas gallinae, Haemoproteus* spp., *Plasmodium* spp.	Belo et al., 2011, Ots & Horak, 1998, Rosenfield et al., 2009

although no pattern was discernible in the one Brazilian host species found in both urban and rural areas. Results for contact-transmitted parasites have been comparably complex. In some cases, parasites are so rare that urbanization effects are either absent or hard to detect statistically, as was the case for *Salmonella* in Chicago, IL (USA; <1% prevalence) and its surrounding suburbs (Hamer et al., 2012). We expect that some parasites may be quite common in cities, especially those that benefit from concentration of host activities in time or space (e.g. *Mycoplasma* sp. (Dhondt et al., 1998) or *Trichomonas* sp. (Robinson et al., 2010)).

9.3 Mechanisms mediating disease risk in cities

More than two thirds of the human population will live in cities by 2040 (Bradley & Altizer, 2007). Subsequently, the threat of spillover (Power & Mitchell, 2004) or spillback (Kelly et al., 2009) from cities will increase as global commerce grows. These events might diversify or genetically invigorate (admix) parasite communities while also encouraging the spread of certain urban-adapted hosts (Carrete & Tella, 2011), many of which are disproportionately able to harbour certain parasites (Dubska et al., 2009; Dubska et al., 2011; Hamer et al., 2009). Below, we review some ecological and organismal mechanisms that probably mediate avian disease risk in cities.

9.3.1 Ecological mechanisms

Risk of infection in cities is likely impacted by host community diversity. In some cases, high biodiversity can elevate infection risk by increasing the chances of exposure to a parasite (Keesing et al., 2006). Most documented cases, however, indicate protective effects of host diversity on infection risk (Keesing et al., 2010). Although many mechanisms are plausible, disease dilution occurs as non-competent hosts lower parasite prevalence by misdirecting parasite or vector attacks (Keesing et al., 2006). As urban avian communities tend to be simple (Chace & Walsh, 2006), dilution effects would likely be weak to absent in most cities. In support, two of the best predictors of American crow (*Corvus brachyrhynchus*) mortality to WNV were the diversity of the passerine community (negatively related to WNV mortality) and the extent of urbanization of a site (positively related to WNV mortality (Koenig et al., 2010)). The most compelling mechanism explaining WNV dilution effects in cities at present is that communities differ in competency for WNV; where competency is low and diversity is high, crows are well-protected (Hamer et al., 2009; Loss et al., 2009). Although it is yet unknown whether risk of other diseases in cities is due to low host diversity, the various host–parasite systems in which dilution has been documented (Keesing et al., 2010) emphasizes the importance of follow-up work.

Still, what mechanisms link changes in host diversity to dilution? One possibility is predation. Although adult crows and other large passerines suffer little predation, given the changes in predation risk observed in cities versus wildlands (McKinney, 2002), dilution effects via predation warrant study. For many birds, nest predation is a significant source of mortality (Martin & Clobert, 1996); however, in cities, it tends to be comparatively low (Ryder et al., 2010). With lower predation, host population densities are likely to increase and become dominated by comparatively older birds, many of which may have generated immune memory to previously encountered parasites (Evans et al., 2009). High host density and a right-shifted age distribution both should impact infection risk (Shochat, 2004). However, as the effects of parasites can differ whether their transmission is frequency or density dependent (Fenton et al., 2002), the effects of high density are difficult to predict, particularly when coupled with an older average population (and potential effects of memory of prior infections). Density- and age-driven effects of avian disease in cities are thus worthy of study. In the American crow study above (Koenig et al., 2010), increases in WNV-driven mortalities were strongly predicted by crow population density prior to WNV arrival in 1999.

Other ecological mechanisms for urban disease risk warrant study too. For the majority of species studied, urban conditions rarely enhance productivity (Chamberlain et al., 2009). With a few exceptions (i.e. urban-exploiters (Stracey & Robinson, 2012)), reproductive output per year is comparable between urban and wildland sites although timing of breeding is often altered (Scheoech & Bowman, 2003). Differences in reproductive timing could be important to disease cycles; breeding seasons in cities tend to start earlier (Partecke et al., 2005) and last longer (Diego Ibañez-Alamo & Soler, 2010), biasing the age-structure of populations towards flushes of immunologically naïve juveniles for most of the year. Adult survival prospects are more strongly impacted than reproduction in cities, although effects can be age dependent (Stracey & Robinson, 2012). The most consistent effect of cities on survival is a reduction in over-winter mortality (Evans et al., 2011). Intriguingly, survival effects are more pronounced for some species than others (Whittaker & Marzluff, 2009); these patterns may arise because species that cope best with urban conditions have an advantage over those forced to move into or remain in cities (Shochat et al., 2006). When increased density is coupled with increased predictability (but not necessarily abundance) of food, individual contact rates increase concurrently with variance in individual quality (Shochat, 2004). This broader distribution of individuals might encourage the transmission of certain parasites (Beldomenico & Begon, 2010), especially if individuals in poor health are disproportionately responsible for infections. The effects of density in cities appear strong within-species, even for urban exploiters like house sparrows (*Passer domesticus*), European blackbirds (*Turdus merula*), and American crows. Among seven populations, urban house sparrows were smaller and in poorer condition than rural birds (Liker et al., 2008), urban European blackbirds had lower fat reserves than rural birds (Partecke et al., 2005), and rural American crows were substantially larger and presumed healthier (as reflected by generic blood chemistry measures) than urban crows (Heiss et al., 2009).

9.3.2 Organismal mechanisms: behaviour

Interspecific variation in avian behaviour, in many forms, is likely to be important to disease in cities. Alterations in nesting height could alter host exposure to vectors (Wang et al., 2008), changes in propensity to roost communally could impact contact rates (Everding & Jones, 2006), and changes in migratory or dispersal behaviour could impact whether city-dwelling individuals move parasites long distances (Partecke & Gwinner, 2007). Even changes in diet could affect exposure, especially for predatory birds that can contract parasites by eating infected hosts (Estes & Mannan, 2003). Importantly, behavioural responses of birds to humans could impact zoonotic risk. Some urban species appear to adjust their activity patterns to human schedules (Ditchkoff et al., 2006), and many can quickly become less fearful of humans (Carrete & Tella, 2011; Evans et al., 2010; Atwell et al., 2012). Whereas all of these mechanisms warrant future study, we focus below on superspreaders (Lloyd-Smith et al., 2005) and dilution hosts (Hawley & Altizer, 2011; Keesing et al., 2006). Superspreaders and dilution hosts represent ends of a continuum with respect to transmission competency: superspreaders are disproportionately responsible for escalating disease risk whereas (less studied) dilution hosts (Keesing et al., 2009a) should quell it.

Contrary to theory underlying many early epidemiological models, it sometimes matters less to how many individuals a host is connected than it does which individuals a host contacts (Drewe, 2010). These ideas are encapsulated by the 20/80 rule, which emphasizes that 80% of infections are caused by 20% of a population (Woolhouse et al., 1997), individuals identified as superspreaders. Determining which organisms are, or can become, superspreaders (Hawley & Altizer, 2011) is important and not just for disease risk in cities (Tompkins et al., 2011). As Galvani and May (2005) wrote, 'heterogeneous infectiousness and its extreme, superspreading, are likely to be general properties of disease transmission in populations'. Data suggest that avian superspreading in cities can occur at species, population, and even individual levels. At the species level for instance, we know that three abundant (migratory) avian synanthropes in Europe are among the most competent (Martin et al., 2010) reservoirs for *Borrelia garini*, a Lyme's disease parasite (Dubska et al., 2009). Also, in the US, urban exploiters and/or adapters are often among the most competent species for WNV (Kilpatrick et al., 2006), and many are preferred for feeding by *Culex* sp. vectors (Hamer et al., 2009). In Britain, bird species with highest urban densities are also those with the broadest geographic distribution; although it is unknown whether a broad geographic distribution influences superspreading potential, previous linkages between species and population life histories and their immune defences emphasize the need for additional study (Martin et al., 2006).

At the individual level, some birds appear more adept at enduring cities than others, but whether this endurance impacts their potential as superspreaders is now unclear. One of the earliest urban ornithology studies found that experimental introductions of European blackbirds to cities in which blackbirds were not then present were more successful when urban than rural blackbirds were used (Graczyk, 1982). A more recent study found that urban blackbird populations in European cites came from repeated colonization by independent populations (Evans et al., 2009). At a broader scale, 'tame' individuals (within several species) were found more successful at colonizing cities (Carrete & Tella, 2011). What these data emphasize is that some individuals are exceptionally flexible, enough to exploit non-natural (urban) conditions. If the capacity to be flexible alters how hosts deal with parasites, urban birds may be more disposed to act as superspreaders (or dilution hosts) than members of natural communities.

Cases in point are common non-native species, which thrive in cities world-wide. Although only a few such species have been studied, their antiparasite coping strategies may make them prone to superspreading (Martin et al., 2010). Although no definitive avian examples of this possibility yet exist, in grey squirrels (*Sciurus carolinensis*), something about the host species enables it to use a paramyxovirus as a novel weapon to outcompete the native red squirrel (*Sciurus vulgaris*) (Tompkins et al., 2003). If introduced species and urban exploiters and/or adapters share similar approaches to dealing with parasites, there is reason to expect that superspreading via introduced species underlies some disease phenomena in cities (Prenter et al., 2004). For one invader however, the house sparrow, enemy release, or the loss of parasites during or prior to introduction, (Torchin et al., 2003), not novel weapons (Callaway & Ridenour, 2004), seems a more viable hypothesis for explaining its global distribution (Marzal et al., 2011). Still, future work on other species and various time scales of invasion are critical to reveal the significance of superspreading in cities.

9.3.3 Organismal mechanisms: physiology

Two physiological mechanisms might be particularly important for superspreading and disease transmission and emergence more generally in urban birds: stress hormones and immune responses. Stress hormones such as corticosterone, the dominant avian stress steroid (Romero, 2004), are notorious for altering immune functions and behaviours relevant to disease. Thus, corticosterone should impact host–parasite interactions in many ways in urban environments. For instance, social conflicts often elevate corticosterone (Hawley et al., 2006), and urban dwelling organisms at high densities might experience such challenges more often, which could affect how they cope with infections. Pollutants, exposure to introduced species, and other factors of urban environments could also impact regulation of corticosterone (Martin et al., 2010), leading to different interactions with parasites than would be observed in natural environments. On the other hand, a recent review provided little evidence that corticosterone is regulated differently in urban versus non-urban birds (Bonier, 2012): some species increase it in urban conditions (Bonier et al., 2007; Zhang et al., 2011) whereas others reduce it (Partecke et al., 2006; Atwell et al., 2012) and in others it is unaffected (Bokony et al., 2012). The effects of urbanization on corticosterone thus appear contingent on the species studied, the larger environmental matrix in which cities occur, and/or the aspect of the corticosterone regulatory process considered (Fokidis et al., 2009). Nevertheless, it would be premature to conclude that such complexity means that stress hormones are not impactful of host–parasite interactions in cities. More likely, organismal responses to urbanization, not to mention urbanization itself, are too heterogeneous for simple generalizations. In other words, we know that stress hormones are integral to the coordination of animal behaviour and physiology, especially immune responses (Martin, 2009); perhaps urban-adapter and urban-exploiter species thrive in urban conditions and only urban-avoiders suffer stress (Bonier, 2012) and its requisite effects on disease.

We expect that such a nuanced perspective will eventually elucidate the role of stress hormones and immunity in cities. So far, most urban avian ecological immunology has been opportunistic, characterizing immune parameters that are comparatively simple to measure (Jacquin et al., 2011) but difficult to interpret (Martin et al., 2006). For instance, urban house sparrow populations can be exposed to more pollutants than rural (Chandler et al., 2004), and pollutants can have profound effects on vertebrate immune functions (Martin et al., 2010). However, one study found that total immunoglobulins (antibodies) did not differ among three populations (urban, suburban, and rural) of house sparrows (Chavez-Zichinelli et al., 2010); whereas one could interpret such results as a lack of urbanization effect on immunity, such a conclusion would be overstated. The immune system is complex. To develop a viable urban ecological immunology, more sophisticated approaches will be necessary.

As available data are so few, we propose the following approach for urban avian ecological immunology (Figure 9.1). This approach emphasizes several key concepts. First, it emphasizes that the host is but one member of a complex interaction, so changes in host phenotype alone may be insufficient to understand why and how urban conditions impact infection risk. Vector choice of hosts, for instance, varies quite dramatically, both among (Simpson et al., 2009) and within (Burkett-Cadena et al., 2011) species, and these vector behaviours can have strong effects on parasite transmission (Simpson et al., 2012). Second, it emphasizes that host interactions with the parasite (and vector) can be fixed or plastic. Specific resistance (i.e. gene-for-gene coevolution) has been the focus of the majority of disease ecology and epidemiology, but a large fraction of host coping mechanisms entails plastic responses with varying levels of specificity, namely induced immune responses. It will be critical to collect data on such responses, over various timescales and incorporate them into new models in conjunction with subtler recognition of the role of immune memory. Third, host-parasite interactions occur in nested stages (Schmid-Hempel & Ebert, 2003); individuals must first encounter a parasite before they resist an infection. If some parasites can be avoided behaviourally, it probably makes little sense to investigate such parasites in cities. After exposure, a host can resist or tolerate an infection (see below), and the outcome of this balancing act likely determines the role of a particular host in a disease cycle. Indeed, the last stage in the interaction involves the host's capacity to transmit the parasite (i.e. competency), but this transmission is the culmination of all upstream processes. The fourth point of emphasis in Figure 9.1 comprises what continues to be a major issue in ecological immunology: what to measure (Boughton, Joop & Armitage, 2011). Each box includes various factors that could reveal the mechanistic points of vulnerability of a host. Fortunately, many of these techniques require very small tissue/blood volumes, have also been measured previously in birds, and can be performed with little training and equipment.

One suite of defences that would seem to be particularly important for disease cycles in cities is sickness behaviour (Adelman & Martin, 2009). These behaviours include traits that influence the extent to which individuals contact each other and expose themselves to parasites and vectors (Hawley & Altizer, 2011). One study found that sick individuals were much less likely to engage in aggression at bird feeders, enabling healthy (and perhaps lower-quality) individuals to feed nearby; such effects of sickness behaviour, if common, could further enhance risk of parasite transmission at feeders (Bouwman & Hawley, 2010), which can be common in cities. Furthermore, many sickness behaviours are impacted by changes in steroid hormones, which if impacted by urban conditions, sickness could be exacerbated or attenuated, altering propensity to transmit disease (Tompkins et al., 2011). Urban dark-eyed juncos (*Junco hyemalis*) tend to be less aggressive, so perhaps for some diseases, attenuated aggression (and its physiological mediators) may reduce risk of transmission in cities (Newman et al., 2006).

Another parasite coping mechanism warranting special attention in cities is tolerance (Raberg et al., 2007). Although sample sizes are yet small, multiple urban-adapting birds, and especially non-native ones, can endure infections with minimal negative effects on performance (Lee et al., 2005). This high parasite tolerance may enable individuals to remain infectious for longer periods than others or spread

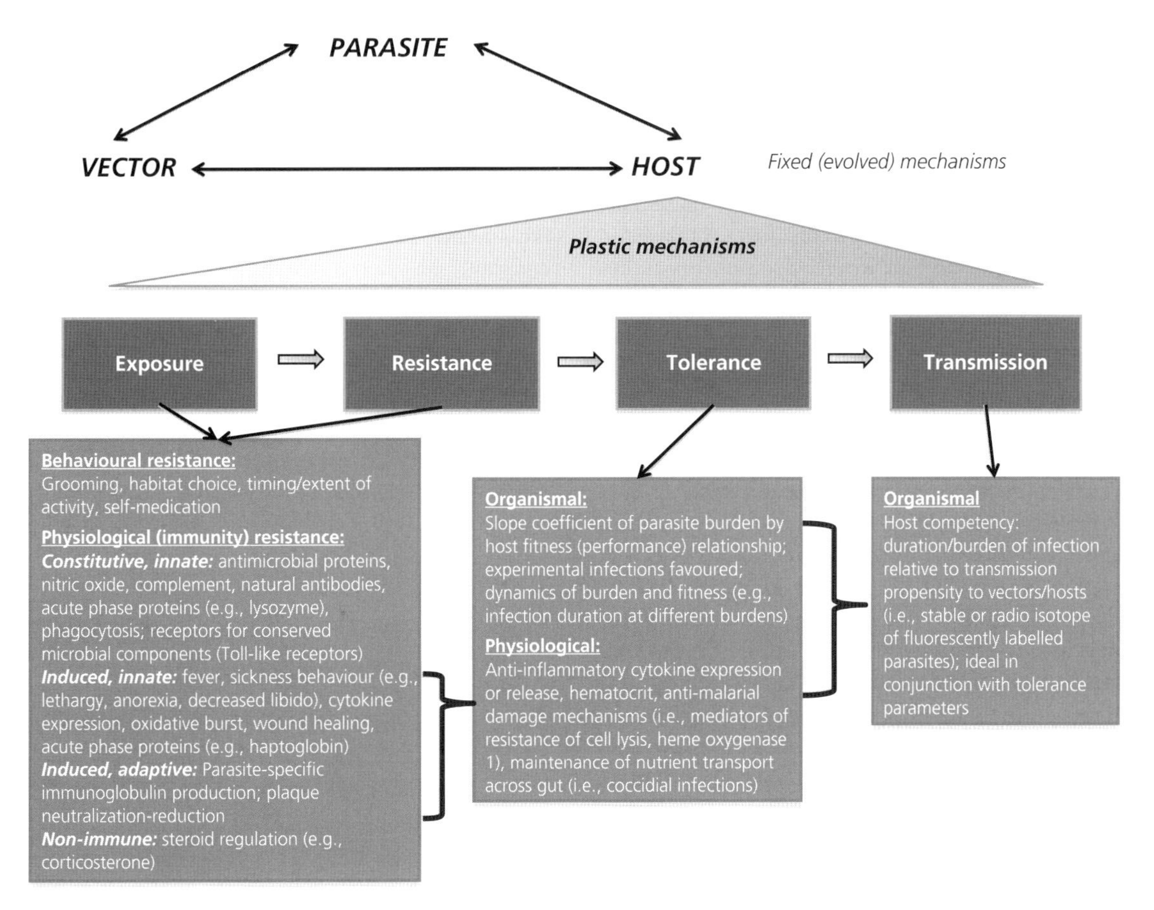

Figure 9.1 Plastic host mechanisms mediating the outcomes of host–parasite interactions. Host traits might involve fixed vulnerabilities to infection (e.g. specific host cell receptor susceptibility to viral infection) or plastic variation (shaded triangle and dark grey boxes below). Plastic variation can occur at multiple levels (dark grey boxes): host exposure to parasites (and/or vectors, including vector feeding choice), host resistance to parasites, host tolerance of successful infections, and host competency to transmit parasite to other hosts. Urbanization could impact hosts at one or more levels, however, arrows depict the nestedness of host-parasite interactions (i.e. that probability of transmission of a parasite between individuals is approximately the product of parasite transition probabilities between stages; Schmid-Hempel and Ebert 2003). Generally, urban factors leading to enhancements of traits in left-most boxes would tend to reduce the suitability of urban hosts for parasites whereas reductions in traits in the left boxes (and/or positive effects on the right side) would increase the importance of urban species/individuals in disease transmission. Light grey boxes on the lower half of the figure list parameters that could be measured in wild birds (summarized in Boughton et al., 2011) to test hypotheses about the impacts of urbanization on host-parasite interactions.

parasites more broadly than hosts that become moribund upon infection. Indeed, a promising area of future ecoimmunology research in general is to reconcile links between host tolerance and competency: differential competency is well-known for WNV among birds (Kilpatrick et al., 2007) and for *Borrelia burdorferi* among birds and mammals (Keesing et al., 2009b). The extent to which competency and tolerance are related, much less whether urban conditions affect the strength of their relationship, is yet unknown. We propose that high tolerance often equates to high competency and underlies superspreading. What would be valuable to learn next is what organismal traits comprise dilution hosts and whether such traits can be encouraged in cities.

9.4 Future studies

We have only scratched the surface in understanding avian host–parasite interactions in cities

(Bradley & Altizer, 2007). One of the most obvious areas for future work is the spatiotemporal scale of studies (Evans et al., 2011); heterogeneity in urban habitats abounds and such variation needs to be addressed (Shochat et al., 2006). Studies should account explicitly for the scales over which specific hosts, parasites and their vectors interact. A second key topic is co-infection; when two or more parasites infect a host, consequences at the individual, population and community levels can change relative to single infections (Pedersen & Fenton, 2007). In cities, parasite diversity (like host diversity) is probably lower than outlying areas, but the impacts of particular suites of coinfecting parasites may have critical implications for the health of avian communities. A third area warranting further study is how urban communities develop: are urban adapters and/or exploiters successful because they adapt genetically, adjust plastically, or both (Carrete & Tella, 2011)? This area is important because parasite-directed defences often co-vary (Brunner & Ostfeld, 2008), so if responses to urbanization are constrained, certain species may be less able to endure cities or more apt to spillover or spillback parasites (Chamberlain et al., 2009). Further, if most heterogeneity in superspreading arises via plasticity, disease risk could fluctuate quite rapidly, often much faster than via evolutionary change.

Among all the factors that impact avian disease in cities, perhaps the most controllable is human provisioning of resources. Although landfills, restaurants, food processing sites, and farms all likely impact avian disease cycles (and epizootic risk), the maintenance and use of bird feeders might be the one that, if modified, could rapidly and most profoundly impact urban health. Provisioning of food by humans has become incredibly popular; 43% of US and ~75% of UK households regularly feed birds (Robb et al., 2008). Bird feeding impacts almost all aspects of avian biology (Robb et al., 2008), including disease risk; feeders have been implicated in outbreaks of salmonellosis, trichomoniasis, aspergillosis, avian pox and avian mange (Brittingham & Temple, 1986; Nesbitt & White, 1974), the recent spread of mycoplasmosis in the US (Fischer et al., 1997), and die-offs of greenfinches (*Carduelis chloris*) and other species in Europe (Robinson et al., 2010). These outcomes have motivated the Garden Bird Health Initiative (UK) and Project Feeder Watch (Cornell Lab of Ornithology, USA) to disseminate guidelines for feeder use and maintenance. Fortunately, learning more about effects of feeders on disease will have both practical and theoretical value (Robb et al., 2008). For instance, feeders seem to have greater impacts on survival than reproductive output (Robb et al., 2008), suggesting that feeder effects on disease are most likely to occur by increasing host contacts with each other, concentrating hosts for vectors, or enabling infected hosts to endure conditions (e.g. winter) that might otherwise cause mortality. Still, as so few feeder studies have been performed, it is as yet unclear how feeders affect avian infection risk in cities.

9.5 Conclusion

Urbanization impacts on avian communities will continue to grow as humans further colonize and expand cities and modify the environment (Bradley & Altizer, 2007). We understand very little at this point about urban avian host-parasite ecology (Delgado & French, 2012), except for a few parasites with implications for human health (Kilpatrick, 2011). Interdisciplinary approaches will likely be most effective to help alleviate (or eliminate) future emergences of diseases circulating in wild birds and sometimes spilling over into humans (Lloyd-Smith et al., 2009).

Acknowledgements

The authors thank Diego Gil and two anonymous referees for comments on a prior draft. L.B. Martin recognizes NSF-IOS grants 0920475 and 0947177 for support during the writing of this chapter.

References

Adelman, J. and Martin, L. (2009). Vertebrate sickness behavior: an adaptive and integrated neuroendocrine immune response. *Integrative and Comparative Biology*, **49**, 202–214.

Arzua, M., Da Silva, M. A. N., Famadas, K. M., et al. (2003). *Amblyomma aureolatum and Ixodes auritulus* (Acari: Ixodidae) on birds in southern Brazil, with notes on their ecology. *Experimental and Applied Acarology*, **31**, 283–296.

Atwell, J. W., Cardoso, G. A. C., Whittaker, D.J., et al. (2012). Boldness behavior and stress physiology in a novel urban environment suggest rapid correlated evolutionary adaptation. *Behavioral Ecology*, **5**, 960–969.

Baucom, R. S. and de Roode, J. C. (2011). Ecological immunology and tolerance in plants and animals. *Functional Ecology*, **25**, 18–28.

Beldomenico, P. M. and Begon, M. (2010). Disease spread, susceptibility and infection intensity: vicious circles? *Trends in Ecology & Evolution (Personal edition)*, **25**, 21–27.

Belo, N. O., Pinheiro, R. T., Reis, E. N. S., et al. (2011). Prevalence and lineage diversity of avian haemosporidians from three distinct cerrado habitats in Brazil. *PLoS One*, **6**, e17654.

Bentz, S. T., Rigaud, T., Barroca, M., et al. (2006). Sensitive measure of prevalence and parasitaemia of haemosporidia from European blackbird (*Turdus merula*) populations: value of PCR-RFLP and quantitative PCR. *Parasitology* **133**, 685–692.

Boal, C. W., Mannan, R. W., and Hudelson, K. S. (1998). Trichomoniasis in Copper's Hawks from Arizona. *Journal of Wildlife Diseases*. **34**, 590–593.

Boal, C. W. and Mannan, R. W. (1999). Comparative breeding ecology of Cooper's hawks in urban and exurban areas of Southeastern Arizona. *The Journal of Wildlife Management*, **63**, 77–84.

Bokony, V., Seress, G., Nagy, S., et al. (2012). Multiple indices of body condition reveal no negative effect of urbanization in adult house sparrows. *Landscape and Urban Planning*, **104**, 75–84.

Bonier, F. (2012). Hormones in the city: Endocrine ecology of urban birds. *Hormones and Behavior*, **61**, 763–772.

Bonier, F., Martin, P. R., Sheldon, K. S., et al. (2007). Sex-specific consequences of life in the city. *Behavioral Ecology*, **18**, 121–129.

Boughton, R. K., Joop, G. and Armitage, S. A. O. (2011). Outdoor immunology: methodological considerations for ecologists. *Functional Ecology*, **25**, 81–100.

Bouwman, K. M. and Hawley, D. M. (2010). Sickness behaviour acting as an evolutionary trap? Male house finches preferentially feed near diseased conspecifics. *Biology Letters*, **6**, 462–465.

Bradley, C. A. and Altizer, S. (2007). Urbanization and the ecology of wildlife diseases. *Trends in Ecology & Evolution*, **22**, 95–102.

Bradley, C. A., Gibbs, S. E. J. and Altizer, S. (2008). Urban land use predicts West Nile Virus exposure in songbirds. *Ecological Applications*, **18**, 1083–1092.

Brittingham, M. C. and Temple, S. A. (1986). A survey of avian mortality at winter feeders. *Wildlife Society Bulletin*, **14**, 445–450.

Brunner, J. L. and Ostfeld, R. S. (2008). Multiple causes of variable tick burdens on small-mammal hosts. *Ecology*, **89**, 2259–2272.

Burkett-Cadena, N. D., McClure, C. J. W., Ligon, R. A., et al. (2011). Host reproductive phenology drives seasonal patterns of host use in mosquitoes. *PLoS One*, **6**, e17681.

Callaway, R. M. and Ridenour, W. M. (2004). Novel weapons: invasive success and the evolution of increased competitive ability. *Frontiers in Ecology and the Environment*, **2**, 436–443.

Carrete, M. and Tella, J.L. (2011). Inter-individual variability in fear of humans and relative brain size of the species are related to contemporary urban invasion in birds. *PLoS One*, **6**, 1–8.

Chace, J. F. and Walsh, J. J. (2006). Urban effects on native avifauna: a review. *Landscape and Urban Planning*, **74**, 46–69.

Chamberlain, D. E., Cannon, A. R., Toms, M. P., et al. (2009). Avian productivity in urban landscapes: a review and meta-analysis. *Ibis*, **151**, 1–18.

Chandler, R. B., Strong, A. M. and Kaufman, C. C. (2004). Elevated lead levels in urban house sparrows: a threat to sharp-shinned hawks and merlins? *Journal of Raptor Research*, **38**, 62–68.

Chavez-Zichinelli, C. A., MacGregor-Fors, I., Talamas Rohana, P., et al. (2010). Stress responses of the House Sparrow (Passer domesticus) to different urban land uses. *Landscape and Urban Planning*, **98**, 183–189.

Croci, S., Butet, A. and Clergeau, P. (2008). Does urbanization filter birds on the basis of their biological traits? *Condor*, **110**, 223–240.

Delgado, C. A. V. and French, K. (2012). Parasite-bird interactions in urban areas: Current evidence and emerging questions. *Landscape and Urban Planning*, **105**, 5–14.

Dhondt, A. A., Tessaglia, D. L. and Slothower, R. L. (1998). Epidemic mycoplasmal conjunctivitis in house finches from Eastern North America. *Journal of Wildlife Diseases*, **34**, 265–280.

Diego Ibanez-Alamo, J. and Soler, M. (2010). Investigator activities reduce nest predation in blackbirds Turdus merula. *Journal of Avian Biology*, **41**, 208–212.

Ditchkoff, S. S., Saalfeld, S. T. and Gibson, C. J. (2006). Animal behavior in urban ecosystems: Modifications due to human-induced stress. *Urban Ecosystems*, **9**, 5–12.

Dobson, A. P. and Hudson, P. J. (1986) Parasites, Disease and the Structure of Ecological Communities. *Trends in Ecology & Evolution*, **1**, 11–15.

Drewe, J. A. (2010). Who infects whom? Social networks and tuberculosis transmission in wild meerkats. *Proceedings of the Royal Society B-Biological Sciences*, **277**, 633–642.

Dubska, L., Literak, I., Kocianova, E., et al. (2011). Synanthropic birds influence the distribution of borrelia species: analysis of ixodes ricinus ticks feeding on passerine birds. *Applied and Environmental Microbiology*, **77**, 1115–1117.

Dubska, L., Literak, I., Kocianova, E., et al. (2009). Differential role of passerine birds in distribution of borrelia spirochetes, based on data from ticks collected from birds during the postbreeding migration period in central Europe. *Applied and Environmental Microbiology*, **75**, 596–602.

Ellis, J.B. (2006). Pharmaceutical and personal care products (PPCPs) in urban receiving waters. *Environmental Pollution*, **144**, 184–189.

Estes, W. A. and Mannan, R. W. (2003). Feeding behavior of Cooper's Hawks at urban and rural nests in southeastern Arizona. *Condor*, **105**, 107–116.

Evans, J., Boudreau, K. and Hyman, J. (2010). Behavioural syndromes in urban and rural populations of song sparrows. *Ethology*, **116**, 588–595.

Evans, K. L., Chamberlain, D. E., Hatchwell, B. J., et al. (2011). What makes an urban bird? *Global Change Biology*, **17**, 32–44.

Evans, K. L., Gaston, K. J., Sharp, S. P., et al. (2009). Effects of urbanisation on disease prevalence and age structure in blackbird Turdus merula populations. *Oikos*, **118**, 774–782.

Everding, S. E. and Jones, D. N. (2006). Communal roosting in a suburban population of Torresian crows (*Corvus orru*). *Landscape and Urban Planning*, **74**, 21–33.

Fenton, A., Fairbairn, J. P., Norman, R., et al. (2002). Parasite transmission: reconciling theory and reality. *Journal of Animal Ecology*, **71**, 893–905.

Fischer, J. R., Stallknecht, D. E., Luttrell, M. P., et al. (1997). Mycoplasmal conjunctivitis in wild songbirds: The spread of a new contagious disease in a mobile host population. *Emerging Infectious Diseases*, **3**, 69–72.

Fokidis, H. B., Greiner E. C., and Deviche, P. (2008). Interspecific variation in avian blood parasites and haematology associated with urbanization in a desert habitat. *Journal of Avian Biology*, **30**, 300–310.

Fokidis, H. B., Orchinik, M. and Deviche, P. (2009). Corticosterone and corticosteroid binding globulin in birds: Relation to urbanization in a desert city. *General and Comparative Endocrinology*, **160**, 259–270.

Francis, R. A. and Chadwick, M. A. (2012). What makes a species synurbic?*Applied Geography*, **32**, 514–521.

Galvani, A. P. and May, R. M. (2005). Epidemiology: dimensions of superspreading. *Nature*, **438**, 293–295.

Geue, D. and Partecke, J. (2008). Reduced parasite infestation in urban Eurasian blackbirds (*Turdus merula*): a factor favoring urbanization? *Canadian Journal of Zoology*, **86**, 1419–1425.

Gibbs, S. E. J., Wimberly, M. C., Madden, M., et al. (2006). Factors affecting the geographic distrubution of West Nile virus in Georgia, USA: 2002–2004. *Vector-borne and Zoonotic Diseases*, **6**, 73–82.

Githeko, A. K., Lindsay, S. W., Confalonieri, U. E., et al. (2000). Climate change and vector-borne diseases: a regional analysis. *Bulletin of the World Health Organization*, **78**, 1136–1147.

Graczyk, R. (1982). Ecological and ethological aspects of synanthropization of birds. *Memorabilia Zoologica*, **37**, 79–91.

Gregoire, A., Faivre, B., Heeb, P., et al. (2002). A comparison of infestation patterns by Ixodes ticks in urban and rural populations of the common blackbird *Turdus merula*. *Ibis*, **144**, 640–645.

Grimm, N. B., Faeth, S. H., Golubiewski, N. E., et al. (2008). Global change and the ecology of cities. *Science*, **319**, 756–760.

Hamer, G. L., Kitron, U. D., Goldberg, T. L., et al. (2009). Host selection by *Culex pipiens* mosquitoes and West Nile Virus amplification. *American Journal of Tropical Medicine and Hygiene*, **80**, 268–278.

Hamer, S. A., Lehrer, E. and Magle, S. B. (2012). Wild birds as sentinels for multiple zoonotic pathogens along an urban to rural gradient in greater Chicago, Illinois. *Zoonoses and Public Health*, **59**, 355–364.

Hawley, D. M. and Altizer, S. M. (2011). Disease ecology meets ecological immunology: understanding the links between organismal immunity and infection dynamics in natural populations. *Functional Ecology*, **25**, 48–60.

Hawley, D. M., Lindstrom, K. and Wikelski, M. (2006). Experimentally increased social competition compromises humoral immune responses in house finches. *Hormones and Behavior*, **49**, 417–424.

Heiss, R. S., Clark, A. B. and McGowan, K. J. (2009). Growth and nutritional state of American Crow nestlings vary between urban and rural habitats. *Ecological Applications*, **19**, 829–839.

Jacquin, L., Lenouvel, P., Haussy, C., et al. (2011). Melanin-based coloration is related to parasite intensity and cellular immune response in an urban free living bird: the feral pigeon Columba livia. *Journal of Avian Biology*, **42**, 11–15.

Keesing, F., Belden, L. K., Daszak, P., et al. (2010). Impacts of biodiversity on the emergence and transmission of infectious diseases. *Nature*, **468**, 647–652.

Keesing, F., Brunner, J., Duerr, S., et al. (2009a). Hosts as ecological traps for the vector of Lyme disease. *Proceedings of the Royal Society B-Biological Sciences*, **276**, 3911–3919.

Keesing, F., Brunner, J., Duerr, S., et al. (2009b). Hosts as ecological traps for the vector of Lyme disease. *Proceedings of the Royal Society B-Biological Sciences*,**1675**, 3911–3919.

Keesing, F., Holt, R.D. and Ostfeld, R.S. (2006). Effects of species diversity on disease risk. *Ecology Letters*, **9**, 485–498.

Kelly, D. W., Paterson, R. A., Townsend, C. R., et al. (2009). Parasite spillback: A neglected concept in invasion ecology? *Ecology*, **90**, 2047–2056.

Kilpatrick, A. M. (2011). Globalization, land use, and the invasion of West Nile virus. *Science*, **334**, 323–327.

Kilpatrick, A. M., Daszak, P., Jones, M. J., et al. (2006). Host heterogeneity dominates West Nile virus transmission. *Proceedings of the Royal Society B-Biological Sciences*, **273**, 2327–2333.

Kilpatrick, A. M., LaDeau, S. L. and Marra, P. P. (2007). Ecology of west nile virus transmission and its impact on birds in the western hemisphere. *Auk*, **124**, 1121–1136.

Koenig, W. D., Hochachka, W. M., Zuckerberg, B., et al. (2010). Ecological determinants of American crow mortality due to West Nile virus during its North American sweep. *Oecologia*, **163**, 903–909.

Kullas, H., Coles, M., Rhyan, J., et al. (2002). Prevalence of *Escherichia coli* serogroups and human virulence factors in faeces of urban Canada geese (*Branta canadensis*). *International Journal of Environmental Health Research*, **12**, 153–162.

LaDeau, S. L., Kilpatrick, A. M. and Marra, P. P. (2007). West Nile virsu emergence and large-scale declines of North American bird populations. *Nature*, **447**, 710–714.

Le Gros, A., Stracey, C. M. and Robinson, S. K. (2011). Associations between northern mockingbirds and the parasite *Philornis porteri* in relation to urbanization. *The Wilson Journal of Ornithology*, **123**, 788–796.

Lee, K. A., Martin, L. B. and Wikelski, M. C. (2005). Responding to inflammatory challenges is less costly for a successful avian invader, the house sparrow (*Passer domesticus*), than its less-invasive congener. *Oecologia*, **145**, 244–251.

Liker, A., Papp, Z., Bokony, V. et al. (2008). Lean birds in the city: body size and condition of house sparrows along the urbanization gradient. *Journal of Animal Ecology*, **77**, 789–795.

Lloyd-Smith, J. O., George, D., Pepin, K. M., et al. (2009). Epidemic dynamics at the human–animal interface. *Science*, **326**, 1362–1367.

Lloyd-Smith, J. O., Schreiber, S. J., Kopp, P. E., et al. (2005). Superspreading and the effect of individual variation on disease emergence. *Nature*, **438**, 355–359.

LoGiudice, K., Ostfeld, R. S., Schmidt, K. A., et al. (2003). The ecology of infectious disease: effects of host diversity and community composition on Lyme disease risk. *Proceedings of the National Academy of Sciences of the United States of America*, **100**, 567–571.

Longcore, T. and Rich, C. (2004). Ecological light pollution. *Frontiers in Ecology and the Environment*, **2**, 191–198.

Loss, S. R., Hamer, G. L., Walker, E. D., et al. (2009). Avian host community structure and prevalence of West Nile virus in Chicago, Illinois. *Oecologia*, **159**, 415–424.

MacGregor-Fors, I. (2011). Misconceptions or misunderstandings? On the standardization of basic terms and definitions in urban ecology. *Landscape and Urban Planning*, **100**, 347–349.

Martin, L.B. (2009). Stress and immunity in wild vertebrates: timing is everything. *General and Comparative Endocrinology*, **163**, 70–76.

Martin, L.B., Han, P., Lewittes, J., et al. (2006). Phytohemagglutinin (PHA) induced skin swelling in birds: histological support for a classic immunoecological technique. *Functional Ecology*, **20**, 290–300.

Martin, L.B., Hasselquist, D. and Wikelski, M. (2006). Immune investments are linked to pace of life in house sparrows. *Oecologia*, **147**, 565–575.

Martin, L. B., Hawley, D. M. and Ardia, D. R. (2011). An introduction to ecological immunology. *Functional Ecology*, **25**, 1–4.

Martin, L. B., Hopkins, W. A., Mydlarz, L., et al. (2010). The effects of anthropogenic global changes on immune functions and disease resistance. *Annals of the New York Academy of Science*, **1195**, 129–148.

Martin, T. E. and Clobert, J. (1996). Nest predation and avian life-history evolution in Europe versus North America: A possible role of humans? *American Naturalist*, **147**, 1028–1046.

Marzal, A., Ricklefs, R. E., Valkiunas, G., et al. (2011). Diversity, loss, and gain of malaria parasites in a globally invasive bird. *PLoS One*, **6**(7), e21905.

McKinney, M. L. (2002). Urbanization, biodiversity, and conservation. *Bioscience*, **52**, 883–890.

McLean, R. G., Shriner, R. B., Kirk, L. J., et al. (1989). Western equine encephalitis in avian populations in North Dakota, 1975. *Journal of Wildlife Diseases*, **25**, 481–489.

Morishita, T. Y., Aye, P. P., Ley, E. C., et al. (1999). Survey of pathogens and blood parasites in free-living passerines. *Avian Diseases*, **43**, 549–552.

Nesbitt, S. A. and White, F. H. (1974). A salmonella-typhimurium outbreak at a bird feeding station. *Florida Field Naturalist*, **2**, 46–47.

Newman, M. M., Yeh, P. J. and Price, T. D. (2006). Reduced territorial responses in dark-eyed juncos following population establishment in a climatically mild environment. *Animal Behaviour*, **71**, 893–899.

Ots, I., and Horak, P. (1998). Health impact of blood parasites in bleeding Great Tits. *Oecologia*, **116**, 441–448.

Partecke, J. and Gwinner, E. (2007). Increased sedentariness in European Blackbirds following urbanization: A consequence of local adaptation?*Ecology*, **88**, 882–890.

Partecke, J., Schwabl, I. and Gwinner, E. (2006). Stress and the city: urbanization and its effects on the stress physiology in European Blackbirds. *Ecology*, **87**, 1945–1952.

Partecke, J., Van't Hof, T. J. and Gwinner, E. (2005). Underlying physiological control of reproduction in urban and forest-dwelling European blackbirds Turdus merula. *Journal of Avian Biology*, **36**, 295–305.

Pedersen, A. B. and Fenton, A. (2007). Emphasizing the ecology in parasite community ecology. *Trends in Ecology & Evolution*, **22**, 133–139.

Power, A. G. and Mitchell, C. E. (2004). Pathogen spillover in disease epidemics. *The American Naturalist*, **164**, S79–S89.

Prenter, J., MacNeil, C., Dick, J. T. A., et al. (2004) Roles of parasites in animal invasions. *Trends in Ecology & Evolution*, **19**, 385–390.

Raberg, L., Sim, D. and Read, A. F. (2007). Disentangling genetic variation for resistance and tolerance to infectious diseases in animals. *Science*, **318**, 812–814.

Raffel, T. R., Martin, L. B. and Rohr, J. R. (2008). Parasites as predators: unifying natural enemy ecology. *Trends in Ecology & Evolution*, **23**, 610–618.

Ringa, A. M., Blitvich, B. J., Hyun-Young, K., et al. (2004). Antibody prevalence of Nest Nile virus in birds, Illinois 2002. *Emerging Infectious Diseases*, **10**, 1120–1124.

Robb, G. N., McDonald, R. A., Chamberlain, D. E., et al. (2008). Food for thought: supplementary feeding as a driver of ecological change in avian populations. *Frontiers in Ecology and the Environment*, **6**, 476–484.

Robinson, R. A., Lawson, B., Toms, M. P., et al. (2010), Emerging infectious disease leads to rapid population declines of common British birds. *PLoS One*, **5**, e12215.

Romero, L. M. (2004), Physiological stress in ecology: lessons from biomedical research. *Trends in Ecology & Evolution*, **19**, 249–255.

Rosenfield, R. N., Bielefeldt, J., Rosenfield, L. J., et al. (2002). Prevalence of *Trichomonas gallinae* in nestling Copper's Hawks among three North American populations. *The Wilson Bulletin*, **144**, 145–147.

Rosenfield, R. N., Taft, S. J., Stout, W. E., et al. (2009). Low prevalence of *Trichomonas gallinae* in urban and migratory Copper's Hawks in northcentral North America. *The Wilson Journal of Ornithology*, **121**, 641–644.

Ruiz, G., Rosenmann, M., Novoa, F. F., et al. (2002). Hematological parameters and stress index in rufous-collared sparrows dwelling in urban environments. *Condor*, **104**, 162–166.

Ryder, T. B., Reitsma, R., Evans, B., et al. (2010). Quantifying avian nest survival along an urbanization gradient using citizen- and scientist-generated data. *Ecological Applications*, **20**, 419–426.

Schmid-Hempel, P. and Ebert, D. (2003). On the evolutionary ecology of specific immune defence. *Trends in Ecology & Evolution*, **18**, 27–32.

Schoech, S. J. and Bowman, R. (2003). Does differential access to protein influence differences in timing of breeding of florida scrub-jays (*Aphelocoma coerulescens*) in suburban and wildland habitats? *Auk*, **120**, 1114–1127.

Shochat, E. (2004). Credit or debit? Resource input changes population dynamics of city-slicker birds. *Oikos*, **106**, 622–626.

Shochat, E., Warren, P. S., Faeth, S. H., et al. (2006). From patterns to emerging processes in mechanistic urban ecology. *Trends in Ecology & Evolution*, **21**, 186–191.

Simpson, J. E., Folsom-O'Keefe, C. M., Childs, J. E., et al. (2009). Avian host-selection by *Culex pipiens* in experimental trials. *PLoS One*, **4**, e7861.

Simpson, J. E., Hurtado, P. J., Medlock, J., et al. (2012). Vector host-feeding preferences drive transmission of multi-host pathogens: West Nile virus as a model system. *Proceedings of the Royal Society B-Biological Sciences*, **279**, 925–933.

Sol, D., Jovani R., and Torres, J. (2000). Geographical variation in blood parasites in feral pigeons: the role of vectors. *Ecography*, **23**, 307–314.

Stracey, C. M. and Robinson, S. K. (2012). Are urban habitats ecological traps for a native songbird? Season-long productivity, apparent survival, and site fidelity in urban and rural habitats. *Journal of Avian Biology*, **43**, 50–60.

Tompkins, D. M., Dunn, A. M., Smith, M. J., et al. (2011). Wildlife diseases: from individuals to ecosystems. *Journal of Animal Ecology*, **80**, 19–38.

Tompkins, D. M., White, A. R. and Boots, M. (2003). Ecological replacement of native red squirrels by invasive greys driven by disease. *Ecology Letters*, **6**, 189–196.

Torchin, M. E., Lafferty, K. D., Dobson, A. P., et al. (2003). Introduced species and their missing parasites. *Nature*, **421**, 628–630.

Wang, Y., Chen, S., Jiang, P., et al. (2008), Black-billed Magpies (*Pica pica*) adjust nest characteristics to adapt to urbanization in Hangzhou, China. *Canadian Journal of Zoology-Revue Canadienne De Zoologie*, **86**, 676–684.

Whittaker, K. A. and Marzluff, J. M. (2009), Species-specific survival and relative habitat use in an urban landscape during the postfledging period. *Auk*, **126**, 288–299.

Woolhouse, M. E. J., Dye, C., Etard, J. F., et al. (1997). Heterogeneities in the transmission of infectious agents: Implications for the design of control programs. *Proceedings of the National Academy of Sciences of the United States of America*, **94**, 338–342.

Zhang, S., Lei, F., Liu, S., et al. (2011). Variation in baseline corticosterone levels of Tree Sparrow (Passer montanus) populations along an urban gradient in Beijing, China. *Journal of Ornithology*, **152**, 801–806.

PART 3

Evolutionary Processes

CHAPTER 10

Mechanisms of phenotypic responses following colonization of urban areas: from plastic to genetic adaptation

Jesko Partecke

10.1 Urbanization and its ecological effects on animals

The urban environment is an ecosystem that significantly differs from nearby non-urban 'natural' habitats in a variety of abiotic and biotic factors. Recent studies show that urban environmental conditions cause changes in ecological processes on a small spatial scale such as alterations of the phenology of plants and animals (Shochat et al., 2006). Blooming and leaf formation occurs earlier and urban bird populations extend their breeding seasons (e.g. Partecke et al., 2005; Sukopp, 1998). Other life history traits in animals are also affected by urban environmental conditions (Klausnitzer, 1989). A shift to increased sedentariness, for instance, has been suggested for several migratory bird species colonizing urban habitats (Adriaensen & Dhondt, 1990; Evans et al., 2012).

Many of the ecological constraints that exist in wild habitats seem to be changed in urban environments. For instance, urban animals may suffer less from climatic stress in temperate cities due to the warmer micro-climate ('heat island effect') but heat stress could be increased by urbanization in tropical areas. (Shochat et al., 2004). Moreover, in addition to natural food resources, urban animals obtain food from feeders, exotic vegetation and human refuse (Shochat et al., 2006). These conditions may increase the carrying capacity of the urban environment, and serve to explain the global pattern of extremely high urban population densities of urban 'exploiters' (species thriving as urban commensals to the point that they become dependent on urban resources) and urban 'adapters' (species that successfully invade urban habitats but still live in non-urban 'natural' settings) (Shochat et al., 2006). On the other hand, animals living in urban areas are confronted with many novel and potentially stressful conditions rarely experienced in their original, 'natural' environments such as unfamiliar food sources, elevated anthropogenic disturbance, permanent presence and high density of humans, dogs and cats, and increased levels of artificial lighting and noise. The combination of these factors creates a new environment that may favour individuals which are able to cope behaviourally and physiologically with these altered and 'novel' conditions in a city.

10.2 Urbanization and its overlooked evolutionary force

Whereas ecological effects of human-induced habitat alteration are receiving increasing attention, evolutionary consequences of urbanization are still largely neglected (Diamond, 1986; Shochat et al., 2006; Sih et al., 2011; Tuomainen & Candolin, 2011). This is insofar remarkable because there is increasing evidence that human ecological impact has enormous evolutionary consequences and can greatly

Avian Urban Ecology. Edited by Diego Gil and Henrik Brumm

accelerate evolutionary change in species around us (Palumbi, 2001; Yeh, 2004). The vast majority of cases of microevolutionary changes in contemporary populations involve human disturbance or activity, such as the introduction of species (Ashley et al., 2003; Svensson & Gosden, 2007). In particular, colonization events and the establishment of populations in novel environments altered by humans are most often associated with rapid evolutionary events (Reznick & Ghalambor, 2001). There have been several demonstrations of rapid evolution after colonization of new habitats in association with the direct influence of climate or novel resources, prey, or predators (e.g. Carroll et al., 1997; Hendry et al., 2000; Huey et al., 2000). Common responses of animals include changes in morphology, behaviour, physiology, and life history traits (e.g. Carroll et al., 1997; Huey et al., 2000; Quinn et al., 2001).

10.2.1 Microevolution via natural selection

Similar to other colonization events, we might expect that invasion of urban areas also leads to microevolutionary changes due to the altered and novel selection regimes in urban settings (Diamond, 1986). The likelihood of an evolutionary response may, however, be influenced by the extent of gene flow from the surrounding areas (urban areas are often located within the pre-existing range of the colonizing species), confounded by genetic drift, opposed by natural selection due to temporal environmental variability, and constrained by the genetic architecture of the underlying traits (Kawecki & Ebert, 2004). Especially for highly mobile organisms such as birds, small-scale genetic differentiation is generally not expected, because of gene flow. If environmental conditions, however, change drastically within short distances, as is more likely in large cities than in the adjacent rural areas, strong selection may act on the phenotype, thus making local differentiation probable.

10.2.2 Immigrant selection

Even if selective forces in urban settings are not strong enough to cause shifts in the genetic composition of urban populations compared to rural populations, immigrant selection could still cause phenotypic and genetic changes between urban and rural populations (Lomolino, 1984). Some individuals of the source population (i.e. rural population) may be particularly pre-adapted to the novel environmental conditions in urban areas and may thus be the only individuals which are able to colonize urban areas. It may be, for example, conceivable that some individuals of the source population with genetically preadapted traits such as higher dispersal propensity or extreme behavioural disposition such as higher boldness are more prone to colonize novel environments such as urban areas and thus have a bigger advantage for successful colonization. This immigrant selection could explain shifts in genotypic frequency distribution following colonization of urban areas without selection acting on urban individuals in the urban environment. Indeed on an inter-species level, it has been shown that species with a broad environmental tolerance (e.g. generalists) may be better predisposed to successfully colonize urban areas than ecological specialists which tolerate only a narrow range of climatic conditions, specialize on few resources or occur in a limited range of habitats (Bonier et al., 2007).

10.2.3 Phenotypic plasticity

Most organisms exhibit phenotypic plasticity as an evolved response to environmental variation (e.g. Pigliucci, 2001). Hence, it is also likely that urban environmental conditions may directly induce phenotypic changes in an individual's morphology, physiology, and behaviour, and hence, do not necessarily elicit a genetic change. As first noted by Baldwin (Baldwin, 1896), these plastic responses may be crucial to colonizers to establish populations and, hence for the successful persistence of the population (Price et al., 2003). The ability of individuals to respond to new ecological challenges by means of phenotypic plasticity may reduce the strength of natural selection on morphological, physiological, or behavioural characters (Price et al., 2003). However, it has been suggested that the optimal urban phenotype will probably require genetic adaptation as well. The requirement for genetic adaptation may arise for four reasons (Evans et al., 2010b). First, even if plasticity results in an adaptive phenotype it may do so in a more costly manner than that possible with genetic adaptation, resulting in selection pressures that ultimately generate genetic change. Second once the population is established and the plastic response approaches the optimum urban phenotype, heritable differences may accumulate by natural selection, so that the adaptive phenotype initially achieved by plasticity becomes genetic (also termed 'genetic accommodation'; West-Eberhard, 2003). Third, the plastic responses induced by arrival in novel environments, such as urban areas, may generate additional selection pressures that drive genetic adaption. Fourth, phenotypic plasticity is often based on reaction norms that are themselves the result of natural selection (e.g. Schlichting & Pigliucci, 1998). Consequently, for the understanding of potential evolutionary processes operating in urban areas it is important to explicitly identify the effects of genetic and environmental sources of phenotypic variation following urbanization.

10.2.4 Maternal and epigenetic effects

The phenotype is not only shaped by the genotype and the environment an individual experiences during development and later life, but it is frequently also influenced by the environmental experience of other individuals such as parents and in particular the mother. These so called maternal effects can modify offspring phenotype, potentially influencing offspring fitness, and therefore potentially playing an important role in evolutionary dynamics (Mousseau & Fox, 1998). One potent mechanism through which mothers can influence offspring phenotype is by developmental programming via maternal hormones. In birds, these maternal hormones are mainly transmitted into the egg yolk (Schwabl, 1993). Variation in exposure to steroids during early life may cause within-sex and among-sibling phenotypic variation in adult life (Groothuis et al., 2005). The effects during early life described above may be a short-term, transient consequence of exposure to maternal steroids and hence activational in their mode of action (Carere & Balthazart, 2007). However, longitudinal studies identified organizational effects of yolk steroids on morphology, physiology and behaviour that last into adulthood. For example, yolk testosterone has been shown to influence aggressive, dominance, and sexual behaviour and plumage characteristics in adults (Groothuis et al., 2005). These data provide evidence that hormone-mediated maternal effects are an epigenetic mechanism causing intrasexual variation in adult phenotype. Thus, when elucidating evolutionary dynamics and adaptive phenotypic adjustments following colonization of novel environments such as cities, we may also need to consider maternal effects as a potential mechanism causing phenotypic changes in the urban setting.

10.3 History and timing of colonization

In order to fully understand evolutionary dynamics of urban life on species one would ideally want to know the history and timing of colonization events of these species at multiple sites across a large geographical region. Evolutionary processes operating in urban areas may depend on three important facts: First, where did the urbanization process of the specific species begin geographically, second, where do colonizing individuals originate from and, third, how did the urbanization events spread across large geographical ranges? The answers to these questions would help to understand the evolutionary and ecological dynamics of the urbanization process. Two distinct colonization scenarios have mainly been discussed which explain the colonization history (Partecke et al., 2006a): (1) repeated and independent colonization of urban areas from surrounding neighbouring non-urban rural populations or (2) establishment via 'leap-frog' mechanism (Figure 10.1.). The leap-frog colonization model assumes that following an initial urban colonization event, subsequent colonization events occurred from already existing and established urban populations. These urban colonizing individuals are thought to be either adapted to the novel habitat or at least imprinted on it which facilitates the colonization success. There are only few studies which discussed these potential colonization processes. A comparison of randomly selected genetic markers via amplified fragment length polymorphism analysis of an urban and a nearby rural population of the European blackbird (*Turdus merula*) showed little or no genetic isolation indicating an absence of genetic differentiation between these two populations (Partecke et al., 2006a).

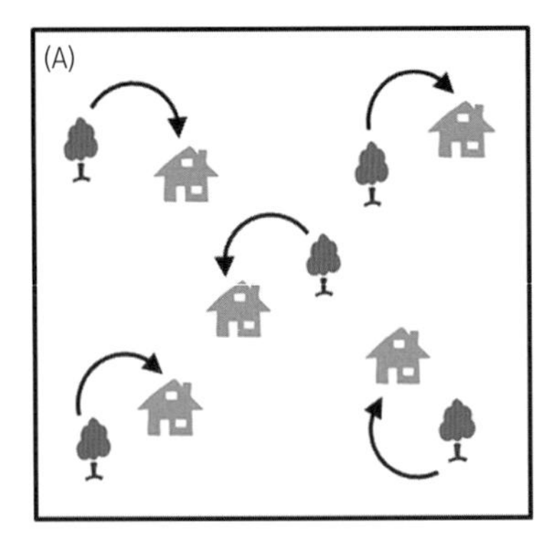

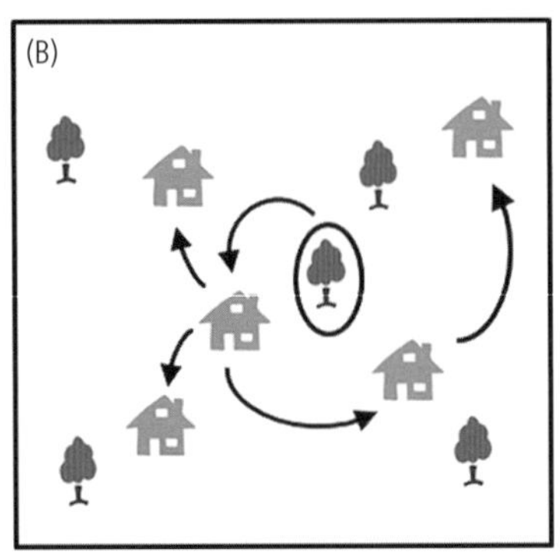

Figure 10.1 Timing and history of colonization events; (A) Repeated and independent colonization of urban areas (houses) from surrounding neighbouring non-urban rural populations (trees). (B) Establishment via 'leap-frog' mechanism. The leap-frog colonization model assumes that following an initial urban colonization event (encircled tree), subsequent colonization events occurred from already existing and established urban populations.

A more elaborated study again on the European blackbird extends these conclusions. Using genetic diversity estimates from microsatellite data of 25 urban and rural populations located across the western Palearctic, this study indicates that urban areas have been mainly colonized by neighbouring forest individuals instead of leap frog colonization of urban individuals (Evans et al., 2009a). However, this study also found overall genetic divergence and loss of genetic diversity was related to timing of colonization. The conclusion of this study is interesting because it suggests that in species where urbanization requires genetic adaptations, colonization of multiple urban areas could be particularly difficult.

10.4 Phenotypic adjustments to the urban environment and their underlying mechanisms: from plastic to genetic adaptations

There is no doubt that ecological differences between urban and rural areas are remarkable and ubiquitous. It is, therefore, conceivable that almost all species which colonize towns and cities from rural sites will require some phenotypic adjustment to the novel urban environment. In fact animals colonizing urban areas have changed a suite of behaviours, physiology, and morphology. In the following paragraph I will review examples of behavioural, physiological, and morphological changes following urbanization and the underlying processes (i.e. plastic and/or genetic responses) that generate these required phenotypic adjustments to the urban environment (Figure 10.2.).

10.4.1 Behaviour

Because colonization of urban areas is principally a behavioural phenomenon, we need to explicitly consider the role of behaviour in shaping adaptive processes in an urban environment. At least one behavioural precondition is thought to be essential for most vertebrate species to settle in cities. Invading individuals have to reduce their timidity in order to establish themselves successfully in urban habitats (Gliwicz et al., 1994; Shochat et al., 2006). Indeed, anecdotal observations and recent studies indicate that urbanized individuals of mammalian and avian species are often tamer than their conspecifics in nearby 'natural' areas (Bowers & Breland, 1996; Cooke, 1980; Gliwicz et al., 1994; Møller, 2008; Shochat et al., 2006). Moreover, changes in population densities in cities may result in either stronger or weaker inter- and intraspecific competition for food resources (Shochat et al., 2006). Changes in competition for food and predation risk may alter foraging behaviour of urban animals. For instance, urban individuals of several bird species seem to be more efficient in exploiting food sources than their rural counterparts (Shochat et al., 2004). For species with higher population densities in urban areas, higher density may also result in higher competition for territories, smaller territory size, and consequently increased aggressive interactions (Luniak et al., 1990). Furthermore, extended daily activity pattern (i.e. shifts of activity into the night or into dawn and dusk phases) in avian species have been attributed to higher levels of artificial lighting and urban noises (Kempenaers et al., 2010). Likewise, urban birds alter the frequency structure or amplitude of

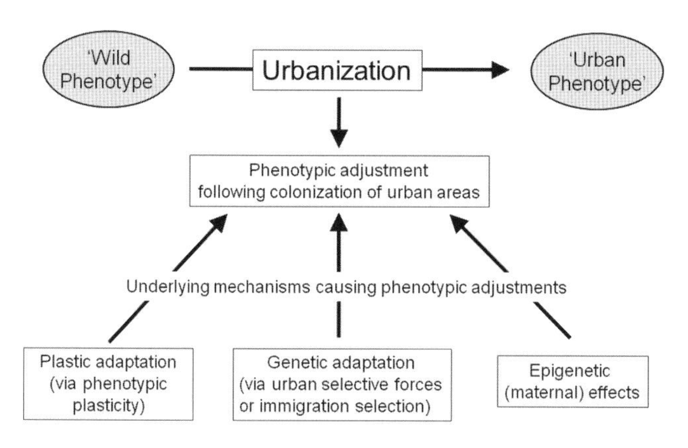

Figure 10.2 In the process of urbanization an individual has to cope with significant different environmental conditions than its former wild conspecific. Given these marked differences between urban and rural habitats species which colonize cities will require some phenotypic adjustments to the novel environment. The potential underlying mechanisms that may generate these adjustments are: phenotypic plasticity, genetic differences (adaptation via micro-evolution), and maternal effects. These different mechanisms are not mutually exclusive but can act together in order to produce an adaptive response of the 'urban phenotype'.

their songs to avoid masking anthropogenic noise (Brumm, 2004; Slabbekoorn & Peet, 2003). However, whether this change in vocal communication following urbanization is adaptive or not is still under debate (Nemeth et al., 2013).

These observations suggest that urbanization may change the general behavioural disposition of individuals thriving in urban areas. For instance, higher premiums may be placed on individuals being 'adventurous' in urban habitats because bold and aggressive individuals would have greater access to food resources in human vicinity than more timid individuals. Thus urbanization may favour organisms with different behavioural characteristics compared to individuals living in 'natural' habitats. However, to what extent these behavioural changes reflect phenotypic plasticity or genetic adaptation is often unknown.

Behavioural flexibility has been often regarded to be high, immediate, and reversible, allowing individuals to maximize their fitness in the many different environments they encounter during life (Sih et al., 2004). Thus, behavioural changes following the colonization of urban areas could very well be plastic responses. However, contrary to the notion of behavioural flexibility as the major adaptive cause of phenotypic variation in behaviour (Dall et al., 2004), animals often show very limited behavioural and physiological flexibility (Sih et al., 2004) and commonly differ consistently in their reaction towards the same environmental stimuli (Wilson et al., 1994). Consistent individual differences in behaviour are, moreover, frequently expressed across time and across a wide range of contexts (e.g. feeding, mating, parental care, and contest). Furthermore, evidence is accumulating that individuals differ in whole suites of functionally-distinct behavioural traits (e.g. Sih et al., 2004). In animal research, the study of behavioural syndromes has gained considerable momentum, largely because the behavioural type of an individual, population, or species can have important ecological and evolutionary implications, including major effects on species distributions, and on the relative tendencies of species to be invasive or to respond well to environmental change (Sih et al., 2004).

In conclusion different 'coping styles' are thought to represent alternative, coherent and adaptive strategies which might do well in different environments. Whereas proactive individuals tend to dominate and outcompete reactive ones in a stable environment, reactive individuals appear to pay close attention to their environment and, as a result, may be favoured in more variable environments (Sih et al., 2004). Although ecologists have shown a renewed interest in the importance of individual variation within species (Sih et al., 2004), the possible role of within-species variation in behavioural types in enabling species to adapt to different environments and more specifically to cope with human-induced rapid environmental change such as urbanization remains poorly studied. For many species, environmental change as a result of anthropogenic influences is likely to be a major factor in shaping their future. The presence of different personality types in populations and the degree of flexibility shown by individuals will presumably determine species' ability to persist (Sih et al., 2004). In particular, successful colonization of urban areas might represent a strong, non-random bottleneck with respect to behavioural types. Bold and neophilic individuals may, for instance, be better in adapting to the novel environmental conditions in urban areas. Thus, bottleneck selection on a behavioural syndrome due to urbanization could have critical, yet underappreciated impacts on the ecology and evolution of species.

In fact evidence is accumulating, that urbanization modifies personalities and even changes behavioural syndromes in urban animals. A recent field study on song sparrows (*Melospiza melodia*) showed that urban individuals were bolder towards humans and exhibited higher levels of territorial aggression than rural conspecifics (Evans et al., 2010a).

Artificial selection experiments demonstrated that personalities in animals are heritable (Drent et al., 2003). To date our knowledge is still very limited to what extent changes in personalities following urbanization are the results of microevolutionary changes or whether behavioural differences are plastic responses to the specific environmental conditions in urban habitats. Recent studies, however, seem to confirm that both plastic and genetic responses may be responsible for shifts in personalities in urban populations. In a recently established urban population of dark-eyed juncos (*Junco hyemalis*) in San Diego, California, free-living individuals exhibited reduced flight initiation distances and captive hand-reared urban individuals showed increased boldness compared to non-urban individuals in an common garden experiment (Atwell et al., 2012). Likewise but contrary to the predictions, hand-reared common blackbirds (*Turdus merula*) were more neophobic and less neophilic than hand-reared conspecifics from a rural population when tested in a common garden experiment (Miranda et al., 2013). Although these intrinsic behavioural differences of hand-reared urban and rural birds argue in favour of genetic differences, maternal effects and early environmental conditions during the nestling phase cannot be rule out at this stage. In addition to these intrinsic behavioural differences, both urban and rural blackbirds were able to change their behavioural responses when repeatedly exposed to different novel objects indicating the ability to habituate to novel situations by plastically adjusting their reactions.

10.4.2 Physiology

Among the physiological mechanisms, hormones are essential components for the development and expression of behaviour. In addition, the endocrine system maintains internal physiological systems, ensuring survival and reproduction are maximized despite environmental perturbations. Our knowledge about the effects of urban life on the endocrine system is, however, still very limited. There are two recent reviews elaborately summarizing in detail how the endocrine system may be affected by urbanization (Bonier, 2012; Buchanan & Partecke, 2012). Here, I will outline one important finding which emphasizes that urbanization can have evolutionary consequences on physiological coping mechanisms, even when flexibility is a key characteristic of the endocrine system.

Glucocorticoid steroid hormones play a major role to ensure survival under adverse environmental conditions. The acute short-term secretion of these hormones is considered beneficial in that it helps to mediate adaptive responses, such as stimulating gluconeogenesis, inhibiting glucose utilization in non-essential tissues, mobilizing fat stores, and redirecting behaviour (Romero, 2004). In prolonged stressful situations, however, chronically elevated levels of circulating glucocorticoids can impair reproductive, immune, and brain functions (Romero, 2004). Hence, if urban environments create novel challenges, glucocorticoids might play an important role mediating behavioural and physiological responses of organisms to those challenges. As mentioned above, urban bird species that thrive in urban areas are often noticeably bold or tame in the presence of humans (Atwell et al., 2012; Evans et al., 2010a; Møller, 2008). In line with these behavioural changes, reduced acute corticosterone stress response in captive urban blackbirds has been interpreted as a result of local adaptation to the urban-specific environmental condition (Partecke et al., 2006b). An attenuated acute hormonal stress response following handling and restraint has been also confirmed in free-living and hand-reared urban dark-eyed juncos (Atwell et al., 2012). This study also reports negative correlation between stressed-induced corticosterone levels and a personality trait—i.e. boldness. Again, early developmental effects cannot be ruled out. Nevertheless, these data suggest contemporary adaptive evolution of correlated hormonal and behavioural traits associated with colonization of urban areas. In contrast, existing field studies on acute corticosterone stress responses including six bird species are quite conflicting. Thus far, no consistent patterns have been revealed. Urban birds have been found to have higher, lower or similar acute corticosterone stress response compared to rural population of the same species (Bonier, 2012). It is conceivable that these inconsistencies between different species may be explained by differences in life histories between species or between urban habitats on large geographical scales (Buchanan & Partecke, 2012).

10.4.3 Phenology

In addition to changes in behavioural and physiological traits, urban life has also been shown to modify the phenology of animals thriving in cities. One of the best studied changes is the timing and duration of reproduction. Most bird species studied so far show an earlier onset and longer duration of the breeding season in urban compared to non-urban habitats (Chamberlain et al., 2009). Analogous to the effects of global climate change on the timing of reproduction, one of the most common arguments for the earlier onset of breeding in urban birds is the warmer microclimate in the cities (Møller, 2012). The urban microclimate is generally warmer than the surrounding natural habitats (Klausnitzer, 1989). Evidence that ambient temperature can have a stimulating effect comes from studies in which birds were exposed to different ambient temperatures (Wingfield et al., 2003). However, urban environments differ from non-urban habitats in several other aspects other than warmer micro-climate that might modify timing of reproduction in urban populations. First, food resources are known to have a proximate influence on reproductive timing (Hahn et al., 1997). Urban birds consume extra anthropogenic food—for instance, offered at feeders during winter and spring—and they also take advantage of all kinds of human waste. In addition, a warmer microclimate in an urban habitat may induce an earlier increase of natural food abundance. Hence, earlier and/or higher food availability may induce an earlier timing of reproduction in urban birds. Urban populations have generally also higher breeding densities than non-urban populations. Such higher breeding density is probably associated with higher frequency of social stimulation, which may advance timing of reproduction in spring (Silverin & Westin, 1995). Another potential factor that may be responsible for the advancement of reproduction in urban birds is artificial light at night. The increase in vernal day-length stimulates the development of the reproductive system. Urban birds, however, are exposed to a combination of natural day-length and light at night. Indeed, a recent study showed that low light intensities of 0.3 lux during night can advance gonadal development during spring for up to 4 weeks (Dominoni et al., 2013). Whatever factors trigger the shift in the timing of reproduction in urban birds, studies demonstrate that the pronounced difference in the timing of breeding between urban and nearby non-urban populations is mainly caused by adaptive phenotypic plasticity (Partecke et al., 2004; Partecke et al., 2005; Yeh & Price, 2004). Especially, during the early stage of colonization phenotypic plasticity in timing of reproduction has been suggested to be crucial for the population persistence (Yeh & Price, 2004). Nevertheless, there are also some indications that differences in the timing of reproduction between urban and non-urban populations are also partly the result of microevolutionary changes. In a common garden experiment hand-reared urban European blackbirds showed different seasonal gonadal cycles than hand-reared forest conspecifics suggesting that micro-evolutionary changes are also involved and could contribute to the variations in the timing of reproduction in the wild (Partecke et al., 2004).

10.4.4 Other life history events

Other annual events such as migration seem to be affected by urban life, too. Field studies revealed a tendency for sedentariness to increase in urban populations of migratory species such as the European blackbird (*Turdus merula*) and the European robin (*Erithacus rubecula*) (Adriaensen & Dhondt, 1990; Evans et al., 2012; Stephan, 1999). This change in migratory behaviour is probably a consequence of increased survival probabilities during the winter due to the milder urban microclimate and is additionally supported by anthropogenic food supply, which enables birds to overwinter on urban breeding grounds in higher numbers. Moreover, urban environmental conditions facilitate earlier breeding. As a consequence, overwintering (i.e. sedentary) individuals may gain a fitness benefit by advancing their reproductive activities before migratory conspecifics arrive at their breeding grounds. The obvious life history differences between sedentary and migratory individuals suggest that the trend to increased sedentariness in urban populations has resulted from natural selection. Studies of captive birds have shown that, in partially migratory species in which some individuals migrate to winter quarters every year whereas others overwinter at the breeding areas, the behavioural dimorphism is under endogenous control and to a large extent inherited (Berthold, 1996). Field and theoretical studies however, proposed that phenotypic plasticity, induced by social and environmental factors, could be an influential modifier of the innate migration programme (Adriaensen & Dhondt, 1990). A common garden experiment with hand-reared urban and rural European blackbirds using nocturnal activity as a proxy for the migratory propensity indicate that urban blackbirds showed a reduced migratory behaviour compared to rural individuals (Partecke & Gwinner, 2007). These data suggest that changes in migratory behaviour following the colonization could be the result of local adaptation. Furthermore, this study confirmed that intrinsic shifts to sedentariness seem to be adaptive in urban habitats, because birds with lower migratory propensity advanced the timing of reproduction. These results corroborate the idea that urbanization may have evolutionary consequences for life history traits such as timing of breeding and migratory behaviour.

10.4.5 Morphology

There is also evidence for morphological changes following urbanization. Most studies so far reported on changes in overall body size, body mass, and bill ratio (Evans et al., 2009b; Liker et al., 2008; Richner, 1989; Ruiz et al., 2002). Another study even reported on changes in sexually selected plumage characteristics (Yeh, 2004). The effect of urbanization on morphological changes was, however, not consistent among species and among different populations of the same species, indicating that morphological changes are species-specific and locality dependent (Evans et al., 2009b). Our knowledge about the underlying mechanisms which generate these morphological differences is still very limited. However, even here both mechanisms, micro-evolution and phenotypic plasticity seem to act. The study on dark-eyed juncos (*Junco hyemalis*) revealed that selection operating in an urban environment has caused rapid local adaptation of a socially selected plumage character in urban individuals (Yeh, 2004). In contrast, a food-supplementation experiment with urban carrion crows (*Corvus corone*) suggests that smaller tarsus length in urban crows presumably results from urban specific conditions such as food shortage during the nestling period (Richner, 1989, 1992). Hence, smaller body size in urban populations may not seem to be adaptive but rather a consequence of malnutrition during ontogeny.

10.5 Conclusion and future directions

Even if our knowledge about the ecological effects of urban life on wild animals is still growing, the preceding survey and other chapters of this book already illustrate that urbanization causes significant changes in behaviour, physiology and morphology on avian species thriving in urban areas. Knowledge about the underlying mechanisms that generate these phenotypic adjustments is, however, still very limited. The few studies that have tried to elucidate the causes of these phenotypic changes suggest that phenotypic plasticity is a key factor causing morphological, physiological, and behavioural changes in urban populations. Clear evidence for contemporary evolution following the colonization of cities is still lacking, although several studies using common garden experiments suggest that intrinsic differences between urban and non-urban populations again in morphology, physiology, and behaviour may be the result of genetic changes.

In order to solve this question, future studies exploring the mechanisms of phenotypic adjustments to urban environment are needed to evaluate the relative contribution of phenotypic plasticity, epigenetic effects, and/or genetic adaptation. Studies using molecular techniques such as candidate gene or whole genomic approaches may be quite promising to examine evolutionary consequences of urbanization (Mueller et al., 2013).

From a reaction-norm perspective, phenotypic variation in morphological, physiological, and behavioural traits reflects both genetic and plastic effects, along with possible genetic variation in this plasticity (Schlichting & Pigliucci, 1998). Hence, most phenotypic responses to urban environmental conditions will potentially involve both plastic and genetic contributions, as has been shown in other studies of phenotypic responses to environmental changes (Phillimore et al., 2010). For future studies in urban ecology and evolution, we need, thus, to consider the concept of reaction-norm using urban–rural gradients or common garden experiments to elucidate both plastic and genetic responses.

In most cases fitness consequences of phenotypic adjustments following the colonization of urban areas are not known. That is, we still do not understand whether phenotypic changes following the colonization of urban areas are adaptive or maladaptive. In order to understand the ecological and evolutionary impact of urbanization there is a need for studies which focus on the link between phenotypic adjustments and their fitness consequences.

If we know that urbanization causes phenotypic changes in natural populations that thrive in cities and if these changes have been shown to be partly driven by genetic adaptation, a future challenge will be to understand the eco-evolutionary dynamics in a rapidly urbanizing world. Urbanization does not only affect phenotypic responses but also population demography and dynamics, for example, higher population density of certain species such as urban adapters. To what extent, however, phenotypic and genetic changes influence population dynamics in urban areas is an open research field. Moreover urbanization also influences the composition of species and, therefore, significantly modifies species communities. An increasing integration of ecology and evolution in urban research will provide key insights into the forces how urban and non-urban ecosystems function and interact with each other.

Acknowledgements

I thank Diego Gil and Henrik Brumm for inviting me to contribute this chapter. I also would like to thank Diego Gil and two anonymous referees for their valuable comments on an earlier draft. Funding was provided by the Volkswagen Foundation (Initiative 'Evolutionary Biology').

References

Adriaensen, F. and Dhondt, A. A. (1990). Population dynamics and partial migration of the European robin (*Erithacus rubecula*) in different habitats. *Journal of Animal Ecology*, **59**, 1077–1090.

Ashley, M. V., Willson, M. F., Pergams, O. R. W., O'Dowd, D. J., Gende, S. M. and Brown, J. S. (2003). Evolutionary enlightened management. *Biological Conservation*, **111**, 115–123.

Atwell, J. W., Cardoso, G. C., Whittaker, D. J., Campbell-Nelson, S., Robertson, K. W. and Ketterson, E. D. (2012). Boldness behavior and stress physiology in a novel urban environment suggest rapid correlated evolutionary adaptation. *Behavioral Ecology*, **23**, 960–969.

Baldwin, J. M. (1896). A new factor in evolution. *The American Naturalist*, **30**, 441–451 536–553.

Berthold, P. (1996). *Control of Bird Migration*. Chapman & Hall, London.

Bonier, F. (2012). Hormones in the city: Endocrine ecology of urban birds. *Hormones and Behavior*, **61**(5), 763–772. doi: 10.1016/j.yhbeh.2012.03.016

Bonier, F., Martin, P. R. and Wingfield, J. C. (2007). Urban birds have broader environmental tolerance. *Biology Letters*, **3**(6), 670–673. doi: 10.1098/rsbl.2007.0349

Bowers, M. A. and Breland, B. (1996). Foraging of Gray squirrels on an urban-rural gradient: use of the GUD to assess anthropogenic impact. *Ecological Applications*, **6**, 1135–1142.

Brumm, H. (2004). The impact of environmental noise on song amplitude in a territorial bird. *Journal of Animal Ecology*, **73**, 434–440.

Buchanan, K. L. and Partecke, J. (2012). The endocrine system: can homeostasis be maintained in a changing world? In U. Candolin and B. B. M. Wong, eds, *Behavioural Responses to a Changing World: Mechanisms and Consequences*, pp. 32–41. Oxford University Press, Oxford.

Carere, C. and Balthazart, J. (2007). Sexual versus individual differentiation: the controversial role of avian maternal hormones. *Trends in Endocrinology and Metabolism*, **18**, 73–80.

Carroll, S. P., Dingle, H. and Klassen, S. P. (1997). Genetic differentiation of fitness-associated traits among rapidly evolving populations of the soapberry bug. *Evolution*, **51**, 1182–1188.

Chamberlain, D. E., Cannon, A. R., Toms, M. P., Leech, D. I., Hatchwell, B. J. and Gaston, K. J. (2009). Avian productivity in urban landscapes: a review and meta-analysis. *Ibis*, **151**(1), 1–18. doi: 10.1111/j.1474-919X.2008.00899.x

Cooke, A. S. (1980). Observations on how close certain passerine species will tolerate an approaching human in rural and suburban areas. *Biological Conservation*, **18**, 85–88.

Dall, S. R. X., Houston, A. I. and McNamara, J. M. (2004). The behavioural ecology of personality: consistent individual differences from an adaptive perspective. *Ecology Letters*, **7**, 734–739.

Diamond, J. M. (1986). Rapid evolution of urban birds. *Nature*, **324**, 107–108.

Dominoni, D., Quetting, M. and Partecke, J. (2013). Artificial light at night advances avian reproductive physiolgy *Proceedings of the Royal Society B: Biological Sciences*, **280**, doi: doi:10.1098/rspb.2012.3017. Available from http://rspb.royalsocietypublishing.org/content/280/1756/20123017.abstract?sid=208e218c-d336-4089-8f5d-884d9fc63bee (accessed 5 July 2013).

Drent, P. J., van Oers, K. and van Noordwijk, A. J. (2003). Realized heritability of personalities in the great tit (*Parus major*). *Proceedings of the Royal Society Biological Sciences Series B*, **270**, 45–51.

Evans, J., Boudreau, K. and Hyman, J. (2010a). Behavioural syndromes in urban and rural populations of song sparrows. *Ethology*, **116**, 588–595.

Evans, K. L., Gaston, K. J., Frantz, A. C., Simeoni, M., Sharp, S. P., McGowan, A., and Hatchwell, B. J. (2009a). Independent colonization of multiple urban centres by a formerly forest specialist bird species. *Proceedings of the Royal Society B-Biological Sciences*, **276**(1666), 2403–2410. doi: 10.1098/rspb.2008.1712

Evans, K. L., Gaston, K. J., Sharp, S. P., McGowan, A. and Hatchwell, B. J. (2009b). The effect of urbanisation on avian morphology and latitudinal gradients in body size. *Oikos*, **118**(2), 251–259. doi: 10.1111/j.1600-0706.2008.17092.x

Evans, K. L., Hatchwell, B. J., Parnell, M. and Gaston, K. J. (2010b). A conceptual framework for the colonisation of urban areas: the blackbird *Turdus merula* as a case study. *Biological Reviews*, **85**(3), 643–667. doi: 10.1111/j.1469-185X.2010.00121.x

Evans, K. L., Newton, J., Gaston, K. J., Sharp, S. P., McGowan, A. and Hatchwell, B. J. (2012). Colonisation of urban environments is associated with reduced migratory behaviour, facilitating divergence from ancestral populations. *Oikos*, **121**(4), 634–640. doi: 10.1111/j.1600-0706.2011.19722.x

Gliwicz, J., Goszczynski, J. and Luniak, M. (1994). Characteristic features of animal populations under synurbanization—the case of the Blackbird and of the Striped Field Mouse. *Memorabilia Zoologica*, **49**, 237–244.

Groothuis, T. G. G., Müller, W., von Engelhardt, N., Carere, C. and Eising, C. (2005). Maternal hormones as a tool to adjust offspring phenotype in avian species. *Neuroscience and Biobehavioral Reviews*, **29**, 329–352.

Hahn, T. P., Boswell, T., Wingfield, J. C. and Ball, G. F. (1997). Temporal flexibility in avian reproduction. In V. J. Nolan and E. D. Ketterson, eds, *Current Ornithology*, pp. 39–80. Plenum, New York.

Hendry, A. P., Wenburg, J. K., Bentzen, P., Volk, E. C. and Quinn, T. P. (2000). Rapid evolution of reproductive isolation in the wild: evidence from introduced salmon. *Science*, **290**, 516–518.

Huey, R. B., Gilchrist, G. W., Carlson, M. L., Berrigan, D. and Serra, L. (2000). Rapid evolution of a geographic cline in size in an introduced fly. *Science*, **287**, 308–309.

Kawecki, T. J. and Ebert, D. (2004). Conceptual issues in local adaptation. *Ecology Letters*, **7**, 1225–1241.

Kempenaers, B., Borgström, P., Loës, P., Schlicht, E. and Valcu, M. (2010). Artificial night lighting affects dawn song, extra-pair siring success, and lay date in songbirds. *Current Biology*, **19**, 1735–1739.

Klausnitzer, B. (1989). *Verstädterung von Tieren*. Die Neue Brehm-Bücherei, Wittenberg Lutherstadt.

Liker, A., Papp, Z., Bókony, V. and Lendvai, Á. Z. (2008). Lean birds in the city: body size and condition of house sparrows along the urbanization gradient. *Journal of Animal Ecology*, **77**(4), 789–795. doi: 10.1111/j.1365-2656.2008.01402.x

Lomolino, M. V. (1984). Immigrant Selection, Predation, and the Distributions of Microtus pennsylvanicus and Blarina brevicauda on Islands. *The American Naturalist*, **123**(4), 468–483. doi: 10.2307/2460993

Luniak, M., Mulsow, R. and Walasz, K. (1990). Urbanization of the European Blackbird—expansion and adaptations of urban population. In M. Luniak, ed., *Urban Ecological Studies in Central and Eastern Europe; international symposium Warsaw, Poland*, pp. 187–198. Polish Academy of Sciences, Warsaw, Poland.

Miranda, A. C., Schielzeth, H., Sonntag, T. and Partecke, J. (2013). Urbanization and its effects on personality traits: a result of microevolution or phenotypic plasticity? *Global Change Biology*, doi: 10.1111/gcb.12258.

Møller, A. P. (2008). Flight distance of urban birds, predation and selection. *Behavioral Ecology and Sociobiology*, **63**, 63–75.

Møller, A. P. (2012). Reproductive behaviour. In U. Candolin and B. B. M. Wong, eds, *Behavioural Responses to a*

Changing World, pp. 106–118. Oxford University Press, Oxford.

Mousseau, T. A. and Fox, C. W. (1998). *Maternal Effects as Adaptations*. Oxford University Press, New York.

Mueller, J. C., Partecke, J., Hatchwell, B. J., Gaston, K. J. and Evans, K. L. (2013). Candidate gene polymorphisms for behavioural adaptations during urbanization in blackbirds. *Molecular Ecology*, **22**, 3629–3637

Nemeth, E., Pieretti, N., Zollinger, S. A., Geberzahn, N., Partecke, J., Miranda, A. C. and Brumm, H. (2013). Bird song and anthropogenic noise: vocal constraints may explain why birds sing higher-frequency songs in cities. *Proceedings of the Royal Society B: Biological Sciences*, **280**(1754). Available from http://rspb.royalsocietypublishing.org/content/280/1754/20122798 (accessed 5 July 2013).

Palumbi, S. R. (2001). Humans as the World's greatest evolutionary force. *Science*, **293**, 1786–1790.

Partecke, J. and Gwinner, E. (2007). Increased sedentariness in European blackbirds following urbanization: A consequence of local adaptation? *Ecology*, **88**, 882–890.

Partecke, J., Gwinner, E. and Bensch, S. (2006a). Is urbanisation of European blackbirds (*Turdus merula*) associated with genetic differentiation. *Journal of Ornithology*, **147**, 549–552.

Partecke, J., Schwabl, I. and Gwinner, E. (2006b). Stress and the city: Urbanization and its effects on the stress physiology in European blackbirds. *Ecology*, **87**, 1945–1952.

Partecke, J., Van't Hof, T. and Gwinner, E. (2004). Differences in the timing of reproduction between urban and forest European blackbirds (*Turdus merula*): result of phenotypic flexibility or genetic differences? *Proceedings of the Royal Society Biological Sciences Series B*, **271**, 1995–2001.

Partecke, J., Van't Hof, T. and Gwinner, E. (2005). Underlying physiological control of reproduction in urban and forest-dwelling European blackbirds *Turdus merula*. *Journal of Avian Biology*, **36**, 295–305.

Phillimore, A. B., Hadfield, J. D., Jones, O. R. and Smithers, R. J. (2010). Differences in spawning date between populations of common frog reveal local adaptation. *Proceedings of the National Academy of Sciences of the United States of America*, **107**, 8292–8297.

Pigliucci, M. (2001). *Phenotypic Plasticity*. The Johns Hopkins University Press, Baltimore.

Price, T. D., Qvarnström, A. and Irwin, D. E. (2003). The role of phenotypic plasticity in driving genetic evolution. *Proceedings of the Royal Society Biological Sciences Series B*, **270**, 1433–1440.

Quinn, T. P., Kinnison, M. T. and Unwin, M. J. (2001). Evolution of chinook salmon (*Oncorhynchus tshawytscha*) populations in New Zealand: pattern, rate, and process. *Genetica*, **112–113**, 493–513.

Reznick, D. N. and Ghalambor, C. K. (2001). The population ecology of contemporary adaptations: what empirical studies reveal about the conditions that promote adaptive evolution. *Genetica*, **112–113**, 183–198.

Richner, H. (1989). Habitat-specific growth and fitness in carrion crows (*Corvus corone corone*). *Journal of Animal Ecology*, **58**, 427–440.

Richner, H. (1992). The effect of extra food on fitness in breeding carrion crows. *Ecology*, **73**(1), 330–335.

Romero, L. M. (2004). Physiological stress in ecology: Lessons from biomedical research. *Trends in Ecology and Evolution*, **19**(5), 249–255.

Ruiz, G., Rosenmann, M., Novoa, F. F. and Sabat, P. (2002). Hematological parameters and stress index in rufous-collared sparrows dwelling in urban environments. *Condor*, **104**, 162–166.

Schlichting, C. D. and Pigliucci, M. (1998). *Phenotypic Evolution: A Reaction Norm Perspective*. Sinauer Associates, Sunderland, MA.

Schwabl, H. (1993). Yolk is a source of maternal testosterone for developing birds. *Proceedings of the National Academy of Sciences of the United States of America*, **90**, 11446–11450.

Shochat, E., Lerman, S. B., Katti, M. and Lewis, D. B. (2004). Linking optimal foraging behavior to bird community structure in an urban-desert landscape: Field experiments with artificial food patches. *The American Naturalist*, **164**, 232–243.

Shochat, E., Warren, P. S., Faeth, S. H., McIntyre, N. E. and Hope, D. (2006). From patterns to emerging processes in mechanistic urban ecology. *Trends in Ecology & Evolution*, **21**, 186–191.

Sih, A., Bell, A. M. and Johnson, J. C. (2004). Behavioral syndromes: an ecological and evolutionary overview. *Trends in Ecology & Evolution*, **19**, 372–378.

Sih, A., Ferrari, M. C. O. and Harris, D. J. (2011). Evolution and behavioural responses to human-induced rapid environmental change. *Evolutionary Applications*, **4**(2), 367–387. doi: 10.1111/j.1752–4571.2010.00166.x

Silverin, B. and Westin, J. (1995). Influence of the opposite sex on photoperiodically induced LH and gonadal cycles in the Willow Tit (*Parus montanus*). *Hormones and Behavior*, **29**, 207–215.

Slabbekoorn, H. and Peet, M. (2003). Birds sing at a higher pitch in urban noise. *Nature*, **424**, 267.

Stephan, B. (1999). *Die Amsel*. Die Neue Brehm-Bücherei Bd. 95, Westarp Wissenschaften, Hohenwarsleben.

Sukopp, H. (1998). Urban ecology—scientific and practical aspects. In J. Breuste, H. Feldmann and O. Uhlmann, eds, *Urban Ecology*, pp. 3–16. Springer, Berlin.

Svensson, E. I. and Gosden, T. P. (2007). Contemporary evolution of secondary sexual traits in the wild. *Functional Ecology*, **21**(3), 422–433. doi: 10.1111/j.1365–2435.2007.01265.x

Tuomainen, U. and Candolin, U. (2011). Behavioural responses to human-induced environmental change. *Biological Reviews*, **86**(3), 640–657. doi: 10.1111/j.1469–185X.2010.00164.x

West-Eberhard, M. J. (2003). *Developmental Plasticity and Evolution*. Oxford University Press, Oxford.

Wilson, D. S., Clark, A. B., Coleman, K. and Dearstyne, T. (1994). Shyness and boldness in humans and other animals. *Trends in Ecology & Evolution*, **9**, 442–446.

Wingfield, J. C., Hahn, T. P., Maney, D. L., Schoech, S. J., Wada, M. and Morton, M. L. (2003). Effects of temperature on photoperiodically induced reproductive development, circulating plasma luteinizing hormone and thyroid hormones, body mass, fat deposition and molt in mountain white-crowned sparrows, *Zonotrichia leucophrys oriantha*. *General & Comparative Endocrinology*, **131**, 143–158.

Yeh, P. J. (2004). Rapid evolution of a sexually selected trait following population establishment in a novel habitat. *Evolution*, 58, 166–174.

Yeh, P. J. and Price, T. D. (2004). Adaptive phenotypic plasticity and the successful colonization of a novel environment. *The American Naturalist*, **164**, 531–542.

CHAPTER 11

Landscape genetics of urban bird populations

Kathleen Semple Delaney

11.1 Introduction

Urbanization leads to habitat fragmentation, and sometimes only leaves small patches of suitable wildlife habitat. Urban sprawl also changes the biodiversity of an area as animals and plants that were once widespread with few barriers to dispersal can become isolated in suitable patches because of inhibited movement through the landscape (Amos et al., 2012; Frankham, 2006; Reed et al., 2003). In particular, the urban development and roads between habitat fragments will be difficult to cross for many terrestrial species, and even for some birds. If urban structures or roads act as barriers to dispersal, then decreased movement across the landscape can have several potentially detrimental effects. Two consequences of decreased movement that I will discuss in this chapter are increases in genetic divergence between isolated populations and the decrease in genetic diversity within isolated populations. Genetic changes can accumulate through genetic drift or adaptation and can increase genetic divergence between populations, effectively increasing the genetic difference between the groups. Because movement is limited between populations, gene flow is also limited. Various life history traits and landscape features can influence the level of gene flow between two populations (Table 11.1). Over time, groups isolated within patches of suitable habitat can become inbred or even go extinct locally (Frankham, 2006; Reed et al., 2003; Brook et al. 2002; Spielman et al., 2004). Novel predators, harmful chemicals and poisons, disease, and invasive species can kill or compete with individuals that are stuck on habitat 'islands'. For birds living in small habitat 'islands', there can be other consequences such as decreased reproductive output (Holmes et al., 1996; Kuitunen et al., 2003; Roberts & Norment, 1999), decreased survival (Holmes et al., 1996), higher nest predation (Hoover et al., 1995), and increased brood parasitism (Burke & Nol, 2000). Decreased survival and reproductive output could lead to fewer individuals available for breeding in small patches. Decreased movement of animals across landscape barriers is assumed to be highest for less mobile species, and this assumption is generally upheld (Table 11.1) (Frankham, 2006). Because of their ability to fly, birds are thought to be highly mobile and should be able to cross urban landscape barriers such as roads or development. In this chapter I will discuss studies that have found genetic changes in bird populations both between and within habitat patches despite the assumption that birds should be able to fly out of harm's way.

What makes an 'urban' landscape? As shown in Chapter 1, there are many definitions of urban landscapes or urbanization. Urban areas can include small and large roads including freeways, low density housing, and high density housing like subdivisions or commercial development. Some urban landscape elements may be less detrimental and even beneficial in some cases to wildlife populations such as city parks, golf courses, and agriculture. However, species with specific habitat requirements, such as forest species, may find agriculture fields as foreign as many birds find parking lots and shopping malls. Deforestation in the tropics has not generally been considered urbanization, but when

Avian Urban Ecology. Edited by Diego Gil and Henrik Brumm

Table 11.1 Geographic and biological factors affecting the level of gene flow (high or low) across a landscape.

Gene flow:	High	Low
	No barriers in the landscape.	Barriers in the landscape: Rivers, mountain ranges, agriculture, development, roads, deforestation.
	Highly mobile species.	Sedentary species.
	Species with short generation times.	Species with long generation times.
	Habitat fragments are geographically close to other fragments or open space.	Habitat fragments are geographically isolated from other fragments or open space.
	Habitat fragments are large.	Habitat fragments are small.
	Species that are generalists.	Species that are habitat specialists (i.e. have adaptive differences).

trees are removed and the forest is destroyed, only patches of intact forest are left. Roads, villages, or cities often surround these patches. Depending on the bird, roads may not be much of a barrier to dispersal for many species as they can simply fly over them. However, for some species, road mortalities may substantially affect population size and density (e.g. Laurance et al., 2004; Mumme et al., 2000; Reijnen et al., 1995; Van Der Zande et al., 1980). In the end, large swaths of poor habitat such as high density housing or commercial development with only small slivers or patches of suitable habitat will present a severe barrier to many species, particularly habitat specialists (e.g. birds that require riparian habitat).

11.2 The question

The studies reviewed in this chapter all ask a similar question: 'How has the urbanized landscape within the range of a focal study bird or birds changed its population genetic structure on a landscape or population or individual level?' It should be noted that this is a fine-scale landscape genetic question and that most, but not all, studies I have highlighted in this chapter are generally done on a city or regional level.

There are many reasons to be concerned about the genetic consequences of urbanization on natural bird populations. Urbanization can decrease biodiversity and bird abundance within an area over a relatively short time (Benitez-Lopez et al., 2010). Urban sprawl is easy to measure and characterize with the use of a geographic information system (GIS) and local land use data. Urbanized areas can represent some of the most prohibitive habitat for birds and other animals to use or pass through. However, because birds are thought to be able to fly over urbanization or other unsuitable habitat, relatively few fine-scale population genetic studies have been done on them. In fact, birds make excellent study subjects because they are generally easily captured, marked, and observed. The ability of birds to simply fly over urbanized habitat to suitable large natural areas could possibly negate the detrimental effects that have been shown for crawling or walking species. However, there are bird species that do not fly long distances and tend to live in a small area for their whole life. It is expected that these species will be the most affected by habitat fragmentation by urbanization.

11.3 Genetic markers

There are several types of genetic markers that are typically used in landscape genetic studies. These markers are used to genotype or sequence small segments of DNA. It is important for these studies that these segments are evolving rapidly and in a 'neutral' or non-adaptive way. Neutral markers track small mutations in the non-coding DNA of an individual. That way small changes due to inbreeding or genetic drift can be detected within a relatively short time scale. If the accumulation of mutations in neutral DNA occurs over relatively short time scales these markers allow us to detect genetic changes that are occurring over the landscape (Bruford & Wayne, 1993). Simple sequence repeats (SSR) or microsatellite DNA markers are used in the majority of studies discussed in this chapter. Microsatellites are segments of DNA made up of strings of repeated motifs (e.g. TATATATA or CCGGCCGGCCGG) and are easy to read based on length differences of alleles generated by the loss or gain of repeats. Microsatellites have the advantage of being codominant, with one copy (allele) inherited from each parent. This way, parentage can be definitively detected and pedigrees can be constructed (Queller et al., 1993). Microsatellites have a disadvantage that they are relatively species-specific, so that microsatellite libraries need to be generated for each species or closely related sister species. However, with the growing availability and reasonable cost of large volume or whole genome sequencing, microsatellite libraries are now relatively easy and cheap to generate for any species. Similarly, amplified fragment length polymorphism or AFLP markers can be generated in large numbers to create multi-locus DNA profiles of individuals that indicate neutral genetic changes (Bensch & Åkesson, 2005). AFLPs have the disadvantage that they are dominant markers so that pedigree analyses cannot be done with them. However, it is relatively inexpensive to generate large numbers of AFLP markers and inference of population genetic structure has improved with better analysis tools (Falush et al., 2007).

The studies highlighted in this chapter use SSR or microsatellite (msat) loci or Amplified fragment length polymorphisms (AFLPs) because they are the appropriate marker for a fine-scale or small landscape question.

11.4 Genetic analysis

11.4.1 Genetic divergence

To understand how habitat fragmentation is affecting the population genetics of avian species, genetic divergence, and its inverse genetic connectivity, must be understood. Assessing genetic divergence or gene flow can be done several ways. First, by direct observation of individuals. However, radio-tracking or colour-banding of individuals can be difficult, laborious, and may not be actually measuring gene flow because individuals that move between habitat fragments may not actually breed in the new patch. Second, gene flow can be assessed indirectly by calculating divergence across a landscape using genetic markers and *F*-statistics. F_{ST}, a measure of genetic divergence, which ranges from 0 to 1, is a measure that compares the amount of genetic variation among populations to the total amount of genetic variation over all populations (Weir & Cockerham, 1984; Wright, 1951).

The most common type of avian urban landscape genetic study to date has focused on single species living in areas that were once thought to be contiguous but at some point have become fragmented by urban development. These studies include samples from birds from across a landscape, often from small patches with roads and development in between. Some studies measure the size, shape, isolation, or age of patches to correlate them with F_{ST}, or similar analog. There are many analogues of F_{ST} that can be used depending on the type of marker, how polymorphic loci are, and the number of samples (see Meirmans & Hedrick, 2011). Tests can be done to determine if F_{ST} values (and analogues) are statistically different from zero; however, it can be difficult to compare raw F_{ST} values computed using different markers, in different studies, or for different taxa. A program called STRUCTURE (Pritchard et al., 2000) is also widely used in landscape genetic studies. The algorithm in STRUCTURE groups individuals using genotype data to infer the presence of distinct population clusters. The resulting clusters can then be correlated with landscape barriers, such as roads, rivers, mountain ranges, etc. Testing for 'isolation by distance' between genetic and geographic distances can be done with this kind of data. Positive isolation by distance is shown when individuals that live in close proximity also have lower genetic distances (i.e. are more closely related). Isolation by distance patterns can be natural, because of small dispersal distances, or can be caused by landscape barriers such as mountain ranges, rivers, or swaths of urban sprawl.

11.4.2 Genetic diversity

Decreased genetic diversity can be a consequence of increased genetic divergence across a landscape. This is because, as isolated populations get smaller, there can be inbreeding effects, loss of genetic diversity, and the reduced ability to adapt to local environmental changes (Frankham, 2006). Inbreeding can increase the prevalence of deleterious alleles which can decrease survivorship and reproduction (MacDougall-Shackleton et al., 2011). Over time, these effects can lead to an increased risk of extinction. Loss of genetic diversity can be the reason that conservation action is or should be taken. Detecting low genetic diversity with microsatellite markers is generally done by looking for a decrease in expected heterozygosity, a decrease in pairwise or population level relatedness, or a low number of alleles at a locus including fixation (one allele). Expected heterozygosity (H_e) can be thought of as a measure of gene diversity (Nei, 1973). There are two commonly used relatedness measures, Lynch and Ritland's r_{LR} (Lynch & Ritland, 1999) and Queller and Goodnight's r_{QG} (Queller & Goodnight, 1989).

11.5 Barriers to movement and genetic divergence in urbanization-sensitive species

11.5.1 Cities

The most intensive type of urbanization can be found in cities where only small areas of natural open space are left intact after development of roads, housing, and commercial property has fragmented a landscape. Often the only open space within a city are landscaped parks; however, sometimes patches of natural vegetation are preserved. Genetic divergence is expected to increase between samples collected in the natural habitat patches where the intervening landscape is urban development for urban-sensitive species.

Only a few studies have shown strong genetic structure in an urbanized landscape, presumably because of the ability of birds to travel over landscape barriers such as roads, housing, or commercial development. Significant genetic divergence (pairwise F_{ST}) between patches of suitable habitat surrounded by urbanization has been shown in: wrentits (*Chamaea fasciatus*), a sedentary bird species in southern California; song sparrows (*Melospiza melodia*) from in and around metropolitan Seattle, Washington (Unfried et al., 2013); and great tits (*Parus major*) sampled in and around Barcelona, Spain (Bjorklund et al., 2010). All three landscapes were highly fragmented by intense urbanization. Wrentits and song sparrows were sampled from a wide variety of habitat patches of different size, age, and isolation, whereas great tits were mainly sampled from 12 small city parks within Barcelona. Both wrentits and song sparrows showed the highest amount of differentiation between sites with older development age surrounding them. For wrentits, the urbanization surrounding a major freeway was identified as a barrier (Delaney et al., 2010). STRUCTURE identified three genetic clusters, with one cluster found on the west side of the freeway and surrounding urban development, and the other two genetic clusters found on the east side. Over the small study area (all patches separated by <20 km), no evidence of isolation by distance was found, suggesting that wrentit population genetic structure was not due to past genetic structure based on geographic distance, but rather was due to the highly fragmented nature of the landscape. It is interesting to note that wrentits, a very sedentary species with fairly specific habitat requirements (see Table 11.1), showed a stronger pattern of differentiation than song sparrows, a fairly mobile species that will use backyard feeders. However, significant genetic structure was found for song sparrows in the urban landscape, particularly older development, which seems to present a significant barrier to movement for this more mobile species (Unfried et al., 2013). In great tits, a fairly sedentary species, genetic divergence (pairwise F_{ST}) between the city parks was significant in about 2/3 of comparisons (Bjorklund et al., 2010). No isolation by distance pattern was found in this study either, suggesting that patterns of genetic divergence were due to the urban landscape and not past genetic structuring. In addition, there was more gene flow from city parks to a natural forested site outside the city, suggesting that urban parks are not dispersal sinks for great tits in Barcelona. Significant patterns of genetic divergence revealed in these papers are particularly interesting given the highly urbanized intervening landscape between habitat patches and the small geographic scale of these city-level studies.

Several other city-level studies have not shown strong genetic divergence, however, the sample sizes and sampling methods of a couple of the studies may have affected results. Two of three recent studies of urban European kestrels (*Falco tinnunculus*) showed significant genetic divergence between urban and rural populations, however, these studies suffered from major sampling problems. Urban European kestrels are found in higher density and have more stable reproductive success than rural birds (Salvati et al., 1999). Because of these differences population histories, we might expect population genetic differences. Rutkowski et al. (2006) sampled two kestrel nestlings (presumed full siblings) from nests in urban Warsaw (N = 14 nests), suburban Warsaw (N = 7 nests), and outside of the city in a rural area (Rutkowski et al., 2006). Samples were genotyped using six microsatellites and a significant F_{ST} value was found between the city and rural samples. Sample sizes were small in this study but the most problematic aspect of sampling was that they genotyped two siblings from each nest they found and treated them as independent samples. In a later paper, Rutkowski et al. (2010) sampled in the same way using two nestlings from each nest they found. This time they sampled from two rural, one urban, and one suburban site. Pairwise F_{ST} between sites were significant except a comparison between one of the rural sites and the urban site (Warsaw). In addition, the optimal number of genetic units identified by the program STRUCTURE was three, and that the suburban population was almost completely isolated from the others. I believe these studies cannot overcome significant temporal and spatial sampling problems, but further studies could be done with better sampling to see if similar patterns hold for kestrels in urban Warsaw. In an urban area of the Czech Republic, Riegert et al. (2009) found no genetic divergence between urban and rural populations of European kestrels despite ecological and demographic differences such as density, reproductive success, diet, and nest site. It is not surprising that such a mobile species would not show strong genetic structure across a small landscape. This makes the results of the Warsaw kestrel studies even more surprising and better sampling more necessary to elucidate landscape genetic patterns in that city.

One study that found no genetic structure in a city landscape was a study in British Columbia that compared population genetic structure of song sparrows (*Melospiza melodia*) found on oceanic islands and habitat fragments ('urban islands') in the city of Victoria (MacDougall-Shackleton et al., 2011). In this study, there was no significant F_{ST} or genetic clustering within the mainland population or 'urban islands'. There was, however, genetic structure between island and mainland populations. There was significant isolation by distance between mainland sites, even on a scale as small as 4 km between sites, suggesting possible historically restricted gene flow. Because they found isolation by distance but no significant current population genetic structure, the results suggest that there was restricted gene flow historically, but that recent habitat fragmentation has not been severe enough to further restrict gene flow in the present (or recent past). The authors also acknowledge that small sample sizes may have affected results. This study is an interesting contrast to the song sparrow study done in urban Seattle (Unfried et al., 2013). There are a few potential explanations for the different results of the studies, despite being done on the same species and in two geographically close cities. For example, birds in Victoria were sampled over a much smaller scale. Also, there is a large difference in the level and intensity of urban fragmentation between a large city like Seattle and a small city like Victoria. I believe that an explicit comparison of urbanization levels, as measured by land use area and demographic data, between the two cities would be interesting to compare to the genetic structure found in the two studies. It would be a potential way to test the prediction that habitat fragmentation by urbanization is more intense and causes higher levels of genetic divergence between patches.

11.5.2 Agriculture and deforestation

Agriculture and deforestation are not typically thought of as 'urbanization'. However, these anthropogenic changes can severely alter the landscape in a way that could be similar to urbanization. Species with very narrow ecological niches, such as a Mediterranean-type scrub specialist or a forest specialist, may find their movements restricted when encountering agricultural fields or deforested patches. In addition, these kinds of anthropogenic disturbances can be seen as a proxy for the effects that a more highly fragmented landscape might have.

Some weak genetic structure was found in two studies with these types of anthropogenically altered landscapes; a study of capercaille (*Tetrao urogallus*) in the Black Forest region of Germany (Segelbacher et al., 2008) and a study of seven bird species in the Taita Hills in Kenya (Callens et al., 2011). Within the Black Forest, capercaillie are restricted to small forest fragments that are surrounded by habitat that is unsuitable due to deforestation for development of roads and towns. In Kenya, species were sampled from a patchy cloud forest archipelago (three main large patches and several smaller patches) where the intervening landscape was agriculture, villages, and roads. Population genetic structure was found in the most sedentary of the seven African species as shown by high pairwise F_{ST} between sites and genetic clustering according to geography (Callens et al., 2011). In the capercaillie, a sedentary forest specialist, genetic clustering analysis with STRUCTURE suggested no differentiation across four subpopulations in the fragmented Black Forest (Segelbacher et al., 2008). However, significant pairwise F_{ST} between subpopulations (five of six comparisons) suggested restricted gene flow. In addition, there was a significant isolation by distance relationship, also indicating restricted gene flow, although only for males. The difference between male and female isolation by distance suggests that females have longer dispersal distances leading to higher gene flow in this population. Another study on capercaillie from the same area examined the isolation by distance relationship of relatedness and various landscape variables while controlling for the effect of geographic distance, a method similar to a partial Mantel test (Braunisch et al., 2010). Landscape structure and permeability were modelled according to Capercaillie behaviour and land use, and genetic distance between each sample was calculated. Gene flow was positively correlated with preferred landscape features, like coniferous forest, and was negatively correlated with unsuitable habitat such as urbanized and agricultural lands. This suggests that landscape connectivity for capercaillie could be maintained in the area by preserving forested corridors between urbanized areas.

Several studies showed no evidence of genetic divergence in anthropogenically modified landscapes. For example, a study of reed buntings (*Emberiza schoeniclus*), a wetland-restricted species, in southern Switzerland showed no population genetic structure in a landscape consisting of wetland habitat fragments surrounded by intensive agriculture (Mayer et al., 2009). Although reed buntings were not using the agricultural areas as part of their territories, the entire landscape represented one patchy genetic population with high dispersal among patches and high immigration into the patchy population from outside areas. In Costa Rica, white-ruffed manakins (*Corapipo altera*) were sampled in seven forest patches that had been isolated by anthropogenic deforestation (Barnett et al., 2008). Genetic structuring as measured with variation at 15 microsatellite loci was found to be very small. This indicated that white-ruffed manakins are moving between forested patches or that deforestation may have been too recent for genetic changes to occur. For reed buntings and white-ruffed manakins, the ability to disperse over unsuitable habitat may be maintaining gene flow between patches of preferred habitat.

A multi-species study done in Australia tested whether dispersal ability differences were associated with population decline in the face of habitat loss due to deforestation and intervening agriculture (Harrisson et al., 2012). Three of the species had declined in response to fragmentation, eastern yellow robin (*Eopsaltria australis*), weebill (*Smicrornis brevirostris*), and spotted pardalote (*Pardalotus punctatus*), whereas one species was tolerant of the habitat fragmentation, the striated pardalote (*Pardalotus striatus*; Amos et al., 2012). In general, there was high connectivity or gene flow in all four species across the landscape (Amos et al., 2012). The most sedentary of the species, the eastern yellow robin, showed the strongest genetic structure, as measured by F_{ST} values across the 12 sampling sites, but this differentiation was still not statistically significant. Population declines for these four species are not strongly associated with dispersal ability because gene flow remains high, however other population level processes may be contributing to their declines.

11.6 Barriers to movement and genetic divergence in urban-adapted species

The previous section describes the conventional wisdom about landscape genetics and fragmentation—that birds and other wildlife prefer natural habitat, and that therefore populations can be isolated in habitat patches with intervening urban matrix that acts as a barrier to dispersal. However, urban-adapted species may show the exact opposite pattern. These species are restricted to urban areas and surrounding natural habitat can act as barriers to dispersal. Similar genetic tests can be used and similar outcomes are expected, with intervening natural habitat acting as barriers to gene flow, which drives genetic clustering within urban centres. This kind of study is possible with urban-adapted birds only and so far few studies have been done but there are many more urban-adapted species that could be studied in a similar way.

In Finland and Italy, the house sparrow (*Passer domesticus*) is only associated with urbanization, and areas of natural habitat possibly act as barriers to dispersal for this sedentary city bird. Over the last 40 years in Finland, populations of house sparrows have been declining. Genotyping of samples collected countrywide from before and after this decline showed that F_{ST} values were three times higher in post-decline samples (Kekkonen et al., 2011). The pre-decline population appeared to be genetically panmictic whereas the post-decline population showed some evidence of genetic structure. However, no isolation by distance relationship was found and STRUCTURE analysis indicated that a single genetic cluster was most likely. Kekkonen et al. concluded that the decline in population size, accompanied by the increase in F_{ST}, warranted conservation attention for house sparrows in Finland. In Italy, house sparrow samples were collected from four locations with variable levels of urbanization (Fulgione et al., 2000). There was higher gene flow between sites if swaths of urbanization existed between them. Therefore, 'urbanization corridors' can act in a similar way for urban-adapted species as 'wildlife corridors' act for many urban-sensitive species.

Blackbirds (*Turdus merula*) in Europe have expanded from the natural forest habitat to cities. A study using samples from one urban and one rural population in Germany found no differences between populations using AFLP markers (Partecke et al., 2006). A subsequent larger study examined the progression of this urban colonization. Evans et al. (2009) sampled and genotyped birds at 24 microsatellite loci from 12 paired urban and rural sites across Europe. Results showed that urban samples were more closely associated with the closest rural site than with other urban sites, suggesting that there is independent colonization of city centres by blackbirds from nearby rural areas. Therefore, blackbirds do not appear to be 'leapfrogging' across Europe to colonize cities from other cities.

11.7 Genetic diversity in isolated fragments

Even though most birds are mobile, habitat fragmentation can significantly increase the genetic distance between local populations of birds separated by barriers such as roads and urban sprawl. These local populations will become more isolated as time progresses or as urbanization becomes more intense (i.e. agricultural land converted to neighbourhoods converted to commercial development). The general expectation from isolation and low gene flow between habitat fragments is that genetic diversity will decrease within those fragments.

Urban populations of European blackbirds (*Turdus merula*) had lower genetic diversity than paired rural populations as measured by the number of alleles per locus, allelic richness per locus (which is the number of alleles corrected for sample size), and expected heterozygosity (Evans et al., 2009). The difference can be explained by the independent colonization of each urban area (see Section 11.6) and the resulting population bottleneck. Another founder event that resulted in a decrease in genetic diversity (mean number of alleles and H_e) of an urban bird population was documented in dark-eyed juncos (*Junco hyemalis thurberi*) that colonized the campus of University of California San Diego in the early 1980s (Rasner et al., 2004; Yeh & Price, 2004; Yeh, 2004). In Wrentits in southern California, expected heterozygosity and the number of effective alleles was significantly lower in more isolated habitat fragments (Delaney et al., 2010).

Average relatedness within isolated populations is expected to increase, as gene flow is restricted between patches. Within 12 city parks in

Barcelona, Spain and one forested area outside of the city, relatedness was calculated for great tits (*Parus major*; Bjorklund et al., 2010). Significant genetic divergence was found between many of the city parks and four of the 12 park populations had relatedness values that were significantly higher than expected by chance. This suggests that low gene flow between parks is increasing inbreeding within parks. In a study of house sparrows (*Passer domesticus*) along an urban–rural gradient in Belgium, 16 microsatellites were used to estimate average pairwise relatedness for comparison between urban and rural birds (Vangestel et al., 2011). Preliminary results showed that there was a genetic difference between urban and rural sparrow populations, so the authors set out to calculate relatedness (r_{QG}). Average relatedness was higher in the urban birds, and birds closest to the highly urbanized city centre had the highest proportion of closely related individuals in their neighbourhood. These urban-adapted species are sedentary and isolated within the city, therefore inbreeding can occur and produce a population with increased average relatedness.

Even though genetic structure was not detected in mainland samples in song sparrows (*Melospiza melodia*) in British Columbia, MacDougall-Shackleton et al. set out to test whether genetic diversity differs between coastal islands and habitat fragments or 'habitat islands' (2011). Using microsatellite loci, they calculated individual inbreeding coefficients (Milligan, 2003) and multilocus heterozygosity (MLH), the proportion of loci at which an individual is heterozygous (Amos et al., 2001). They also calculated relatedness between mates and overall relatedness within islands and urban fragments. In the urban fragments there was lower heterozygosity (MLH), higher inbreeding coefficients, and higher average relatedness (r_{LR}), even though there was no reported population size differences between urban or oceanic islands. Therefore, human-caused landscape disturbance and fragmentation may have caused a faster decline in genetic diversity than the naturally fragmented landscape of coastal islands. This result highlights the finding that habitat fragmentation as induced by urbanization can cause stronger genetic patterns than other causes of fragmentation.

11.8 Evolutionary impacts

11.8.1 Functional effects of urbanization

Population structure as revealed by neutral genetic markers does not necessarily suggest that populations of isolated birds are adapting to their local environment. However, the two studies I present here have shown local adaptation in response to anthropogenic changes in their environment. A study of house finches (*Carpodacus mexicanus*) in Arizona used samples from two locations, one urban and one a natural desert site (Badyaev et al., 2008). Urban birds had higher bite force and larger bills, presumably because of their dependence on introduced food sources, specifically larger and harder seeds, such as sunflower seeds provided at bird feeders, than seeds used by birds in natural populations. Adaptive evolutionary changes in bill size were also associated with genetic divergence between populations. Using 12 microsatellite loci, Badyaev et al. found significant genetic differentiation between the urban and desert populations, suggesting that divergent selection on bill traits maintained population genetic divergence even in this highly mobile species. In Africa, the little greenbul (*Andropadus virens*) was found in anthropogenically disturbed coffee or cacao plantations and in undisturbed forest sites (Smith et al., 2008). Little greenbuls had significant morphometric, plumage, and song differences between two kinds of sites despite high gene flow and proximity of habitat types. Microsatellite and most AFLP genetic markers showed modest divergence between disturbed and pristine sites, but several outlier markers showed highly significant divergence, suggesting that those markers may be under selection. Human disturbed sites were generally more open (less canopy) and supported a higher density of little greenbuls. Differences in habitat and density, caused by anthropogenic disturbance, may explain these microevolutionary changes and suggests that agriculture and deforestation can not only be a barrier to dispersal for some birds but can initiate adaptive change.

11.8.2 Behavioural differences

There are many studies, and chapters in this book, which associate behavioural changes in birds with urbanization or habitat fragmentation (see Chapters 2, 5, and 9). The following studies, however, have used microsatellite DNA genotyping to try to correlate differences in behaviour (mating behaviour and song dialect) with the levels of habitat fragmentation in the study landscape.

Least flycatchers (*Empidonax minimus*) in Ontario, Canada, form clusters of territories with contiguous borders and unused open space surrounding the clusters. These territory clusters have high rates of extra-pair paternity (EPP; Tarof et al., 2005). Least flycatcher EPP rates were compared between the contiguous population and a nearby fragmented population (Kasumovic et al., 2009); males in the contiguous habitat should have more opportunity for extra-pair copulations. Sex ratio, pairing success, territory size, and clutch size, did not differ between contiguous and fragmented clusters. However, only 11% (1/8) of the nests in the fragmented cluster had extra-pair young compared to 50% (5/10) of the nests within the continuous cluster, a significant difference. These results suggest that birds in more fragmented areas don't have as many opportunities to 'sneak' copulations because nearest neighbours are too far away.

Genetic monogamy or EPP rate was also studied in Florida scrub-jays (*Aphelocoma coerulescens*) over three landscapes with different fragmentation levels (Townsend et al., 2011). In the most fragmented landscape, where only 6% of the habitat is suitable, birds were more inbred. It was thought that EPP rate could increase in this population because mated pairs who were related to each other could increase heterozygosity of their offspring by pairing with unrelated neighbours. However, Florida scrub-jays are highly genetically monogamous, and there were no differences found in EPP rate between fragmented or contiguous habitat.

In a dense urban neighbourhood in Ramat-Aviv, Israel, Leader et al. (2008) investigated the relationship between song dialect and population genetics in orange-tufted sunbirds (*Nectarinia osea*) Birds in this small neighbourhood have two different song dialects, with a few males being 'bilingual'. Mitochondrial DNA haplotypes and typing five microsatellite loci did not reveal any differences between birds of different dialects. This suggests that there are behavioural mechanisms maintaining dialect structure in the neighbourhood despite high gene flow.

11.9 Conclusions

Urban encroachment is degrading habitat and affecting the diversity and abundance of bird communities in many areas of the world (Chace & Walsh, 2006; McKinney, 2002; Miller & Hobbs, 2002; Radeloff et al., 2005). Critical habitat loss can lower population sizes within habitat fragments, which can affect population dynamics, colonization probabilities, and reproductive output. As shown in this chapter, even mobile species such as birds can be genetically affected by urban sprawl and other anthropogenic habitat disturbances, however it is far from a universal pattern. Species with reduced dispersal ability should show the strongest effects of habitat fragmentation (Table 11.1). In general, the studies that found significant and strong genetic structure were about birds that had the most sedentary habits, such as the wrentit (Delaney et al, 2010) and great tit (Bjorklund et al., 2010). Even where strong genetic effects of anthropogenic habitat fragmentation were not found, the most sedentary species were often most affected by urbanization (e.g. Amos et al., 2012; Callens et al., 2011). In addition to dispersal ability, the strength of genetic structuring seemed to be associated with the type of intervening landscape. The studies that showed the strongest effects not only had birds with decreased dispersal ability but the most intensive type of urbanization (i.e. cities and freeways; Bjorklund et al., 2010; Delaney et al., 2010; Unfried et al., 2013). Although many of the studies I have presented have showed moderate to weak genetic structure, I feel that this is an important area for future research. By thoughtfully choosing study species with low dispersal ability and study landscapes where urbanization is intense, strong patterns of genetic divergence may emerge.

Several studies with strong genetic structure also showed genetic diversity decreases associated with fragmentation (Bjorklund et al., 2010; Delaney et al., 2010) and founder events (Evans et al., 2009; Rasner et al., 2004). This decrease in genetic diversity is particularly troubling because when populations decline to critically low numbers, local extirpations extinctions can occur. Most studies in this chapter have used common and widespread species, making it easier to obtain samples from across a landscape. It is troubling to understand that studies cannot be done on rare species because local extinctions have already happened.

Urban habitat fragmentation is expected to increase in the future as the planet becomes more populated. This change in land use will accompany the changing climate. Ranges in bird species are predicted to shift with global temperature rises and changing ecological conditions (Tingley et al., 2009). Bird species that occur in highly fragmented habitat will find limited options for refugia from warm temperatures and suboptimal habitat. In addition, ranges may shift in unpredictable ways, possibly creating new assemblages of birds with no evolutionary cohistory (Stralberg et al., 2009). This could introduce new competition and predator–prey dynamics that could detrimentally affect species. Also, habitat fragmentation will decrease the amount of suitable habitat for birds to colonize during range shifts.

As whole genome sequencing becomes less expensive, researchers will be able to identify and type many more markers or even sequence whole genomes of individuals. This kind of information will provide extensive detail about population genetic structure, pedigree reconstruction, and possibly regions of local genetic adaptation. Identifying local genetic adaptation will be extremely important to develop effective conservation strategies (Calsbeek et al., 2003; Thomassen et al., 2011; Vandergast et al., 2008). Protecting habitat of high ecological value and evolutionary potential will be a priority. Protecting high quality habitat will be necessary also for possible translocations of avian populations to more suitable habitat as climate changes. Multiple study areas that provide replicates of habitat fragmentation will make future studies stronger. If similar patterns are seen across replicate urban areas, the evidence will be strong that urbanization is the cause of the effects observed.

Acknowledgements

My sincere thanks goes Diego Gil, Seth Riley, Borja Milá, and two anonymous reviewers for reviewing earlier versions of this chapter and making critical and helpful suggestions.

References

Amos, J. N., Bennett, A. F., Mac Nally, R., et al. (2012). Predicting landscape-genetic consequences of habitat loss, fragmentation and mobility for multiple species of woodland birds. *PLoS ONE*, **7**, p.e30888.

Amos, W. Worthington, W., Fullard, K., et al. (2001). The influence of parental relatedness on reproductive success. *Proceedings of the Royal Society B: Biological Sciences*, **268**, 2021–2027.

Badyaev, A. V., Young, R. L., Oh, K. P., Addison, C., and Pfenning, D. (2008). Evolution on a local scale: Developmental, functional, and genetic basis of divergence in bill form and associated changes in song structure between adjacent habitats. *Evolution*, **62**, 1951–1964.

Barnett, J. R., Ruiz-Gutierrez, V., Coulon, A., and Lovette, I. J. (2008). Weak genetic structuring indicates ongoing gene flow across White-ruffed Manakin (*Corapipo altera*) populations in a highly fragmented Costa Rica landscape. *Conservation Genetics*, **9**, 1403–1412.

Benitez-Lopez, A., Alkemade, R., and Verweij, P. A. (2010). The impacts of roads and other infrastructure on mammal and bird populations: A meta-analysis, *Biological Conservation*, **143**, 1307–1316.

Bensch, S. and Åkesson, M. (2005). Ten years of AFLP in ecology and evolution: why so few animals? *Molecular Ecology*, **14**, 2899–2914.

Bjorklund, M., Ruiz, I., and Senar, J. C. (2010). Genetic differentiation in the urban habitat: the great tits (*Parus major*) of the parks of Barcelona city. *Biological Journal of the Linnean Society*, **99**, 9–19.

Braunisch, V., Segelbacher, G., and Hirzel, A. H. (2010). Modelling functional landscape connectivity from genetic population structure: a new spatially explicit approach. *Molecular ecology*, **19**, 3664–3678.

Brook, B. W., Tonkyn, D. W., Q'Grady, J. J., and Frankham, R. (2002). Contribution of inbreeding to extinction risk in threatened species. *Conservation Ecology*, **6**, 16. http://www.consecol.org/vol6/iss1/art16/.

Bruford, M. W. and Wayne, R. K. (1993). Microsatellites and their application to population genetic studies. *Current Opinion in Genetics and Development*, **3**, 939–943.

Burke, D. M. and Nol, E. (2000). Landscape and fragment size effects on reproductive success of forest-breeding birds in Ontario. *Ecological Applications*, **10**, 1749–1761.

Callens, T., Galbusera, P., Matthysen, E., et al. (2011). Genetic signature of population fragmentation varies with mobility in seven bird species of a fragmented Kenyan cloud forest. *Molecular Ecology*, **20**, 1829–1844.

Calsbeek, R., Thompson, J. N., and Richardson, J. E. (2003). Patterns of molecular evolution and diversification in a biodiversity hotspot: the California Floristic Province. *Molecular Ecology*, **12**, 1021–1029.

Chace, J. F. and Walsh, J. J. (2006). Urban effects on native avifauna: a review. *Landscape and Urban Planning*, **74**, 46–69.

Delaney, K. S., Riley, S. P. D., and Fisher, R. N. (2010). A rapid, strong, and convergent genetic response to urban habitat fragmentation in four divergent and widespread vertebrates. *PLoS ONE*, **5**, p.e12767. doi:10.1371/journal.pone.0012767.

Evans, K. L., Gaston, K. J., Frantz, A. C., et al.(2009). Independent colonization of multiple urban centres by a formerly forest specialist bird species. *Proceedings of the Royal Society B: Biological Sciences*, **276**, 2403–2410.

Falush, D., Stephens, M. and Pritchard, J. K. (2007). Inference of population structure using multilocus genotype data: dominant markers and null alleles. *Molecular Ecology Notes*, **7**, 574–578.

Frankham, R. (2006). Genetics and landscape connectivity. In K. R. Crooks and M. Sanjayan, eds. *Connectivity Conservation*, pp. 72–96. Cambridge University Press, Cambridge.

Fulgione, D., Procaccini, G., and Milone, M. (2000). Urbanisation and the genetic structure of *Passer italiae* (Vieillot 1817) populations in the South of Italy. *Ethology Ecology & Evolution*, **12**, 123–130.

Harrisson, K., Pavlova, A., Amos, J., et al. (2012). Fine-scale effects of habitat loss and fragmentation despite large-scale gene flow for some regionally declining woodland bird species. *Landscape Ecology*, **27**, 813–827.

Holmes, R. T., Marra, P. P., and Sherry, T. W. (1996). Habitat-specific demography of breeding black-throated blue warblers (*Dendroica caerulescens*): Implications for population dynamics. *Journal of Animal Ecology*, **65**, 183–195.

Hoover, J. P., Brittingham, M. C., and Goodrich, L. J. (1995). Effects of forest patch size on nesting success of wood thrushes. *Auk*, **112**, 146–155.

Kasumovic, M. M., Ratcliffe, L. M., and Boag, P. T. (2009). Habitat fragmentation and paternity in least flycatchers. *The Wilson Journal of Ornithology*, **121**, 306–313.

Kekkonen, J., Hanski, I. K., Jensen, H., Väisänen, R. A., and Brommer, J. E.2011. Increased genetic differentiation in house sparrows after a strong population decline: From panmixia towards structure in a common bird. *Biological Conservation*, 144, 2931–2940.

Kuitunen, M. T., Viljanen, J., Rossi, E., and Stenroos, A. (2003). Impact of busy roads on breeding success in pied flycatchers *Ficedula hypoleuca*. *Environmental Management*, **31**, 79–85.

Laurance, S. G. W., Stouffer, P. C., and Laurance, W. F. (2004). Effects of road clearings on movement patterns of understory rainforest birds in Central Amazonia. *Conservation Biology*, **18**, 1099–1109.

Leader, N., Geffen, E., Mokady, O., and Yom-Tov, Y. (2008). Song dialects do not restrict gene flow in an urban population of the orange-tufted sunbird, *Nectarinia osea*. *Behavioral Ecology and Sociobiology*, **62**, 1299–1305.

Lynch, M. and Ritland, K. (1999). Estimation of pairwise relatedness with molecular markers. *Genetics*, **152**, 1753–1766.

MacDougall-Shackleton, E. A., Clinchy, M., Zanette, L., and Neff, B. D. (2011). Songbird genetic diversity is lower in anthropogenically versus naturally fragmented landscapes. *Conservation Genetics*, **12**, 1195–1203.

Mayer, C., Schiegg, K., and Pasinelli, G. (2009). Patchy population structure in a short-distance migrant: evidence from genetic and demographic data. *Molecular Ecology*, **18**, 2353–2364.

McKinney, M. L. (2002). Urbanization, biodiversity, and conservation. *Bioscience*, **52**, 883–890.

Meirmans, P. G. and Hedrick, P. W. (2011). Assessing population structure: FST and related measures. *Molecular Ecology Resources*, **11**, 5–18.

Miller, J. R. and Hobbs, R. J. (2002). Conservation where people live and work. *Conservation Biology*, **16**, 330–337.

Milligan, B. G. (2003). Maximum-Likelihood Estimation of Relatedness. *Genetics*, **163**, 1153–1167.

Mumme, R. L., Schoech, S. J, Woolfenden, G. E. and J. W. Fitzpatrick. (2000). Life and death in the fast lane: Demographic consequences of road mortality in the Florida scrub-jay. *Conservation Biology*, **14**, 501–512.

Nei, M. (1973). Analysis of gene diversity in subdivided populations. *Proceedings of the National Academy of Sciences*, **70**, 3321–3323.

Partecke, J., Gwinner, E., and Bensch, S. (2006). Is urbanisation of European blackbirds (*Turdus merula*) associated with genetic differentiation? *Journal of Ornithology*, **147**, 549–552.

Pritchard, J. K., Stefens, M., and Donnelly, P. (2000). Inference of population structure using multilocus genotype data. *Genetics*, **155**, 945–959.

Queller, D. and Goodnight, K. (1989). Estimating relatedness using genetic markers. *Evolution*, **43**, 258–275.

Queller, D., Strassmann, J., and Hughes, C. (1993). Microsatellites and kinship. *Trends in Ecology and Evolution*, **8**, 285–292.

Radeloff, V. C., Hammer, R. B., Stewart, S. I., Fried, J. S., Holcomb, S. S., and McKeefry, J. F. (2005). The wildland-urban interface in the United States. *Ecological Applications*, **15**, 799–805.

Rasner, C. A., Yeh, P., Eggert, L. S., Hunt, K. E., Woodruff, D. S. and Price, T. D. (2004). Genetic and morphological evolution following a founder event in the dark-eyed junco, *Junco hyemalis thurberi*. *Molecular Ecology*, **13**, 671–681.

Reed, D. H., Lowe, E. H., Briscoe, D. A., and Frankham, R. (2003). Inbreeding and extinction: Effects of rate of inbreeding. *Conservation Genetics*, **4**, 405–410.

Reijnen, R., Foppen, R., Braak, C. T., and Thissen, J. (1995). The effects of car traffic on breeding bird populations in woodland. III. Reduction of density in relation to the proximity of main roads. *Journal of Applied Ecology*, **32**, 187–202.

Riegert, J., Fainová, D., and Bystřická, D. (2009). Genetic variability, body characteristics and reproductive parameters of neighbouring rural and urban common kestrel (*Falco tinnuculus*) populations. *Population Ecology*, **52**, 73–79.

Roberts, C. and Norment, C.J. (1999). Effects of plot size and habitat characteristics on breeding success of scarlet tanagers. *Auk*, **116**, 73–82.

Rutkowski, R., Rejt, L., and Szczuka, A. (2006). Analysis of microsatellite polymorphism and genetic differentiation in urban and rural kestrels *Falco tinnunculus* (L.). *Polish Journal of Ecology*, **54**, 473–480.

Rutkowski, R., Rejt, Ł., Tereba, A., Gryczyńska-Siemiątkowska, A., and Janic, B. (2010). Population genetic structure of the European kestrel *Falco tinnunculus* in Central Poland. *European Journal of Wildlife Research*, **56**, 297–305.

Salvati, L., Manganaro, A., Fattorini, S., and Piattella, E. (1999). Population features of kestrels *Falco tinnunculus* in urban, suburban and rural areas in Central Italy. *Acta Ornithologica*, **34**, 53–58.

Segelbacher, G., Manel, S., and Tomiuk, J. (2008). Temporal and spatial analyses disclose consequences of habitat fragmentation on the genetic diversity in capercaillie (*Tetrao urogallus*). *Molecular Ecology*, **17**, 2356–2367.

Smith, T. B., Milá, B., Grether, G. F., Slabbekoorn, H., Sepil, I., Buermann, W., Saatchi, S., and Pollinger, J. (2008). Evolutionary consequences of human disturbance in a rainforest bird species from Central Africa. *Molecular Ecology*, **17**, 58–71.

Spielman, D., Brook, B. W., and Frankham, R. (2004). Most species are not driven to extinction before genetic factors impact them. *Proceedings of the National Academy of Sciences of the United States of America*, **101**, 15261–15264.

Stralberg, D., Jongsomjit, D., Howell, C. A., et al. (2009). Re-shuffling of species with climate disruption: a no-analog future for California birds? *PLoS One*, **4**, p.e6825.

Tarof, S. A., Ratcliffe, L. M., Kasumovic, M. M., and Boag, P. T. (2005). Are least flycatcher (*Empidonax minimus*) clusters hidden leks? *Behavioral Ecology*, **16**, 207–217.

Thomassen, H. A., Fuller, T., Buermann, W., et al. (2011). Mapping evolutionary process: a multi-taxa approach to conservation prioritization. *Evolutionary Applications*, **4**, 397–413.

Tingley, M. W., Monahan, W. B., Beissinger, S. R., and Moritz, C. (2009). Birds track their Grinnellian niche through a century of climate change. *Proceedings of the National Academy of Sciences of the United States of America*, **106**, 19637–19643.

Townsend, A. K., Bowman, R., Fitzpatrick, J. W., Dent, M., and Lovette, I. J. (2011). Genetic monogamy across variable demographic landscapes in cooperatively breeding Florida scrub-jays. *Behavioral Ecology*, **22**, 464–470.

Unfried, T. M., Hauser, L. and Marzluff, J. M.(2013). Effects of urbanization on Song Sparrow (Melospiza melodia) population connectivity. *Conservation Genetics*, **14**(1), pp. 41–53.

Van Der Zande, A. N., Ter Keurs, W. J., and Van Der Weijden, W. J. (1980). The impact of roads on the densities of four bird species in an open field habitat-evidence of a long-distance effect. *Biological Conservation*, **18**, 299–321.

Vandergast, A. G., Bohonak, A. J., Hathaway, S. A., Boys, J., and Fisher, R. N. (2008). Are hotspots of evolutionary potential adequately protected in southern California? *Biological Conservation*, **141**, 1648–1664.

Vangestel, C., Mergeay, J., Dawson, D. A., Vandomme, V. and Lens, L. (2011). Spatial heterogeneity in genetic relatedness among house sparrows along an urban—rural gradient as revealed by individual-based analysis. *Molecular Ecology*, **20**, 4643–4653.

Weir, B. S. and Cockerham, C. C. (1984). Estimating F-statistics for the analysis of population structure. *Evolution*, **38**, 1358–1370.

Wright, S. (1951). The genetical structure of populations. *Annals of Eugenics*, **15**, 323–354.

Yeh, P. J. (2004). Rapid evolution of a sexually selected trait following population establishment in a novel habitat. *Evolution*, **58**, 166–174.

Yeh, P. J. and Price, T. D. (2004). Adaptive phenotypic plasticity and the successful colonization of a novel environment. *American Naturalist*, **164**, 531–542.

CHAPTER 12

Reconciling innovation and adaptation during recurrent colonization of urban environments: molecular, genetic, and developmental bases

Alexander V. Badyaev

12.1 Introduction

Evolution of innovations requires not only incorporation of novel genetic or environmental inputs, but also maintenance of functionality of already evolved structures. What is the structure of organismal systems that enables such coexistence? How can newly arising modifications spread and evolve before acquiring full functionality? How do novelties arise developmentally? Answers to these questions require direct analysis of development during origination of novel adaptations, during the window when beneficial novelty can be produced by developmental and genetic systems, but organismal survival still relies on existing adaptations.

Here I address these questions by studying morphological evolution in an invasive species that over the course of colonization of novel environments repeatedly evolves adaptations to urban habitats. Whereas colonization of diverse environments in this species was associated with evolution of novel adaptations to new food sources, recurrent colonization of urban environments resulted in convergent evolution of morphology most beneficial in urban settings. I was particularly interested in the interaction between historical contingency and current selection in the evolution of morphologies adaptive in urban settings.

From a few California birds released from a pet store in New York City in 1939, and through natural expansion of its western range, in just a few decades the house finch (*Carpodacus mexicanus*) came to occupy one of the widest ecological ranges of any extant bird (Badyaev et al., 2012). Originally a bird of hot deserts and dry open habitats of the southwest, it now occurs in nearly all types of landscapes and climates in North America, from edges of northern taiga to ocean coasts, to metropolitan areas. Frequently overlooked, however, is the fact that during colonization of their tremendously diverse ecological range, house finches also repeatedly colonized, from different starting points, a more uniform urban environment.

Here I specifically focus on rapid evolution of beak morphology. In seed-eating birds, beak adaptations reflect not only competing demands of seed handling and crushing (Boag & Grant, 1981; Bowman, 1961; Willson, 1971), but also influences vocal tract configurations and vocalizations (Podos et al., 2004; Slabbekoorn & Smith, 2002). For example, larger beaks are less suited for rapid sound modulations, such as trills, whereas longer beaks and wider gape limit the effective length of the vocal tract and corresponding song frequencies (Nowicki, 1987).

There are two reasons why colonization of urban environments has a particularly strong effect on house finch beak morphology. First, the finches in recently established urban populations depend closely on feeders; all range expansion to urban

Avian Urban Ecology. Edited by Diego Gil and Henrik Brumm

areas, especially at the extreme northern parts of the range (e.g. in south-western Alaska) was enabled by association with feeders (Badyaev et al., 2012). Second, feeders typically supply seeds (such as sunflower or large millet seeds) that are much larger and harder than those constituting the bulk of house finch diet in natural populations (Badyaev et al., 2008; Hensley, 1954). Such contrast is particularly evident in comparisons of urban and rural population in the species' ancestral range in south-western United States: small seeds of grasses and cacti, that constitute the bulk of seed diet in the desert, have average diameter in the direction of cracking of 1.2 ± 0.4mm (n = 377 seeds) and average hardness of 2.7 ± 0.08 Newtons (N), whereas sunflower seeds which are the main part of seed diet in the urban areas have average diameter in the direction of cracking of 6.8 ± 2.1 mm (n = 120) and average hardness of 8.11 ± 0.21 N (Badyaev et al., 2008; Mills et al., 1989).

In this chapter, I will first describe morphological divergence in and natural selection on beak configurations across urban and natural populations. Second, I will examine ways in which innovation and adaptation can be reconciled during adaptive diversifications of beaks. Third, I will discuss developmental and genetic basis of such cycles of divergence and convergence to urban adaptations and the ways by which ontogenetic mechanisms can accomplish precise, diverse, and yet reversible adaptations in beak configurations. I suggest that both urban adaptations and population divergence were facilitated by modular organization of beak development, where small regulatory changes in conserved molecular growth factors and significant functional redundancy in resulting configurations produce a wide range of adaptive beak modifications. I will conclude with a preliminary analysis of molecular basis behind such adaptive evolution and direction for future studies.

12.2 Recurrent urban adaptations during range expansion

Rapid and reversible evolution of local adaptations that accompanied range expansion of house finches across North America over the last 70 years provide an excellent opportunity to directly investigate the nature of developmental change during crucial, but commonly missed stage of evolution—when the initial phenotypic adaptability is converted into local adaptation and when existing adaptations have to coexist with novel modifications. In the northern, recently established parts of the range, house finch populations provide an opportunity to study replicated evolution of adaptations to urban environments from different starting points along different historical contingencies (Figure 12.1), and during the time when the need for rapid reorganization of complex beaks coincides with the need to maintain some stability once the optimal configuration is found. In the southern, ancestral parts of the range, house finch differentiation between urban and rural environments gives us insight into long-term consequences of adaptation for urban environments in morphological, cultural, and developmental systems (Badyaev et al., 2008). Both time scales will be explored here.

12.2.1 Different ecological starting points of urban adaptations

In the house finch, the routes of colonization of novel environments do not correspond to the locations of highest species-specific habitat suitability (Figure 12.2). To establish this, we created detailed geographic information system (GIS) maps with cell resolution = 1 km for >300 locations across newly established populations in northwestern United States, using a cost distance approach as implemented in ESRI's ARC/INFO spatial analysis (Badyaev and Krebs, unpublished data). Briefly, we incorporated measures of species-specific resistance to movement and dispersal by weighing the distance between a source and a target cell based on resistance values of the intervening cells (Koen et al., 2012; Nikolakaki, 2004) and specifically focused on parameters that describe anthropogenic developments and vegetation types, because both determine major changes in seed diet of the house finches (e.g. uniform and large sunflower seeds from feeders versus small and diverse native seeds in rural environments). We calculated ecological similarity of populations based on GIS layers of roads and dwellings, population

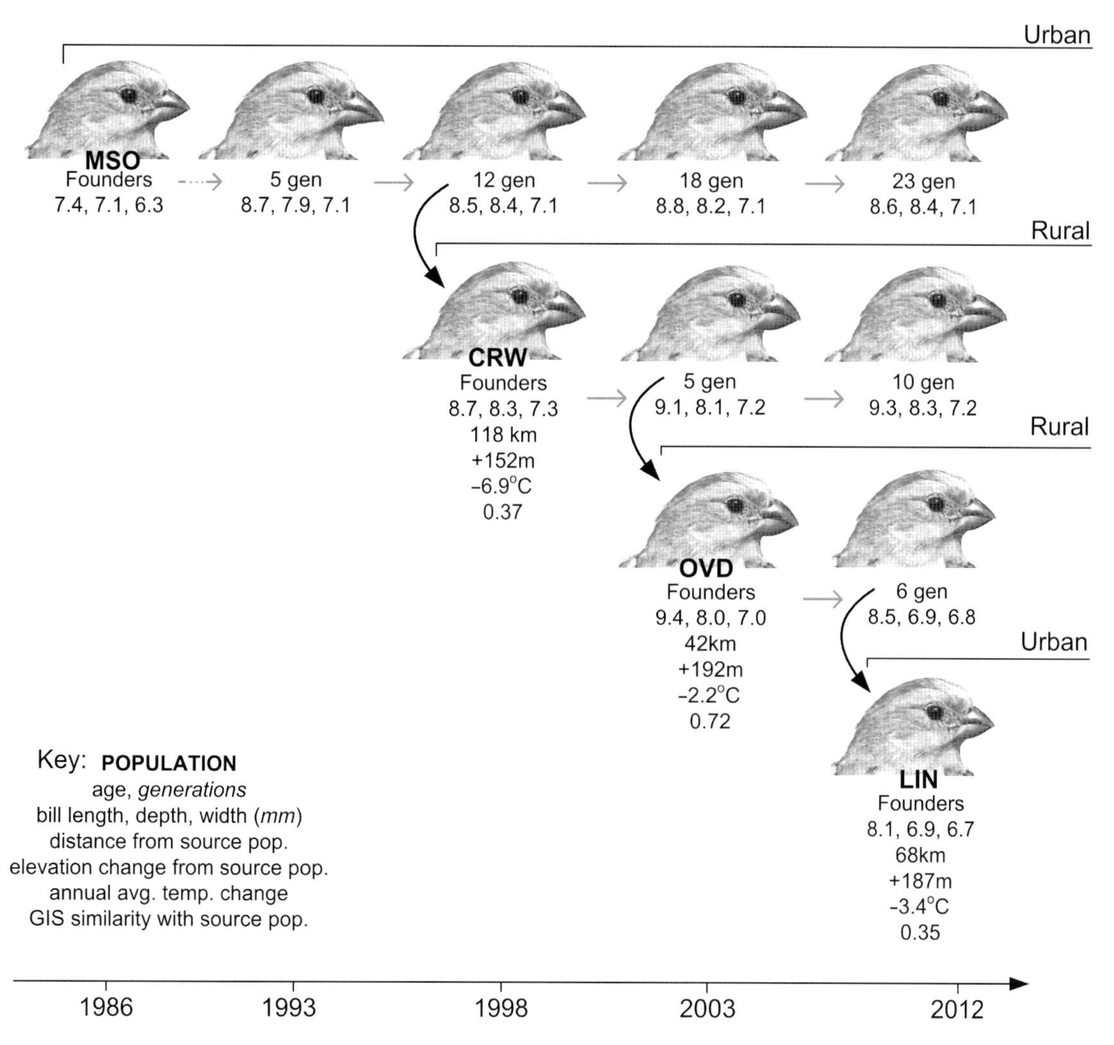

Figure 12.1 An example of urban and rural population sequence established by 'dispersal leaps' following source population establishment (e.g. rural CRW is established by 12 generation-old urban MSO population) in north-western United States. Landscape data and GIS similarity (cost resistance across land cover layers) are from Badyaev and Krebs (unpublished). Climate data from 1948 to 2010. Morphometric-based figures are from Badyaev (2010).

density (the 2000 United States census), elevation, land cover and vegetation types (50 habitat types, output cell resolution = 90 m) to establish that the populations within a typical colonization sequence (e.g. Figure 12.1) experience exceptionally diverse ecological conditions over just a few generations (Figure 12.2). The significance of this finding is that colonization of urban locations en route of such invasions proceeds from widely distinct ecological starting points.

12.2.2 Rapid and reversible evolution of locally adaptive beak morphologies

Colonization of ecologically distinct areas during the range expansion was associated with rapid reorganization in beak morphologies, both within and between newly established populations (Badyaev & Hill, 2000; Badyaev & Martin, 2000). Ultimately, these changes were driven by close covariation of seed diet composition and bite force, which, in turn,

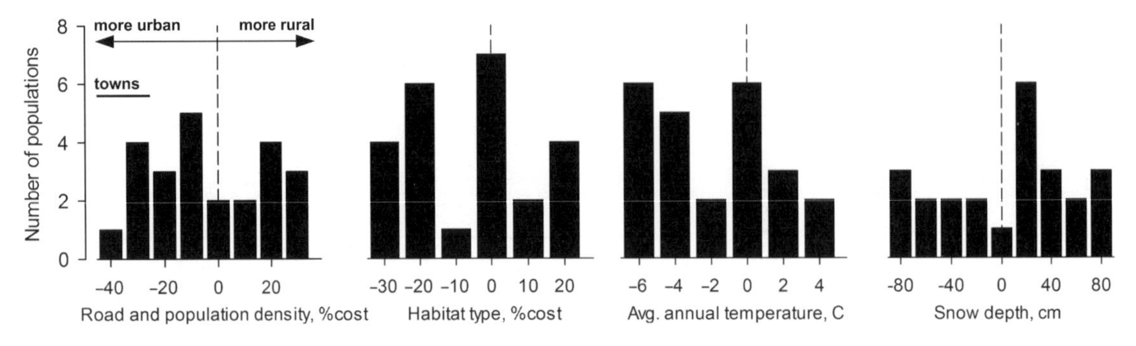

Figure 12.2 Diverse ecological conditions that precede and succeed urban adaptations across 24 recently established house finch populations in north-western United States. Shown are differences from the source populations (dashed lines) in human population density (proxy for feeder abundance), habitat type (% difference in suitability of habitat type for house finches from GIS layers, % cost), average annual temperature, and snow depth (Badyaev and Krebs, unpublished).

are closely linked to beak's bone and muscle development in this species (Badyaev, 2010, 2011a).

Across populations, colonization of urban environments was associated with the evolution of longer and deeper beaks whereas adaptation to rural environments was associated with the evolution of shorter and markedly narrower beaks (Badyaev, 2010; Badyaev et al., 2008). Although these trends were recurrent among populations, during the process of such evolution, finches commonly expressed an unusually wide spectrum of beak morphologies and configurations (see Section 12.2.3).

Long-term study of beak evolution in urban environments (Badyaev, 2010) revealed three main principles. First, the evolution of precise local beak configurations did not diminish future modifications; distinct and locally stable configurations remained reversible, often many generations in the future. Second, functional equivalence of distinct beak configurations enabled initial survival and afforded wide exploration of beak morphospace before the most appropriate configuration was found and stabilized by natural selection. Third, during the first few generations in a new ecological locale, finches showed rapid reorganization and diversity in beak configurations followed by stabilizations of locally appropriate beak configuration (Figure 12.3).

Importantly, the stabilization and onset of local adaptation coincided temporarily with recurrent patterns of dispersal at the edge of house finch geographic range, when groups of resident adults undertake post-breeding 'dispersal leaps' to a new location, often hundreds of kilometres away and become year-round residents there—the main pattern of range expansion in this species (Badyaev et al., 2012). There was no consistent sex-bias in such dispersing flocks which consist mostly of adult pairs (Badyaev et al., 2001); these 'dispersal leaps' occur every 5–10 generations after population establishment. When such 'dispersal leaps' coincide temporarily with origination and stabilization of locally adaptive beak configuration in source populations, they enable the direct study of the effects of the starting developmental configuration on subsequent evolution (Figure 12.3).

12.2.3 Developmental and functional redundancy of beak configurations

Overall, greater bite force (contributed primarily by beak width) and longer beaks were favoured in urban areas where house finch's seed diet consists almost exclusively of sunflower seeds (Badyaev et al., 2008). However, despite a strong trend in bite force in both rural and urban adaptations, there was not a single optimal beak configuration that was favoured in each environment (Figure 12.3c). In urban populations, fitness contribution of proportional expression of beak length and width (under fluctuating directional selection) differed between generations, whilst beak depth was most commonly under stabilizing selection (Figure 12.3).

Three findings illustrate adaptive equivalence of distinct beak configurations during evolution and

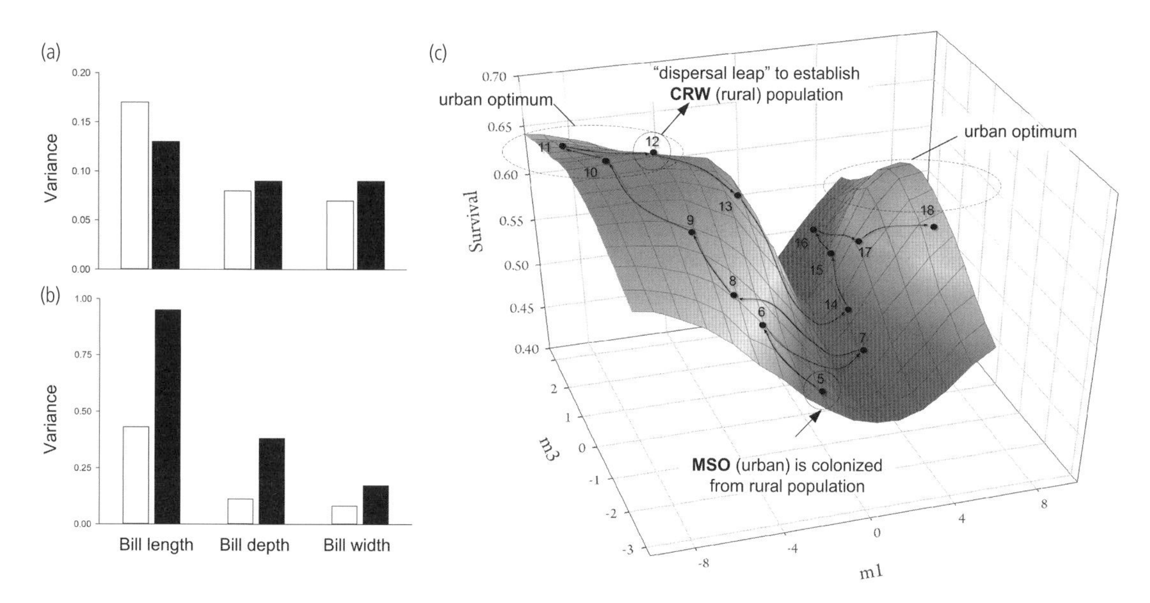

Figure 12.3 Beak morphology of finches in urban populations is more uniform, regardless of the founding population and local climatic conditions, than beak morphology of finches in native/rural populations. (a) Divergence between urban (white bars, $n = 1024$) and rural/native (black bars, $n = 221$) populations in ancestral range in southeastern Arizona. (b) Recent divergence between urban (white bars, $n = 8740$, five populations) and rural populations (black bars, $n = 9130$, 11 populations) in northwestern Montana. (c) Fitness surface (survival selection on bite force) of MSO population over 20 years of adaptation to urban environment, defined by the canonical axes m1 (beak depth) and m3 (relative length and width) (after Badyaev, 2010). Lines connect subsequent generations. Dashed ellipsoids show two distinct areas of high fitness—'urban optimums'—in this urban population. 'Dispersal leap' and founding of rural CRW population coincides with 12th generation of evolution of adaptation in urban MSO population.

prevalence of compensatory developmental and functional interactions among beak components (Badyaev, 2010). First, despite large microevolutionary changes during adaptation to the urban environment, the morphological integration—a measure of developmental and functional correlation between beak components—remained similar. Second, despite widely distinct beak phenotypes across generations during adaptive sequences (e.g. Figure 12.1), the across generation change in overall covariance structure was gradual between adjacent generations, the pattern expected when consistent microevolutionary change in one beak component (e.g. beak depth) is associated with compensatory growth in the other components. Finally, a detailed analysis of phenotypic covariance structure of beaks during evolution revealed its high dimensionality (Badyaev, 2010), implying that different beak components show compensatory variation in different generations.

The major evolutionary significance of compensatory developmental interactions and functional versatility in configurations of beak components is that they maintain substantial developmental and genetic variance in individual components, thereby enabling rapid microevolutionary changes during population establishment. Such organization not only can enable rapid assembly of locally appropriate morphology and exploitation of resources in novel environments, but can also reconcile the necessity for specialization and close functional integration in beaks with extensive evolutionary diversification following changes in natural selection.

12.2.4 Consequences for song production

Combination of local foraging adaptation, effects on the vocal apparatus and cultural inheritance of songs can strongly reinforce local adaptation in beak configurations (Grant & Grant, 1996; MacDougall-Shackleton & MacDougall-Shackleton, 2001; Mayr, 1939). In house finches such reinforcement might be particularly effective for three reasons. First, the beak traits that are most affected by beak modifications in urban environments—beak length and overall size—are most important for production of buzz and trill notes—the song elements that are the target of mate choice in this species (Bitterbaum & Baptista, 1979; Mennill et al., 2006). Second, background noise of urban environments most affects vocalization at maximum frequencies (Patricelli & Blickley, 2006; Slabbekoorn & Smith, 2002) that are, in turn, most affected by urban environment-specific beak modifications. Finally, house finches have unusually small song-learning neighbourhoods—less than 5 km in some environments (Bitterbaum & Baptista, 1979; Tracy & Baker, 1999).

Overall, across their range, urban finches had fewer notes in their songs and had fewer trill rates than did rural finches (Badyaev et al., 2008). The relationship between beak morphology and song characteristics varied among populations because of variation in beak morphology (Badyaev & Hill, 2000), and also because of local cultural inheritance of songs. For example, in ancestral populations in south-eastern Arizona, urban house finches sang fewer notes and had slower trill rates over wider frequency range than did the desert finches—differences attributed mostly to their beaks being longer, deeper and wider (Badyaev et al., 2008). Similarly, in an urban population in north-western United States, house finches sang songs without buzz notes—the song element that is affected the most by urban noise (Mennill et al., 2006). The link between ecological conditions and sexual signalling reinforces, through reproductive isolation, selection for local adaptation in beak configurations and ultimately contributes to faster population divergence.

12.3 Developmental bases of recurrent and reversible urban adaptations

12.3.1 How do adaptive traits arise in development? General principles

Evolution proceeds by combining conserved and modular generative processes into hierarchical configurations at different times, places, and contexts by regulatory changes (Davidson, 2006; King & Wilson, 1975; Wagner, 2011; Wilkins, 2001). The processes arising from fundamental biochemical and physical properties of biological tissues and their interaction with the environment provide the basis for innovations, exploratory behaviours, and developmental plasticity (Forgacs & Newman, 2005; Gerhart & Kirschner, 2007; Müller & Newman, 2005; Newman & Müller, 2000; West-Eberhard, 2003). The extent to which natural selection or genetic drift stabilize these processes versus create and direct them is a debated question (Badyaev, 2005; 2011b; Lynch, 2007; Müller, 2007; Reid, 2007). Regardless, investigation of links between the processes that produce adaptive modifications and processes that maintain and modify them is particularly relevant for the evolution of composite traits in organisms, such as birds, where generation times are often too long and population sizes too small to accomplish gradual step-wise coordinated evolution of every component, each with incremental (not yet functional) stages, as is conventionally assumed (Lynch & Abegg, 2010; Lynch, 2010). Further, the extraordinary conservation of basic molecular and cellular mechanisms over vast phylogenetic distances and organismal systems (Davidson, 2006; Koonin & Wolf, 2009; Shubin et al., 2009) not only brings forth the question of how can natural selection accomplish tremendous phenotypic diversity with such a limited set of ingredients, but, most importantly, gives us a powerful insight into how evolutionary change actually proceeds. The central emerging theme is that novelties, diversification, and adaptation are best understood in terms of conserved developmental processes being stabilized and arranged, secondarily, by natural selection or genetic drift.

12.3.2 Developmental dynamics of beak development

How do developmental processes enable precise local adaptation in beaks (that requires diminishing of developmental variation) with innovation (that requires maintenance of abundant developmental variation) during recurrent colonization of the urban environment? A potential resolution is that many of the emergent and self-regulatory processes that comprise a composite beak (Figure 12.4) are not themselves subject of natural selection acting on beak functioning (e.g. seed handling and bite force).

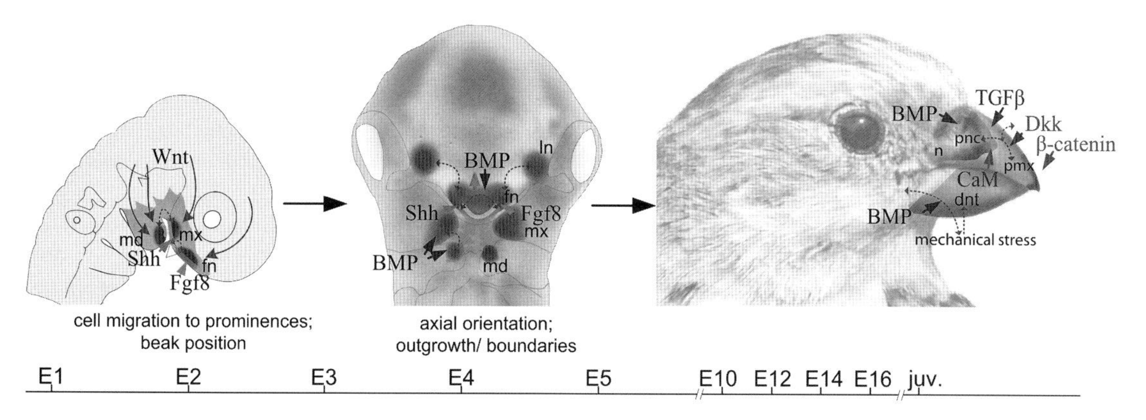

Figure 12.4 Modular core processes are regulated by conserved growth factors in the house finch beak development (Badyaev, 2011a). Main genes are wingless type (*Wnt*), fibroblast growth factor 8 (*Fgf8*), sonic hedgehog (*Shh*), bone morphogenetic proteins (BMP), transforming growth factor beta (*TGFβ*), calmodulin (*CaM*), Dickkopf (*Dkk*), and *β-catenin*. Additional 741 highly expressed genes associated with cell proliferation, ossification, and stress response during establishment of new urban and rural populations are identified (see Figure 12.6). Facial prominences (shown in dark), formed by proliferation of neural crest cells, are frontonasal (fn), lateral nasal (ln), mandibular (md), and maxillary (mx). Cartilage and bone areas arising in late development are prenasal cartilage (pnc), premaxillary bone (pmx), nasal bone (n), and dentary bone (dnt).

Instead the role of natural selection can be largely confined to eliminating or stabilizing post-production configurations of conserved developmental processes and to fine-tuning these configurations to the most recurrent or locally appropriate ecological context. Ubiquitous reuse of conserved regulatory elements throughout beak ontogeny (Figure 12.4) gives selection an opportunity to rapidly accomplish observed adaptive diversifications (Figure 12.1) without excessive waiting time and population sizes needed for incremental evolution of complex beaks by coordinated evolution of each of the developmental components. In the last section of this chapter I will present evidence that such patterns are evident in house finch urban adaptations.

Beak morphogenesis starts with migration of neural crest cells into the embryo's facial region and mandibular arch (Helms & Schneider, 2003). Neural crest cells apparently do not have affinity for a final placement until they aggregate into four to five facial prominences that will ultimately merge into a beak (Geetha-Loganathan et al., 2009; Helms & Schneider, 2003; Wu et al., 2006). Activation–inhibition interactions and compensatory growth of rapidly dividing mesodermal cells in adjacent prominences and their ectodermal envelopes determine the juxtaposition of prominences and delineate the placement of a future beak (Figure 12.4). In turn, continuing cell proliferation induces reciprocal regulatory feedback between the mesenchyme cells and epithelium boundaries within each prominence, with the epithelium layer providing boundary effects and axial orientation directing cell outgrowth (Eames & Schneider, 2008; Hu & Marcucio, 2009; Wu et al., 2006). Frontonasal, two lateral nasal and two maxillary prominences (Figure 12.4) then merge and, directed by activation-inhibition feedback from their epithelium layer, form the upper beak. Merging of two mandible prominences produces the lower beak (Figure 12.4).

In later embryonic stages, interactions between proliferating cartilage and bone areas of the upper beak activates local factors, fine-tuning beak length and curvature, with many shapes emerging as a straightforward geometric consequences of cell proliferation and juxtaposition and the ratio of cartilage–bone cell mass (Abzhanov et al., 2007; Campàs et al., 2010; Mallarino et al., 2011; Wu et al., 2006). Importantly, in later nestling and juvenile stages, when house finches start to consume seeds unmanipulated by adults, mechanical stress associated with muscle attachment and force leads to local production of bone morphogenetic protein (BMP) factors and host of other skeletogenic signals that are commonly responsible for fine-tuning of functional beaks and beak adaptive plasticity (Young & Badyaev, 2007).

The entire process is regulated by conserved regulatory factors that have distinct effects depending on time and place of their expression (Figure 12.4). For example, *BMP4* regulates early commitment of neural crest cells to bone formation at E-2 stage (Abzhanov et al., 2007), later cell proliferation within prominences at E-4 (Wu et al., 2004), species-specific outgrowth of cartilage-bone areas in upper beak at E-6 (Abzhanov et al., 2004), mechanical stress-induced bone formation in both upper and lower beaks, and adaptive plasticity in beak size at E-10 and juvenile stages (Figure 12.4; Badyaev et al., 2008). Thus, regulation of the same molecular factor (or mutation in the same factor) can accomplish beak reorganizations at different points in ontogeny.

12.3.3 Two evolutionary scenarios

Can such developmental organization, with conserved and redundant signalling (Figure 12.4), produce rapid adaptations to urban environments, whilst maintaining extensive evolutionary diversification in other parts of the species' geographic range? Especially considering small sizes of founding populations and short time periods of colonization? There are two general views.

Under conventional population genetics scenario, the evolutionary change in complex beaks depends on mutational input during development of each of beak's components at each evolutionary stage, resulting in gradual, and coordinated microevolutionary change, where modifications are small enough not to disrupt beak functionality, but large enough to maintain directional and continuous change. The conservatism and redundancy of growth factors (Figure 12.4) greatly facilitate such coordinated and directional change. Although developmental and genetic redundancy of pathways by which beaks are produced (Section 12.4) likely provides a large mutational target, the major impediment is likely to be weak selection in small populations over short periods of time to account for large beak modifications during population establishments (Figure 12.1) and the necessity of incremental fitness of intermediate, but not yet functional configurations. Rapid reversals in adaptive beak morphologies and apparently undiminished genetic variation in beak component during adaptive evolution (Badyaev, 2010) are problematic under this scenario. Overall, the hypothesis predicts gradual and continuous microevolutionary change within population sequences, retention and improvement of an optimal beak configuration, and incremental coordinated changes in all developmental components.

Alternatively, beak modification can arise by regulation and stabilization of conserved core developmental components and processes that are not themselves a subject to contemporary natural selection (although they might be shaped, partially, by past selection). This view emphasizes the modularity of beak development, where many phenotypic outcomes arise from emergent, self-regulatory processes of cell division, adhesion, and competition being secondarily stabilized by regulatory factors (Forgacs & Newman, 2005; Müller, 2007; Newman, 2012). Under this view, directionality of evolutionary change and functionality of novel phenotypes are direct products of within-organism developmental dynamics and homeostasis (Baldwin, 1902; Schmalhausen, 1938). Such that novel inputs (genetic or environmental) are accommodated by beak developmental processes; directionality of evolutionary change comes from such integration of external stresses into developmental processes (Waddington, 1959). This process can operate in small populations and produce large, distinct, but functional, modifications of beaks by few regulatory mutations. Overall, the hypothesis predicts rapid reorganizations and abrupt microevolutionary changes between populations, adaptive equivalence of distinct configurations, and suggests that regulatory elements are main targets of heritable change in beak development.

Thus, under the conventional view, arising genetic variation enables evolutionary change, with directionality provided by natural selection. Under the alternative scenario, arising stabilization and accommodation of external stresses is aligned with evolutionary change, the directionality and functionality comes from developmental dynamics of cells and tissues. Data on large-scale patterns of phenotypic and genetic evolution in several house finch populations (Badyaev, 2009; 2010) are in general accordance with predictions of the second hypothesis. Here I examine the existing molecular evidence for these mechanisms.

12.4 Evidence: molecular genetic basis of urban adaptations

12.4.1 Histological analysis of urban adaptations

In urban and rural environments within an ancestral range of house finches, we compared histological changes associated with beak divergence and local adaptation. To assess *BMP* activity and developmental sequence of cartilaginous bone origination, growth, and maturation, we analysed sections of embryos using an anti-phosphorylated Smad1/Smad5/Smad8 (Cell Signalling Technology, Beverly, MA; Ahn et al., 2001). Binding of BMPs to Ser/Thr kinase receptors on the cell membrane activates receptors and subsequently leads to phosphorylation of Smad proteins (Hogan, 1996; Young & Badyaev, 2007).

The mandibular primordia of the large-beaked urban finches expressed BMP earlier and at a higher level than those of the desert finches; urban and rural birds were distinct in BMP activity by embryonic day 5 (7 days prior to hatching, E26 stage in Figure 12.5a) and difference continued to increase throughout the ontogeny. Overall, embryos in the urban population had accelerated ontogenetic transformations of mandibles (zone I; Figure 12.5b) compared to the native desert population (Badyaev et al., 2008). Interestingly, there were no differences in skeletal tissues between the populations. The most significant finding was a major shift in timing of *BMP4* expression between urban and rural populations (Figure 12.5). BMP is a ubiquitous regulator of bone remodelling in vertebrates (Young & Badyaev, 2007); however, its involvement at such earlier developmental stages of supposedly recent population differentiation was surprising. Later studies, however, documented ubiquitous reuse of conserved regulatory factors, including BMP4, throughout beak ontogeny (reviewed in Badyaev, 2011a). Such exceptional differentiation in timing of expression of conserved regulators can enable rapid reorganization of beak development. However, to definitively establish whether this pattern is common in other growth factors and whether gene functions remain constant, we needed to sequence the transcriptome of house finch beak development during initial stages of adaptation to urban environments (Young and Badyaev, unpublished work).

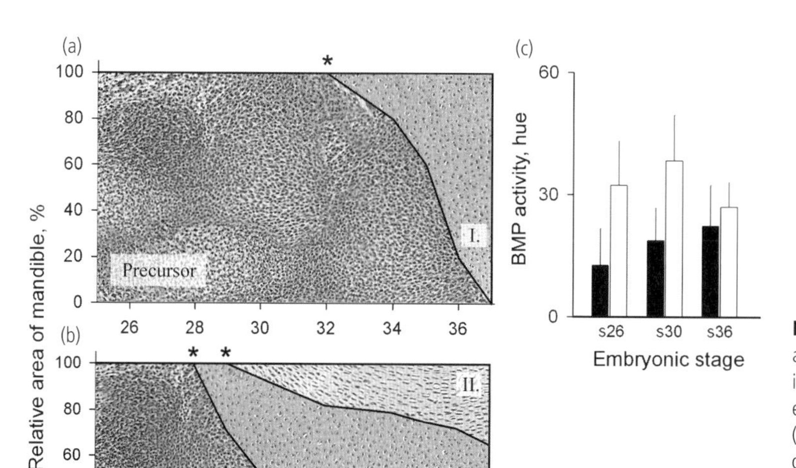

Figure 12.5 Ossification is strongly accelerated in urban environments. Shown is ontogenetic tissue transformation in embryonic lower mandible in (a) rural and (b) urban populations and relative areas occupied by tissues at successive stages of ossification (zone I and zone II by H&H embryonic stages). Asterisk shows the earliest age of appearance of a tissue. (c) Strong overexpression of *Bmp4* (mean ± 1SE) in urban (white bars) versus rural (black bars) populations. After Badyaev et al. (2008).

12.4.2 Genomic analysis of urban adaptations: preliminary results

We extracted and sequenced RNA from embryos along the population sequences, focusing on urban–rural transitions (e.g. Figure 12.1). In RNA-seq analyses it is not always clear whether genes with low read counts come from low expressed genes or from transcriptional noise (Hebenstreit et al., 2011; Wang et al., 2009)—e.g. how to separate expressed and non-expressed genes in low values of frequency distribution (Figure 12.6). Thus, we used a model that considers the number of transcripts detected in a RNA-seq study as a mixture of two distributions—a Poisson distribution for transcripts from inactive genes and a negative Binomial distribution for transcribed genes (Wagner et al., 2012). Application of this model showed that genes with more than one transcript per million transcripts (TPM > 1) come from highly expressed genes (Figure 12.6). We used this criterion to assess and compare gene expression across urban and rural environments (Figures 12.6 and 12.7).

We found that, first, for genes expressed in both urban and rural populations, the expression in the urban population was earlier (corroborating accelerated development of larger beaks there, see Section 12.4.1). Second, controlling for developmental stage, many more genes were expressed in new rural population than in the source urban populations and these genes were associated with a wider range of functions. For example, in the urban environment, $n = 6302$ genes were expressed at stage 36 compared to $n = 5607$ genes in a rural population at stage 30

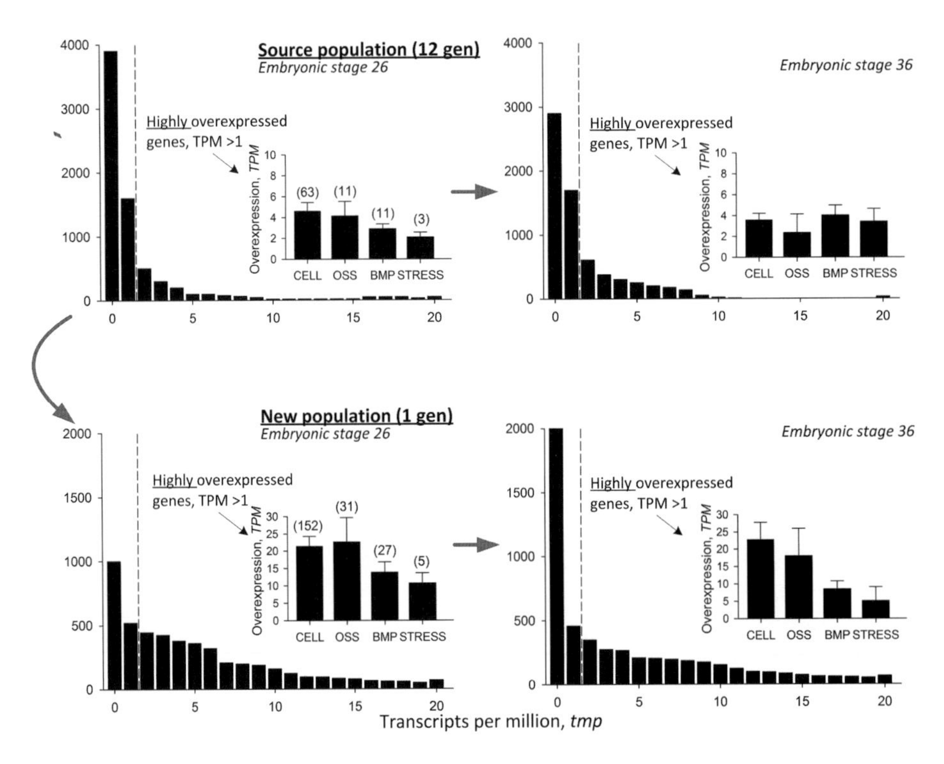

Figure 12.6 Highly distinct RNA-seq embryonic profiles during establishment of a new rural population from an urban population (across rows) and ontogenetic changes within urban (upper row) and rural (lower row) populations. Shown is frequency distribution of transcribed and non-transcribed genes and magnitude of expression in the subset of highly expressed genes involved in cell proliferation (CELL), ossification (OSS), BMP signalling (BMP), and skeletogenic response to corticosterone (STRESS). In parentheses are numbers of genes in each category. Note distinct scales of upper and lower graphs. Tpm-values are normalized to both gene length (kbp) and to number of reads mapped (sum of raw reads mapped) making them comparable across genes, populations, and developmental stages.

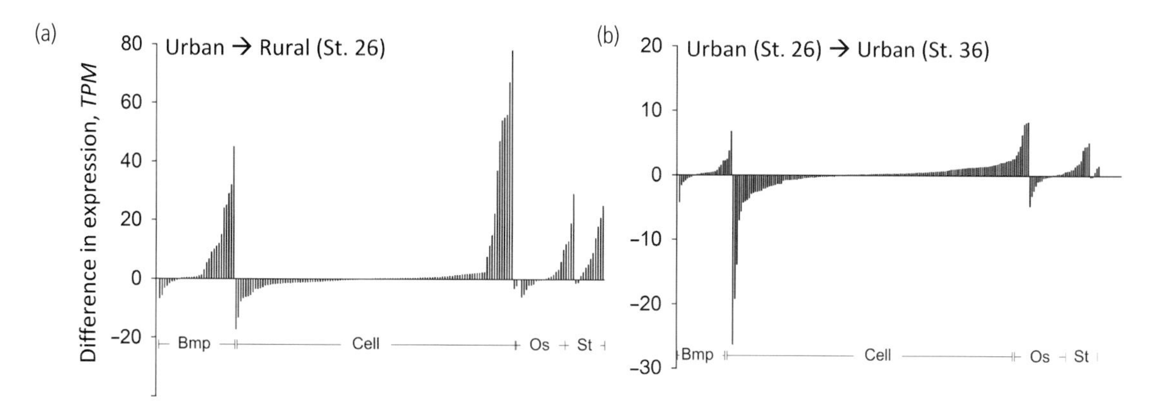

Figure 12.7 Differences in gene expression between source (urban) and new (rural) populations at the same embryonic stage (a) and between embryonic stages within a source urban population (b) for the subset of highly expressed genes involved in cell proliferation (CELL), ossification (OSS), BMP signalling (BMP), and skeletogenic response to corticosterone (STRESS) (based on Figure 12.6). Note manifold changes of gene expression in switches between urban and rural environments. Within groups genes are sorted by magnitude of expression.

and n = 4618 genes in rural population at stage 25, including n = 741 highly expressed genes associated with ossification, cell proliferation, *Bmp* signalling and response to corticosterone (Figure 12.6 and Figure 12.7). Finally, colonization of novel urban environments and onset of novel local adaptation was associated with significantly greater transcriptional noise than in the source population (Figure 12.6, Young and Badyaev, unpublished work).

Importantly, the development of novel beak configuration in urban environments involved not only known skeletogenic genes, but also a cooption of new genes, that often differed between the populations (Young and Badyaev, unpublished work). Two groups of genes offer particularly interesting insights. First, the genes that regulate cartilage formation did not differ between the environments [e.g. in beak and skeleton, the expression of *COL1A1* (fribrillar collagen), *EXT1* (growth factor signalling), *Ihh* (signalling during chondrocyte differentiation), *MYOG* (skeletal muscle development), *BMP1*(cartilage formation)]. Whereas the genes that were closely associated with ossification differed the most [e.g. *BMP4* (endochondral osteogenesis, onset of ossification), *CDH11* (expressed in osteoblast cells, upregulated during osteoblast differentiation), *RARG* (chondrocyte differentiation), *TWIST1* (osteoblast differentiation), and *SPARC* (osteonectin-excreted by osteoblasts, induces mineralization)].

Second, the genes associated with response to corticosterone—a stress hormone (STRESS in Figures 12.6 and 12.7) were overexpressed during initial stages of colonization of urban areas. This finding is particularly significant because baseline corticosterone was elevated in newly established urban populations and under novel environmental conditions in general (Young and Badyaev, unpublished work). Further, because of the shared regulatory pathways between corticosterone synthesis and BMP production, response to stress can mediate ossification and growth under novel ecological conditions (Badyaev et al., 2006).

In sum, our preliminary RNA sequencing results corroborated previous observations of surprising involvement of conserved regulatory factors into earliest stages of divergence and in cooption of gene functions. We uncovered significant interchangeability of gene functions under novel conditions in accounting for abrupt modifications of beaks during evolution of adaptations to urban environments.

References

Abzhanov, A., Protas, M., Grant, B. R., Grant, P. R., and Tabin, C. J. (2004). *Bmp4* and morphological variation of beaks in Darwin's finches. *Science*, **305**, 1462–1464.

Abzhanov, A., Rodda, S. J., Mcmahon, A. P., and Tabin, C. J. (2007). Regulation of skeletogenic differentiation in cranial dermal bone. *Development*, **134**, 3133–3144.

Ahn, K., Mishina, Y., Hanks, M. C., Behringer, R. R., and Crenshaw, E. B. (2001). BMPR-IA signaling is required for the formation of the apical ectodermal ridge and dorsal-ventral patterning of the limb *Development*, **128**, 4449–4461.

Badyaev, A. V. (2005). Stress-induced variation in evolution: from behavioural plasticity to genetic assimilation. *Proceedings of Royal Society London: Biological Sciences*, **272**, 877–886.

Badyaev, A. V. (2009). Evolutionary significance of phenotypic accommodation in novel environments: An empirical test of the Baldwin effect. *Philosophical Transactions of the Royal Society*, **364**, 1125–1141.

Badyaev, A. V. (2010). The beak of the other finch: Coevolution of genetic covariance structure and developmental modularity during adaptive evolution. *Philosophical Transactions of the Royal Society, Biological Sciences*, **365**, 1111–1126.

Badyaev, A. V. (2011a). How do precise adaptive features arise in development? Examples with evolution of context-specific sex-ratios and perfect beaks. *Auk*, **128**, 467–474.

Badyaev, A. V. (2011b). Origin of the fittest: Link between emergent variation and evolutionary change as a critical question in evolutionary biology. *Proceedings of Royal Society London: Biological Sciences*, **278**, 1921–1929.

Badyaev, A. V. and Hill, G. E. (2000). The evolution of sexual dimorphism in the house finch. I. Population divergence in morphological covariance structure. *Evolution*, **54**, 1784–94.

Badyaev, A. V. and Martin, T. E. (2000). Sexual dimorphism in relation to current selection in the house finch. *Evolution*, **54**, 987–997.

Badyaev, A. V., Whittingham, L. A., and Hill, G. E. (2001). The evolution of sexual size dimorphism in the house finch: III. Developmental basis. *Evolution*, **55**, 176–189.

Badyaev, A. V., Hamstra, T. L., Oh, K. P., and Acevedo Seaman, D. (2006). Sex-biased maternal effects reduce ectoparasite-induced mortality in a passerine bird. *Proceedings of the National Academy of Sciences of the United States of America*, **103**, 14406–14411.

Badyaev, A. V., Young, R. L., Oh, K. P., and Addison, C. (2008). Evolution on a local scale: Developmental, functional, and genetic bases of divergence in bill form and associated changes in song structure between adjacent habitats. *Evolution*, **62**, 1951–1964.

Badyaev, A. V., Belloni, V., and Hill, G. E. (2012). House finch (Haemorhous mexicanus). In A. Poole, ed. The Birds of North America Online. Ithaca, Cornell Laboratory of Ornithology; http://bna.birds.cornell.edu/bna/species/046: doi:10.2173/bna.46

Baldwin, J. M. (1902). *Development and Evolution*. Macmillan, New York.

Bitterbaum, E. and Baptista, L. F. (1979). Geographical variation in songs of California House Finches (*Carpodacus mexicanus*). *Auk*, **96**, 462–474.

Boag, P. T. and Grant, P. R. (1981). Intense natural selection in a population of Darwin's finches (Geospizinae) in the Galapagos. *Science*, **214**, 82–85.

Bowman, R. I. (1961). *Morphological differentiation and adaptation in the Galapagos finches*. University of California Press, Berkeley, CA.

Campàs, O., Mallarino, R., Herrel, A., Abzhanov, A., and Brenner, M. P. (2010). Scaling and shear transformations capture beak shape variation in Darwin's finches. *Proceedings of the National Academy of Sciences of the United States of America*, **107**, 3356–3360.

Davidson, E. H. (2006). *The Regulatory Genome: Gene Regulatory Networks in Development and Evolution*. Academic Press, San Diego,CA.

Eames, B. F. and Schneider, R. A. (2008). The genesis of cartilage size and shape during development and evolution. *Development*, **135**, 3947–3958.

Forgacs, G. and Newman, S. A. (2005). *Biological Physics of the Developing Embryo*, Cambridge University Press, Cambridge.

Geetha-Loganathan, P., Nimmagadda, S., Antoni, L., Fu, K., Whiting, C. J., Francis-West, P., and Ruichman, J. M. (2009). Expression of WNT signaling pathway genes during chicken craniofacial delevelopment. *Developmental Dynamics*, **238**, 1150–1165.

Gerhart, J. and Kirschner, M. (2007). The theory of facilitated variation. *Proceedings of the National Academy of Sciences of the United States of America*, **104**, 8582–8589.

Grant, B. R. and Grant, P. R. (1996). Cultural inheritance of song and its role in the evolution of Darwin's finches. *Evolution*, **50**, 2471–2487.

Hebenstreit, D., Fang, M., Gu, M., Charoensawan, V., Oudenaarden, A. V., and Teichmann, S. A. (2011). RNA sequencing reveals two major classes of gene expression levels in metazoan cells. *Molecular Systems Biology*, **7**, 497.

Helms, J. A. and Schneider, R. A. (2003). Cranial skeletal biology. *Nature*, **423**, 326–331.

Hensley, M. M. (1954). Ecological relations of the breeding bird population of the desert biome in Arizona. *Ecological Monographs*, **24**, 185–207.

Hogan, B. L. M. (1996). Bone morphogenetic proteins: multifunctional regulators of vertebrate development *Genes & Development*, **10**, 1580–1594.

Hu, D. and Marcucio, R. S. (2009). Unique organization of the frontonasal ectodermal zone in birds and mammals. *Developmental Biology*, **325**, 200–210.

King, M. C. and Wilson, A. C. (1975). Evolution at two levels in humans and chimpanzees. *Science*, **188**, 107–116.

Koen, E. L., Bowman, J., and Walpole, A. A. (2012). The effect of cost surface parameterization on landscape resistance estimates. *Molecular Ecology Resources*, **12**, 686–696.

Koonin, E. V. and Wolf, Y. I. (2009). The fundamental units, processes and patterns in evolution, and the Tree of Life conundrum. *Biology Direct*, **4**, 33.

Lynch, M. (2007). *The Origins of Genome Architecture*. Sinauer Associates, Sunderland, MA.

Lynch, M. (2010). Scaling expectations for the time to establishment of complex adaptations. *Proceedings of the National Academy of Sciences of the United States of America*, **107**, 16577–16582.

Lynch, M. and Abegg, A. (2010). The rate of establishment of complex adaptations. *Molecular Biology and Evolution*, **27**, 1404–1414.

MacDougall-Shackleton, E. A. and MacDougall-Shackleton, S. A. (2001). Cultural and genetic evolution in mountain white-crowned sparrows: song dialects are associated with population structure. *Evolution*, **55**, 2568–2575.

Mallarino, R., Grant, P. R., Grant, B. R., Herrel, A., Kuoa, W. P., and Abzhanov, A. (2011). Two developmental modules establish 3D beak-shape variation in Darwin's finches. *Proceedings of the National Academy of Sciences of the United States of America*, **108**, 4057–4062.

Mayr, E. (1939). Speciation phenomena in birds. *American Naturalist*, **752**, 249–278.

Mennill, D. J., Badyaev, A. V., Jonart, L. M., and Hill, G. E. (2006). Male house finches with elaborate songs have higher reproductive performance. *Ethology*,**112** 174–180.

Mills, G. S., Dunning, J. J. B., and Bates, J. M. (1989). Effects of urbanization on breeding bird community structure in southwestern desert habitats. *Condor*, **91**, 416–428.

Müller, G. B. (2007). Evo-devo: Extending the evolutionary synthesis. *Nature Reviews Genetics*, **8**, 943–949.

Müller, G. B. and Newman, S. A. (2005). The innovation triad: an EvoDevo agenda. *Journal of Experimental Zoology (Mol Dev Evol)*, **304B**, 487–503.

Newman, S. A. (2012). Physico-genetic determinants in the evolution of development. *Science*, **338**, 217–219.

Newman, S. A. and Müller, G. B. (2000). Epigenetic mechanisms of character origination. *Journal of Experimental Zoology*, **288**, 304–314.

Nikolakaki, P. (2004). A GIS site-selection process for habitat creation: estimating connectivity of habitat patches. *Landscape and Urban Planning*, **68**, 77–94.

Nowicki, S. (1987). Vocal tract resonances in oscine bird sound production: evidence from birdsongs in a helium atmosphere. *Nature*, **325**, 53–55.

Patricelli, G. L. and Blickley, J. L. (2006). Avian communication in urban noise: causes and consequences of vocal adjustment. *Auk*, **123**, 639–649.

Podos, J., Huber, S. K., and Taft, B. (2004). Bird song: the interface of evolution and mechanism. *Annual Reviews of Ecology, Evolution and Systematics*, **35**, 55–87.

Reid, R. G. B. (2007). *Biological Emergences: Evolution by Natural Experiment*. MIT Press, Cambridge, MA.

Schmalhausen, I. I. (1938). *Organism as a Whole in Individual Development and History*. Academy of Sciences, Leningrad, USSR.

Shubin, N., Tabin, C. J. and Carroll, S. (2009). Deep homology and the origins of evolutionary novelty. *Nature*, **457**, 818–823.

Slabbekoorn, H. and Smith, T. B. (2002). Bird song, ecology and speciation. *Philosophical Transactions of the Royal Society of London B Biological Sciences*, **357**, 493–503.

Tracy, T. T. and Baker, M. C. (1999). Geographic variation in syllables of House Finch songs. *Auk*, **116**, 666–676.

Waddington, C. H. (1959). Canalization of development and genetic assimilation of acquired characters. *Nature*, **183**, 1654–1655.

Wagner, A. (2011). *The Origins of Evolutionary Innovations: A theory of transformative change in living systems*. Oxford University Press, Oxford.

Wagner, G. P., Kin, K. and Lynch, V. (2012). Measurement of mRNA abundance using RNA-seq data: RPKM measure is inconsistent among samples. *Theory in Biosciences*, **131**, 281–285.

Wang, Z., Gerstein, M. and Snyder, M. (2009). RNA-Seq: a revolutionary tool for transcriptomics. *Nature Reviews Genetics*, **10**, 57–63.

West-Eberhard, M. J. (2003). *Developmental Plasticity and Evolution*. Oxford University Press, Oxford.

Wilkins, A. S. (2001). *The Evolution of Developmental Pathways*. Sinauer Associates, Sunderland, MA.

Willson, M. F. (1971). Seed selection in some North American finches. *Condor*, **73**, 415–429.

Wu, P., Jiang, T., Suksaweang, S., Widelitz, R. B., and Chuong, C. (2004). Molecular shaping of the beak. *Science*, **305**, 1465–1466.

Wu, P., Jiang, T. X., Shen, J.-Y., Widelitz, R. B., and Chuong, C. M. (2006). Morphoregulation of avian beaks: Comparative mapping of growth zone activities and morphological evolution. *Developmental Dynamics*, **235**, 1400–1412.

Young, R. L. and Badyaev, A. V. (2007). Evolution of ontogeny: linking epigenetic remodeling and genetic adaptation in skeletal structures. *Integrative and Comparative Biology*, **47**, 234–244.

PART 4

Case Studies

CHAPTER 13

Acoustic, morphological, and genetic adaptations to urban habitats in the silvereye (*Zosterops lateralis*)

Dominique A. Potvin, Raoul A. Mulder, and Kirsten M. Parris

13.1 Introduction

Avian species that maintain populations in urban areas are diverse and do not conform to one typical 'urban adaptor' phenotype, even though they experience similar selective pressures and constraints typical of the urban environment. One example of such a pressure is urban noise, which can disrupt the detection and discrimination of acoustic signals essential to avian communication. As they travel from sender to receiver through the environment, acoustic signals are subject to processes of attenuation (the loss of intensity over distance) and degradation (the distortion of sound, e.g. through reverberation or scattering; Richards & Wiley, 1980). In addition, acoustic masking occurs when background noise reduces the hearing threshold of a receiver, thereby decreasing the distance over which a signal can be detected and/or discriminated (Kang, 2007). An individual sending a signal through a masking, reverberating or attenuating environment should therefore adjust characteristics of this signal to increase the likelihood of it reaching the receiver with maximum fidelity. This hypothesis—that animals should adjust their signal patterns according to their surrounding acoustic environment—is known as the acoustic adaptation hypothesis (AAH), and was first proposed by Morton (1975). In birds, the AAH predicts changes—in some cases flexible, in others microevolutionary—in song that are consistent with those observed in urban areas, including frequency shifts (Slabbekoorn & Peet, 2003); timing (Brumm, 2006; Fuller et al., 2007), meme replacement (Cardoso & Atwell, 2011), and tempo (Slabbekoorn & den Boer-Visser, 2006).

Although there is growing evidence for acoustic adaptation in birds, urban adjustments to vocalizations may not be universal. Some authors have hypothesized that certain species are preadapted to communicate in noisy areas (Rheindt, 2003), which might reduce or even eliminate the need to adjust vocalizations. Conversely, the morphology, neurology, and physiology of an individual's song learning and production apparatus, as shaped by genetic and early environmental or developmental factors, may constrain changes to vocalizations. Additionally, if song changes that improve signal transmission in an environment result in any loss of crucial information, then the change is not necessarily adaptive (Boncoraglio & Saino, 2007).

The role of flexibility has recently played a large part in the urban acoustic adaptation literature, with one hypothesis stating that passerines may be well-suited to noisy urban environments largely because of the plasticity of their song, and their ability to learn and therefore to adjust to new noise environments (Bermudez-Cuamatzin et al., 2011; Francis et al., 2011; Verzijden et al., 2010). This flexibility may manifest itself in immediate changes to song frequency, tempo output, or amplitude with fluctuating noise levels, or otherwise as a selective learning mechanism by which song dialects in noisy habitats change over time. It is unclear, however, how less plastic vocalizations such as calls might be

Avian Urban Ecology. Edited by Diego Gil and Henrik Brumm

affected by urban noise and if this selective pressure is shaping calls via genetic, physiological, or developmental means (Leonard & Horn, 2008).

The silvereye (*Zosterops lateralis*) is an ideal species in which to investigate the many facets of urban vocal adjustment, and to test some of the hypotheses regarding urban acoustic adaptation. Silvereyes are native Australian birds that thrive in both cities and rural areas, with a diverse repertoire of songs and calls that overlap with the frequency range of urban noise. Moreover, they are widely distributed around the continent, enabling a study that incorporates multiple populations on a large geographic scale. Within this chapter, we will explore some of the differences between urban and rural silvereyes, and attempt to address several of the issues introduced above. We ask the following questions: Do songs and calls differ between rural and urban silvereye populations, and if so, how are they different? Are any differences potentially adaptive? What evidence is there of signal flexibility? Are any vocalization changes accompanied by morphological or genetic changes? We will also discuss the potential mechanisms and consequences of any adaptations or differences between city and rural birds. To do this, we will be reporting on a case study that compared morphology, genotypes, songs, alarm calls, and contact calls of seven urban populations with seven rural populations of silvereyes across 1 million kilometres of eastern Australia.

13.2 Vocalizations

13.2.1 Song

Birdsong, a learned trait, is usually a sexually selected melodious and complex signal containing individual-specific information produced by passerines (Catchpole & Slater, 2008). Songs lend themselves to acoustic adaptation. Since they are learned, there is the potential for the use of specific memes or syllables to vary between generations, with some that are well-suited to the environment being possibly advantageous and perhaps also more easily learned, and therefore retained or their use increased—an example of cultural evolution. Additionally, species with a number of songs or syllables may be able to respond immediately to the current noise environment by singing more loudly (increasing amplitude; Brumm & Todt, 2002) or by electing to sing those songs or syllables most suited to their immediate acoustic situation (Halfwerk & Slabbekoorn, 2009). Finally, songs may also be shifted temporally—the duration of a song and the number of notes are often plastic, with individuals able to space out notes or syllables or slow down songs in order to keep the information intact (Sakata et al., 2008). The flexibility of songs, therefore, may be an important reason that some passerines are able to communicate effectively in a variety of habitats.

In the silvereye, songs are sung only by males, probably to attract females (Barnett & Briskie, 2007). A single silvereye song consists of 5–20 syllables sung in random order, and silvereye repertoires may contain dozens of syllables in total (Barnett & Briskie, 2007, pers. obs.; Figure 13.1a). Across Australia, silvereyes in urban areas sing songs with minimum frequencies around 200 Hz higher than rural birds, and this is correlated with the level of background noise present (Potvin et al., 2011; Figure 13.2). Urban silvereye songs are also slower, with urban birds singing 0.5 syllables per second slower and 27% fewer syllables per song (Potvin & Parris, 2013). Additionally, urban silvereyes sing a higher percentage of trills (fast, repetitive syllables) in their songs. When entire repertoires are compared for similarity, we find that noise and urban habitat type are strong predictors of repertoire similarity between populations, alongside geographic location (Potvin & Parris, 2012). These multiple and varied differences between urban and rural silvereye song might reveal a number of different effects of the urban acoustic environment, and strategies for adapting to this environment.

In many avian species, upward shifts in the frequency of song have been attributed to the avoidance of lower notes that are likely to be masked by anthropogenic noise, thus improving signal transmission (Bermudez-Cuamatzin et al., 2011; Halfwerk & Slabbekoorn, 2009; Nemeth & Brumm, 2010). Although there is evidence that this modification does not necessarily improve transmission significantly, and that frequency shifts upwards may be due simply to singing louder (the Lombard effect: Nemeth & Brumm, 2010), this adjustment may nevertheless represent a flexible form of acoustic

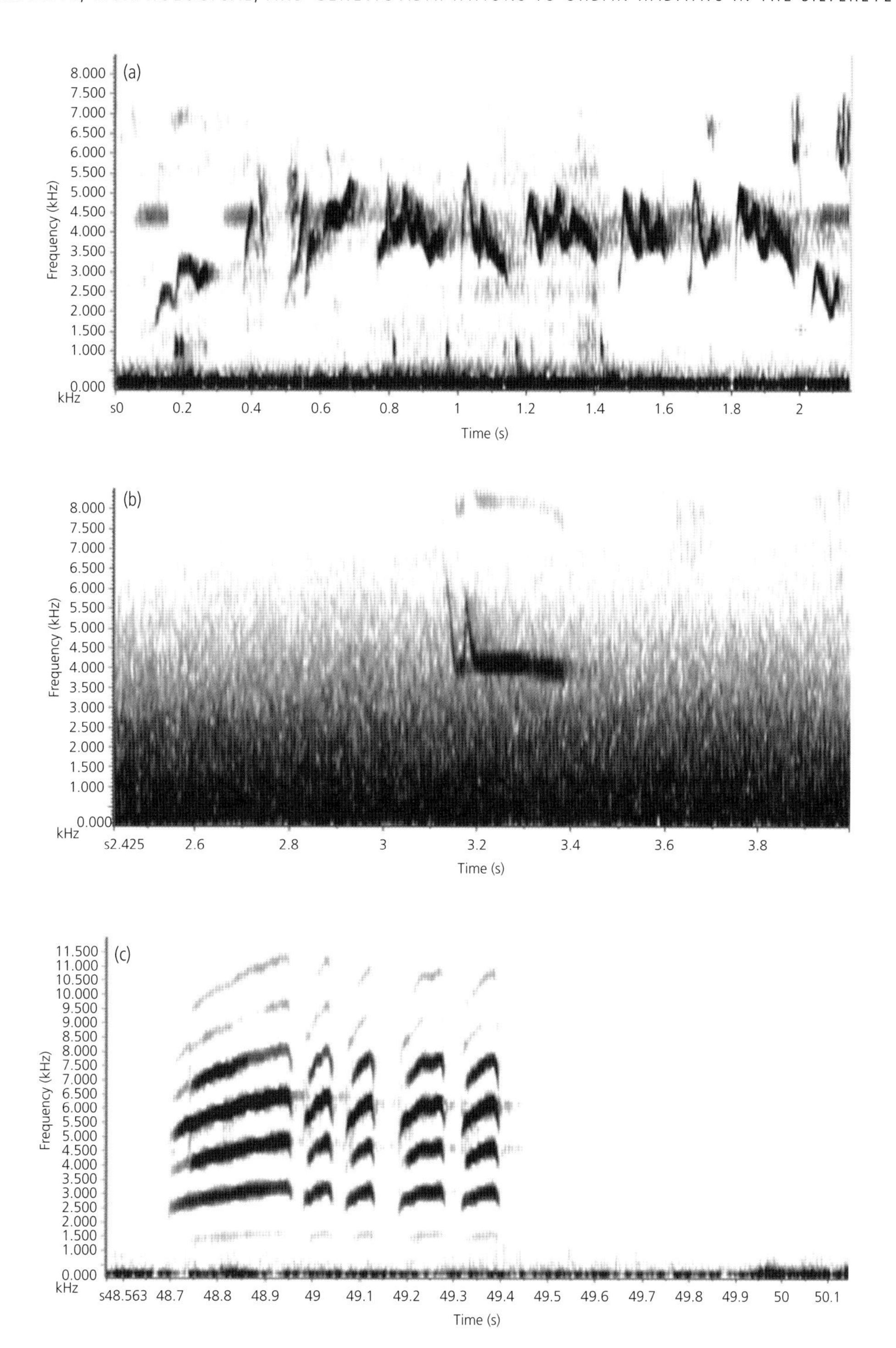

Figure 13.1 An example of a silvereye (a) song, (b) contact call, and (c) alarm call. Pictured are spectrograms, or visual representations of sounds showing frequency versus time.

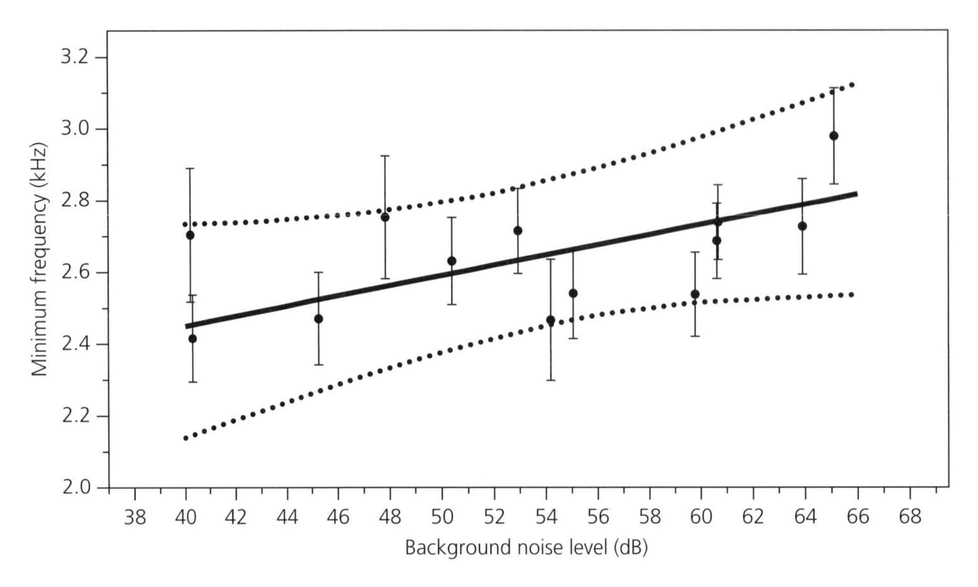

Figure 13.2 Minimum frequency of silvereye song versus the background noise level at a site with 95% credible intervals (dashed lines). Data from Potvin et al., 2011.

adaptation. The use of fewer syllables in city songs may also be a flexible response to noise, because it reduces reverberation or perhaps simply increases signal efficacy by emphasizing each syllable more distinctly (Slabbekoorn et al., 2007). These changes are therefore consistent with the framework of acoustic adaptation, and support for the AAH in silvereyes increases when considering the additional observed changes to song, for example selective meme use.

The increased use of memes such as trills in cities may help to punctuate songs. Hence, their use may augment the likelihood of long-distance detection through noisy acoustic environments. This shift in syllable use may indicate a pattern of cultural evolution (i.e. the process of retaining or discarding particular learned memes across a population). Support for urban populations undergoing cultural evolution and replacing memes has also been found in the dark-eyed junco *Junco hyemalis* (Cardoso & Atwell, 2011) and white-crowned sparrow *Zonotrichia leucophrys* (Cardoso & Atwell, 2011; Luther & Baptista, 2010), and is an intergenerational example of acoustic adaptation. The cultural evolution hypothesis as an explanation for differences in urban and rural song is also supported by the fact that urban variables are highly predictive of repertoire content (Potvin & Parris, 2013).

13.2.2 Contact calls

Both male and female silvereyes use contact calls to communicate in various contexts including foraging, flocking and perching. Contact calls consist of one frequency-modulated tonal note, with one of four identifiable motifs; simple calls, linear calls, variable calls and chips (Figure 13.1b). The minimum frequencies of urban contact calls are around 90 Hz higher than those of rural contact calls, regardless of call type or geographic location. Using captive silvereyes, we found that in the presence of high frequency noise, silvereyes lowered the minimum frequencies of these calls and increased both the amplitude and the duration of the call (Potvin & Mulder, in press). The adjustment of minimum frequencies upwards in urban areas is therefore most likely an immediate response to avoid masking in areas with urban noise.

Contact calls, being only one short syllable, are ultimately less malleable than songs. The probable innate development of the vocalization and the information contained therein probably further contribute to this lack of flexibility. The frequency differences, as supported by captive studies, are likely to be plastic responses to noise levels, designed to reduce masking by low-frequency urban

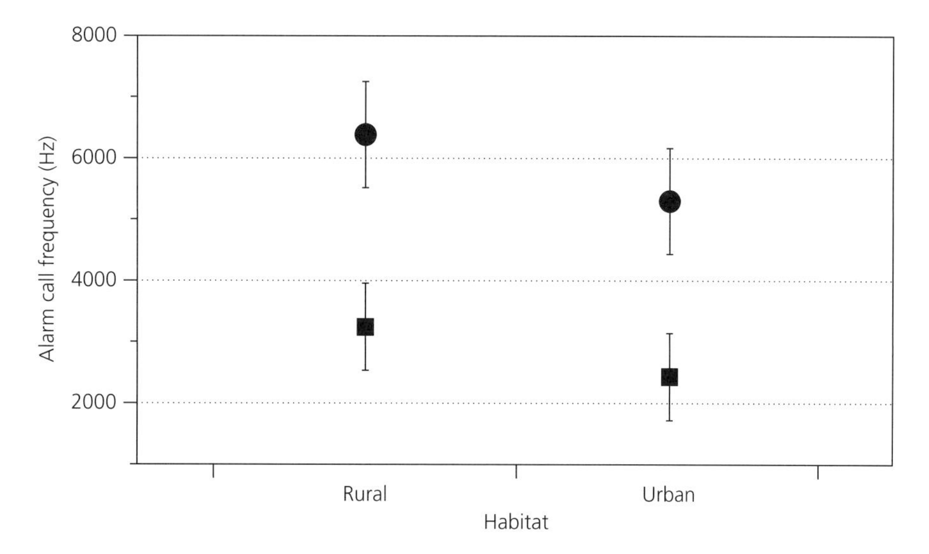

Figure 13.3 Mean minimum (square) and peak (circle) frequencies of silvereye alarm calls from seven pairs of rural and urban habitats. Error bars denote 95% credible intervals.

noise and improve the transmission of the signal. We have not identified any other differences in the structure or use of contact calls in cities. Urban or rural silvereyes do not preferentially use one call type, and the basic structure of the calls (frequency modulation, tempo, and duration) is also similar.

One explanation for why silvereyes do not modify call structure may be that structure contains important information about the sender. Previous studies have demonstrated that silvereye contact calls show individual repeatability and convey recognizable, individual-specific information to conspecifics (Bruce & Kikkawa, 1988; Robertson, 1996). Since changes that increase signal transmission but consequently affect efficacy would be maladaptive and thus should be avoided, silvereyes might be limited to altering frequency (as observed) and amplitude (volume). Adjusting amplitude is a response to noisy environments found in the songs and calls of other species (Brumm & Todt, 2002; Cynx et al., 1998). While we have not yet measured call and song amplitude explicitly in this species in the field, it is possibly that silvereyes may be increasing the volume of vocalizations under noisy conditions independent of, or alongside, frequency shifts.

13.2.3 Alarm calls

Alarm calls are used to warn of danger and elicit mobbing behaviour, and are very similar to the human ear to fledgling begging calls (D.A.P., personal observation). Alarm calls differ from songs and contact calls in that they consist of a repeated note and contain a number of harmonics or formants (Figure 13.1c). Urban silvereyes across Australia use alarm calls with minimum, average, and peak frequencies around 1 kHz lower than rural silvereyes, with slightly increased maximum frequencies, giving a larger frequency range (Figure 13.3).

The direction of these changes is opposite both to that predicted by the AAH (if birds are attempting to reduce the potential for background masking), and also to the higher frequencies observed in urban contact calls and song. Hence, the low frequency of urban alarm calls may be maladaptive. Developmental stresses experienced by birds early in life can negatively affect brain development or learning (Nowicki et al., 2002). Noise is a known stressor (Wright et al., 2007), and if it disrupts brain development or learning in unpredictable ways, this could explain why alarm calls alone appear to be affected in this way. An alternative possibility is that urban stressors are directly disrupting normal development of the vocal tract itself. Alarm calls contain formants, which are produced in the vocal tract using mechanisms different from those involved in tonal (song and contact call) note production (Nowicki & Marler, 1988), so it is possible that subtle changes to vocal apparatus could affect one vocalization type and not others. Nevertheless, there is at present no evidence that developmental stress affects the development of vocal structures (Brumm et al., 2009), and it seems likely that developmental stress might result in gross changes to vocal apparatus that would affect all vocalizations, so this explanation seems less plausible.

Because we do not know what exact information is encoded in silvereye alarm calls, or how these changes might affect other individuals, we cannot exclude a second, adaptive possibility for the changes. We know that nesting success of urban silvereyes is similar to that of rural silvereyes (van Heezik et al., 2008), suggesting that changes in alarm calls in this context are likely still effective, or that they may not be needed if the predators in urban locations differ. The frequency or bandwidth of an alarm call could therefore also potentially signal information about the type of predator found in an urban environment (Klump & Shalter, 1984) or make them less detectable by urban predators. Further research would be required to determine whether dangers do differ for silvereyes in urban areas and if alarm calls are predator or danger-specific.

13.3 Morphology

Given that the ability to vocalize is directly influenced by morphological structures (Derryberry, 2009), one might expect that species demonstrating acoustic differences in urban habitats may also show differences in body structure or size. Urban silvereyes, however, show no evidence of basic morphological change. Structural features (head-bill length, wing length, tarsus length) and mass do not seem to be affected by habitat (Potvin et al., 2013; Figure 13.4). Although there is clearly a connection between morphology and song in many species, including silvereyes (e.g. Martin et al., 2011; Potvin & Parris, 2013, Christensen et al., 2006; Podos et al., 2004), this correlation has only been found in one species where changes in the morphology of urban birds were correlated with changes in song. Urban house sparrows *Passer domesticus* eat different seeds to those in rural areas, and as a consequence, urban birds sing differently due to new constraints of bill morphology (Badyaev et al., 2008). In silvereyes, differences in the dimensions of bills between urban and rural populations due to dietary changes seem unlikely, as silvereyes are mainly frugivorous and food sources tend to be consistent between rural and urban habitats in similar geographic areas. In the absence of obvious differences in bill size, song differences between urban and rural silvereyes may be due to other structural changes, such as changes to beak gape (Podos et al., 2004) or the song apparatus. Alternatively, and more likely, they may be due to physiological changes (Cynx et al., 2005) or to cultural changes such as dialect formation (as above). Although it is difficult to reach firm conclusions about the connection between the morphology of urban and rural silvereyes and their song or calls, it is clear that phenotypic changes observed in urban songs have changed more than the outward appearance of the species in urban populations. Hence, rather than urban environments selecting for morphologies that in turn affect vocal production, it is more probable that changes are due to the plasticity of song (and to some extent, calls), either through cultural evolution or immediate, plastic changes, and the possible physiological or stress effects described on alarm calls.

13.4 Genetics

Urbanization has the potential to segregate populations, either by disrupting dispersal or, more likely, by potentially facilitating assortative mating, especially if song differences between urban and rural populations are distinct (MacDougall-Shackleton & MacDougall-Shackleton, 2001; Patten et al., 2004; Slabbekoorn & Smith, 2002). Findings from avian studies investigating genetic differentiation between urban and rural populations of birds have so far been inconsistent, with some studies finding evidence for genetic divergence between urban and rural populations (Badyaev et al., 2008; Evans et al.,

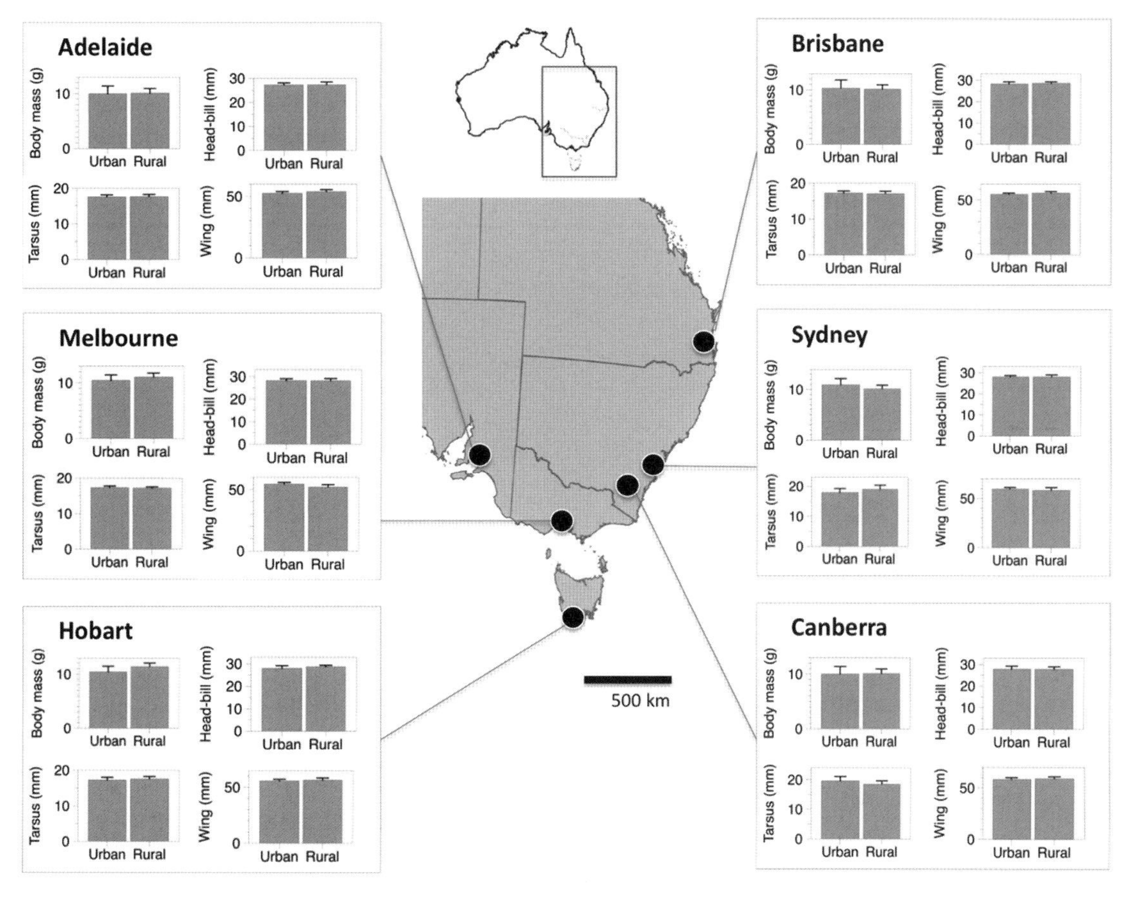

Figure 13.4 Mean measurements of head-bill length, wing length, tarsus length, and mass of rural and urban silvereyes, grouped by geographic area (indicated on map). Error bars represent standard deviation.

2012; Partecke et al., 2004) or decreased genetic diversity in urban areas (Evans et al., 2009), and others failing to identify such differences (Leader et al., 2008). If urbanization is restricting gene flow between populations, segregation and urban inbreeding might be detected using neutral markers or microsatellites. However, urban habitats do not seem to promote higher levels of inbreeding or lower dispersal rates in silvereyes (Potvin et al., 2013). Silvereyes are partial migrants, and have shown a strong ability to move across various landscapes (Chan & Kikkawa, 1997). It therefore seems implausible that dispersal of such a mobile species would be restricted by habitat type, though differentiation of song and call repertoires could limit dispersal between populations.

Through F_{st} analyses, we have found that urban populations are not subject to higher rates of genetic divergence from nearby rural populations than would be expected based on the geographic distance between them, and urban populations do not demonstrate consistent patterns of genetic isolation (Potvin et al., 2013). Consistent with this, we found that habitat type is not a strong predictor of genetic differentiation, based on autocorrelation analyses of microsatellites (Potvin et al., 2013). Of course, we cannot conclude from these negative findings that silvereyes are not adapting genetically to urban habitats. If urban adaptation is a recent phenomenon, genetic differentiation between city and rural populations may take more time to manifest, and detecting functional changes using microsatellite (neutral) markers can be difficult. Nevertheless, it seems more likely that differences in vocalizations result from vocal plasticity, cultural evolution, or physiological effects of noise on song development, rather than genetic isolation.

13.5 Conclusion

Silvereyes are one of the most successful urban adaptors on the Australian continent. Part of the reason they are able to succeed in the city may be the numerous acoustic adaptations they have made in urban areas. Adjusting the frequency of both songs and contact calls provides evidence for real-time adaptive responses to urban noise, and demonstrates the advantage of phenotypic plasticity in a continually fluctuating and noisy environment. Furthermore, urban areas seem to be promoting cultural evolution, and around Australia silvereyes show evidence of preferring the same memes in urban environments. The lack of obvious morphological or genetic changes in urban populations of this species further indicates that adjustments are likely to be due to acoustic plasticity rather than microevolution. However, we have identified potentially maladaptive changes to alarm calls, which may be the result of physiological or developmental impacts of urbanization.

Unfortunately, we have yet to determine all the implications of these changes to acoustic signals in urban silvereyes. Currently, the potential for assortative mating based on habitat-specific song, while not apparent in microsatellite analysis, is still unknown. Additionally, whether any of the observed changes to songs or calls reduce the effectiveness of these signals—especially the urban alarm call—is yet to be ascertained. However, acoustic flexibility in urban populations of the silvereye is probably not restricted to the sender of the signal. If receivers are equally flexible in interpreting and responding to an altered signal, then any negative effect of urbanization on vocalizations may be compensated for. Therefore, being an effective urban adaptor probably relies on flexibility, not just with respect to acoustic communication, but to other aspects of life history as well. The flexibility of silvereyes in acoustic communication supports this theory.

References

Badyaev, A. V., Young, R. L., Oh, K. P., and Addison, C. (2008). Evolution on a local scale: developmental, functional, and genetic bases of divergence in bill form and associated changes in song structure between adjacent habitats. *Evolution*, **62**, 1951–64.

Barnett, C. and Briskie, J. (2007). Energetic state and the performance of dawn chorus in silvereyes (*Zosterops lateralis*). *Behavioral Ecology and Sociobiology*, **61**, 579–587.

Bermudez-Cuamatzin, E., Rios-Chelen, A. A., Gil, D., and Garcia, C. M. (2011). Experimental evidence for real-time song frequency shift in response to urban noise in a passerine bird. *Biology Letters*, **7**, 36–8.

Boncoraglio, G. and Saino, N. (2007). Habitat structure and the evolution of bird song: a meta-analysis of the evidence for the acoustic adaptation hypothesis. *Functional Ecology*, **21**, 134–142.

Bruce, P. and Kikkawa, J. (1988). A sexual difference in the contact calls of silvereyes. *Emu*, **88**, 188–190.

Brumm, H. (2006). Signalling through acoustic windows: Nightingales avoid interspecific competition by short-term adjustment of song timing. *Journal of Comparative Physiology A: Neuroethology, Sensory, Neural Behaviour and Physiology*, **192**, 1279–85.

Brumm, H. and Todt, D. (2002). Noise-dependent song amplitude regulation in a territorial songbird. *Animal Behaviour*, **63**, 891–897.

Brumm, H., Zollinger, S., and Slater, P. B. (2009). Developmental stress affects song learning but not song complexity and vocal amplitude in zebra finches. *Behavioral Ecology and Sociobiology*, **63**, 1387–1395.

Cardoso, G. C. and Atwell, J. W. (2011). Directional cultural change by modification and replacement of memes. *Evolution*, **65**, 295–300.

Catchpole, C. K. and Slater, P. J. B. (2008). *Bird Song*, 2nd ed. Cambridge University Press, Cambridge.

Chan, K. and Kikkawa, J. (1997). Short communication: a silvereye dilemma: to migrate or not to migrate? *Emu*, **97**, 91–93.

Christensen, R., Kleindorfer, S., and Robertson, J. (2006). Song is a reliable signal of bill morphology in Darwin's small tree finch *Camarhynchus parvulus*, and vocal performance predicts male pairing success. *Journal of Avian Biology*, **37**, 617–624.

Cynx, J., Bean, N. J., and Rossman, I. (2005). Testosterone implants alter the frequency range of zebra finch songs. *Hormones and Behavior*, **47**, 446–451.

Cynx, J., Lewis, R., Tavel, B., and Tse, H. (1998). Amplitude regulation of vocalizations in noise by a songbird, *Taeniopygia guttata*. *Animal Behaviour*, **56**, 107–13.

Derryberry, E. P. (2009). Ecology shapes birdsong evolution: variation in morphology and habitat explains variation in white-crowned sparrow song. *American Naturalist*, **174**, 24–33.

Evans, K. L., Gaston, K. J., Frantz, A. C., Simeoni, M., Sharp, S. P., McGowan, A., Dawson, D. A., Walasz, K., Partecke, J., Burke, T., and Hatchwell, B. J. (2009). Independent colonization of multiple urban centres by a formerly forest specialist bird species. *Proceedings of the Royal Society B: Biological Sciences*, **276**, 2403–2410.

Evans, K. L., Newton, J., Gaston, K. J., Sharp, S. P., Mcgowan, A. and Hatchwell, B. J. (2012). Colonisation of urban environments is associated with reduced migratory behaviour, facilitating divergence from ancestral populations. *Oikos*, **121**, 634–640.

Francis, C. D., Ortega, C. P., and Cruz, A. (2011). Vocal frequency change reflects different responses to anthropogenic noise in two suboscine tyrant flycatchers. *Proceedings of the Royal Society B: Biological Sciences*, **278**, 2025–2031.

Fuller, R. A., Warren, P. H., and Gaston, K. J. (2007). Daytime noise predicts nocturnal singing in urban robins. *Biology Letters*, **3**, 368–70.

Halfwerk, W. and Slabbekoorn, H. (2009). A behavioural mechanism explaining noise-dependent frequency use in urban birdsong. *Animal Behaviour*, **78**, 1301–1307.

Kang, J. (2007). *Urban Sound Environment*. Taylor & Francis New York.

Klump, G. M. and Shalter, M. D. (1984). Acoustic behaviour of birds and mammals in the predator context; I. Factors affecting the structure of alarm signals. ii. the functional significance and evolution of alarm signals. *Zeitschrift für Tierpsychologie*, **66**, 189–226.

Leader, N., Geffen, E., Mokady, O., and Yom-Tov, Y. (2008). Song dialects do not restrict gene flow in an urban population of the orange-tufted sunbird, *Nectarinia osea*. *Behavioral Ecology and Sociobiology*, **62**, 1299–1305.

Leonard, M. L. and Horn, A. G. (2008). Does ambient noise affect growth and begging call structure in nestling birds? *Behavioral Ecology*, **19**, 502–507.

Luther, D. and Baptista, L. (2010). Urban noise and the cultural evolution of bird songs. *Proceedings of the Royal Society of London B*, **277**, 469–73.

MacDougall-Shackleton, E. A. and MacDougall-Shackleton, S. A. (2001). Cultural and genetic evolution in mountain white-crowned sparrows: song dialects are associated with population structure. *Evolution*, **55**, 2568–75.

Martin, J. P., Doucet, S. M., Knox, R. C., and Mennill, D. J. (2011). Body size correlates negatively with the frequency of distress calls and songs of Neotropical birds. *Journal of Field Ornithology*, **82**, 259–268.

Morton, E. S. (1975). Ecological sources of selection on avian sounds. *American Naturalist*, **109**, 17.

Nemeth, E. and Brumm, H. (2010). Birds and anthropogenic noise: are urban songs adaptive? *American Naturalist*, **176**, 465–75.

Nowicki, S. and Marler, P. (1988). How do birds sing? *Music Perception: An Interdisciplinary Journal*, **5**, 391–426.

Nowicki, S. N., Searcy, W. S., and Peters, S. P. (2002). Brain development, song learning and mate choice in birds: a review and experimental test of the 'nutritional stress hypothesis'. *Journal of Comparative Physiology A: Neuroethology, Sensory, Neural, and Behavioral Physiology*, **188**, 1003–1014.

Partecke, J., Van't Hof, T., and Gwinner, E. (2004). Differences in the timing of reproduction between urban and forest European blackbirds (*Turdus merula*): result of phenotypic flexibility or genetic differences? *Proceeding of the Royal Society B*, **271**, 1995–2001.

Patten, M. A., Rotenberry, J. T., and Zuk, M. (2004). Habitat selection, acoustic adaptation, and the evolution of reproductive isolation. *Evolution*, **58**, 2144–55.

Podos, J., Southall, J. A., and Rossi-Santos, M. R. (2004). Vocal mechanics in Darwin's finches: correlation of beak gape and song frequency. *Journal of Experimental Biology*, **207**, 607–19.

Potvin, D. A. (2013). Larger body size on islands affects silvereye *Zosterops lateralis* song and call frequency. *Journal of Avian Biology*, **44**, 221–225.

Potvin, D. A. and Parris, K. M. (2013). Song convergence in multiple urban populations of silvereyes (*Zosterops lateralis*). *Ecology and Evolution*, **2**, 1977–1984.

Potvin, D. A. and Mulder, R. A. Immediate, independent adjustment of call pitch and amplitude in response to varying background noise by silvereyes (*Zosterops lateralis*). *Behavioral Ecology*, in press.

Potvin, D. A., Parris, K. M., and Mulder, R. A. (2011). Geographically pervasive effects of urban noise on frequency and syllable rate of songs and calls in silvereyes (*Zosterops lateralis*). *Proceedings of the Royal Society B: Biological Sciences*, **278**, 2464–2469.

Potvin, D. A., Parris, K. M., and Mulder, R. A. (2013). Limited genetic differentiation between acoustically divergent populations of urban and rural silvereyes (*Zosterops lateralis*). *Evolutionary Ecology*, **27**, 381–391.

Rheindt, F. E. (2003). The impact of roads on birds: Does song frequency play a role in determining susceptibility to noise pollution? [Die Auswirkungen von Straßen auf Vögel: Ist Gesangsfrequenz ein Faktor für Lärmempfindlichkeit?] *Journal für Ornithologie*, **144**, 295–306.

Richards, D. G. and Wiley, R. H. (1980). Reverberations and amplitude fluctuations in the propagation of sound in a forest: implications for animal communication. *American Naturalist*, **115**, 381–399.

Robertson, B. C. (1996). Vocal mate recognition in a monogamous, flock-forming bird, the silvereye, *Zosterops lateralis*. *Animal Behaviour*, **51**, 303–311.

Sakata, J. T., Hampton, C. M., and Brainard, M. S. (2008). Social modulation of sequence and syllable variability in adult birdsong. *Journal of Neurophysiology*, **99**, 1700–1711.

Slabbekoorn, H. and Den Boer-Visser, A. (2006). Cities change the songs of birds. *Current Biology*, **16**, 2326–31.

Slabbekoorn, H. and Peet, M. (2003). Ecology: Birds sing at a higher pitch in urban noise. *Nature*, **424**, 267.

Slabbekoorn, H. and Smith, T. B. (2002). Habitat-dependent song divergence in the little greenbul: an analysis of environmental selection pressures on acoustic signals. *Evolution*, **56**, 1849–58.

Slabbekoorn, H., Yeh, P., and Hunt, K. (2007). Sound transmission and song divergence: a comparison of urban and forest acoustics *Condor*, **109**, 67–78.

Van Heezik, Y., Ludwig, K., Whitwell, S., and Mclean, I. G. (2008). Nest survival of birds in an urban environment in New Zealand. *New Zealand Journal of Ecology*, **32**, 155–165.

Verzijden, M. N., Ripmeester, E. A., Ohms, V. R., Snelderwaard, P., and Slabbekoorn, H. (2010). Immediate spectral flexibility in singing chiffchaffs during experimental exposure to highway noise. *Journal of Experimental Biology*, **213**, 2575–81.

Wright, A. J., Soto, N. A., Baldwin, A. L., Bateson, M., Beale, C. M., Clark, C., Deak, T., Edwards, E. F., Fernandez, A., Godinho, A., Hatch, L. T., Kakuschke, A., Lusseau, D., Martineau, D., Romero, M. L., Weilgart, L. S., Wintle, B. A., Notarbartolo-Di-Sciara, G., and Martin, V. (2007). Anthropogenic noise as a stressor in animals: a multidisciplinary perspective. *International Journal of Comparative Psychology*, **20**, 250–273.

CHAPTER 14

Human-induced changes in the dynamics of species coexistence: an example with two sister species

Renée A. Duckworth

14.1 Introduction

Urbanization has eliminated habitat and resources of many avian species and this has had devastating consequences for their populations (Marzluff, 2001; Shochat et al., 2006). In response to this, humans have implemented conservation programmes, some of which have been extremely successful. For example, in many suburban areas as well as urban parks, the placement of nest boxes has led to species recovery and has promoted stable breeding populations (Griffith et al., 2008; Newton, 1994). However, these successes are not without consequences. By altering resource availability, humans can profoundly affect the evolution of adaptations for colonizing new habitat and can also alter the dynamics of species coexistence. At the same time, species that depend on man-made resources present a unique opportunity to gain insight into fundamental problems in ecology and evolution. By exercising fine control over the location, density and stability of these resources, we can use these systems to gain a detailed understanding of species replacement, evolution of colonization strategies, and rapid adaptation to human-altered landscapes. In this chapter, I will use long-term studies of the evolutionary ecology of western and mountain bluebirds as an empirical example to show how human alteration of the density and stability of nest cavities, through the placement of artificial nest boxes, has influenced these species' behaviour, population dynamics and ultimately, their coexistence.

14.2 Mechanisms of species coexistence

Identifying the mechanisms of species coexistence is one of the most difficult problems in ecology (Chesson, 2000). Trade-offs in performance are thought to underlie patterns of species coexistence (Baraloto et al., 2005; Kneitel & Chase, 2004; Tilman, 2004) as superior competitive ability often comes at the expense of dispersal ability, abiotic tolerances, reproductive investment or efficiency of resource use (Cadotte, 2007; Hughes et al., 2003; Pfennig & Pfennig, 2005; Tilman, 1994). Such trade-offs underlie niche differences between species which can allow for stable coexistence locally by increasing the strength of intraspecific competition relative to interspecific competition (Chesson, 2000).

In combination with trade-offs acting at a local scale, regional-scale processes also have the potential to strongly influence patterns of species coexistence (Amarasekare, 2003; Cadotte, 2007; Kneitel & Chase, 2004; Leibold et al., 2004; Tilman, 1994). For example, the competition–colonization trade-off, where species that are superior colonizers are inferior competitors (Amarasekare, 2003; Cadotte, 2007; Yu & Wilson, 2001), can impact regional diversity as species alternately colonize and become extinct from habitat patches over time. Colonizing species are often highly dispersive and arrive at new habitat patches first, but are then displaced from these patches once less dispersive, better competitors arrive. The long-term coexistence of such species depends on disturbance such that new habitat patches

Avian Urban Ecology. Edited by Diego Gil and Henrik Brumm

must continually become available to enable the persistence of the colonizing species (Brawn et al., 2001; Levin & Paine, 1974). Thus, resource and habitat stability is an essential component underlying species coexistence.

In addition to competition–colonization tradeoffs, regional patterns of diversity are also influenced by environmental heterogeneity. If resource availability or abiotic conditions vary across the landscape, species that dominate in one area may be at a disadvantage in another, ultimately fostering regional coexistence (Amarasekare & Nisbet, 2001; Kneitel & Chase, 2004; Snyder & Chesson, 2003). For example, if species vary in their ability to exploit resource-rich versus resource-poor habitat patches, this can alter competitive dominance across patches that vary in resource density (Palmer, 2003; Tessier & Woodruff, 2002). Thus, resource patchiness, stability and density are all key factors that influence patterns of species coexistence and urbanization can have both subtle and dramatic effects on all of these factors.

14.3 Urbanization and its impact on resource distribution and stability

In densely populated areas humans typically try to minimize the effects of natural disturbance. For example, in coastal areas humans erect bulkheads and other structures to stabilize a disturbance-prone beach habitat, in forest and grassland regions, fire suppression is used to protect local communities and forest and agricultural resources, and in major river corridors, levees, dams, and reservoirs are erected to control water levels to protect historically flood-prone areas from damage.

Such suppression of ecological disturbance means that habitat and resource stability have been fundamentally altered in many human-dominated ecosystems and this can have strong negative impacts on species adapted to disturbance-mediated habitats (Brawn et al., 2001). Interestingly, this frequently overlooked result of urbanization—habitat stabilization—is the opposite of the massive disturbance typically associated with urbanization. Yet, while it is true that urbanization causes major disturbance to many stable communities, it also simultaneously results in stabilization of habitats that were previously disturbance prone as humans create a novel environment and then seek to protect and maintain it. How such habitat stabilization has influenced community composition and species adaptations for colonizing new habitat has been a largely overlooked question in the context of urban ecology research.

Secondary cavity nesting birds provide a clear example of species whose resource distribution, density and stability have been fundamentally altered by human intervention. The initial impacts of urbanization on these species are usually negative—factors such as deforestation, fire suppression, and the introduction of non-native cavity nesting species, typically decrease the availability of their most limiting resource—nest cavities—by reducing the number of dead trees where natural nesting holes are usually found or by adding novel competitors to the environment (e.g. Prescott, 1982). On the other hand, in more recent years, many secondary cavity nesting species have become the focus of conservation efforts and this has resulted in widespread implementation of nest-box programmes to provide an abundance of nesting sites for these highly charismatic species. Across the world, nest boxes have been used to maintain healthy populations [e.g. bluebirds, *Sialia* spp. (Gowaty & Plissner, 1998; Guinan et al., 2000) and wood ducks, *Aix sponsa*, in North America (Semel & Sherman, 1995)] and to rescue threatened populations [e.g. Eurasian rollers, *Coracias garrulus*, in Spain (Rodriguez et al., 2011); Gouldian finch, *Erythrura gouldiae*, in western Australia (Brazill-Boast et al., 2012)] of secondary cavity nesters. Even beyond formal conservation efforts, in many areas, the placement of nest boxes is common among the general public, not just in rural areas, but also in urban parks and suburbs as a way of attracting favoured avian species. In all of these cases, nest boxes are placed to maximize breeding densities of a particular species and not to mimic natural resource distributions and dynamics.

The distribution and persistence of natural tree cavities varies depending on habitat characteristics, climate and the source of cavity creation [e.g. by cavity excavators such as woodpeckers versus fungal or insect damage (Cockle et al., 2011)]. In North America, new habitat for cavity nesters is often created as a result of forest fire, which produces an abundance of dead snags and is quickly colonized by primary excavators that make new cavities which pave the way for colonization of secondary cavity nesters. The density of nest cavities in these burned forests can be quite high, but their distribution is often clumped and any particular nest cavity may last only for a few years (Chambers & Mast, 2005; Lehmkuhl et al., 2003). Nest boxes, on the other hand, are usually distributed evenly across the landscape and will last for as long as humans actively clean, replace and repair them. Moreover, nest-box trails are usually placed near human settlements and roadways while naturally occurring habitat occurs most frequently in rural or wilderness areas. Consequently, the placement of nest boxes is likely to drastically alter the density, stability and distribution of a key resource that is known to be limited and to impact both intra- and interspecific competition in birds (Newton, 1994).

In the next section, I describe how manipulation of this resource has produced insights into competitive dynamics of two sister species, western and mountain bluebirds (*Sialia mexicana* and *S. currucoides*, respectively) and how human conservation programmes in response to urbanization may be altering the coexistence and evolutionary trajectory of many secondary cavity nesting bird species. I then discuss more generally how the effects of urbanization on resource stability may be an overlooked but potentially important factor in understanding the diversity of species in urban environments.

14.4 Empirical example: disruption of bluebird colonization cycles

14.4.1 Historical context for species coexistence and evolution of dispersal strategies

Before the widespread placement of nest boxes, new bluebird habitat was largely created by forest fire which generates suitable habitat by opening up understory vegetation and creating dead snags. Eventually, as the forest regrows, bluebirds are no longer able to breed in these habitat patches because snag density decreases and regrowth of the forest eliminates the open meadows bluebirds depend on to forage for insect prey (Guinan et al., 2000; Power & Lombardo, 1996). Recent studies on the distribution, density and stability of nest cavities in post-fire forests of North America indicate that nest cavities occur in densities of 10–20 per hectare, are usually clumped in their distribution, and can last for up to 25 years (Lehmkuhl et al., 2003; Remm & Lõhmus, 2011; Saab et al., 2002). Most cavities, however, have a much shorter longevity with a median life span of 14 years (Cockle, Martin & Wesolowski, 2011). Thus, historically, bluebirds' habitat and main limiting resource—nest cavities—were patchily distributed and relatively ephemeral.

The successional nature of post-fire habitat meant that the persistence of bluebirds depended on their ability to continually recolonize new habitat patches (Schieck & Song, 2006). In general, the two species have evolved distinct strategies for finding and settling in new habitat. Mountain bluebirds are more dispersive than western bluebirds (Guinan et al., 2000; Power & Lombardo, 1996) and are frequently among the earliest colonizers following forest fires (Hutto, 1995; Schieck & Song, 2006), whereas, western bluebirds often show delayed patterns of colonization (Kotliar et al., 2007; Saab et al., 2007). However, once new habitat is found, competition for nest cavities among these and other secondary cavity nesting species is intense and often involves aggressive displacement (Brawn & Balda, 1988; Duckworth, 2006b). Western bluebirds, while less dispersive and slower to find new habitat, are on average, more aggressive than mountain bluebirds and rapidly displace them when they colonize newly available habitat (Duckworth & Badyaev, 2007). Thus, the coexistence of these two species is at least partly explained by regional scale processes in the form of a competition–colonization trade-off where mountain bluebirds are the superior colonizers because of their higher dispersiveness and western bluebirds are the superior competitors due to their higher aggressiveness.

The apparent trade-off between aggression and dispersal seen at the species level in this system is not upheld at the intraspecific level as western bluebirds have evolved two distinct dispersal strategies in which there is a positive association between dispersal and aggression. Highly aggressive western bluebird males tend to leave their natal populations and disperse to new areas to breed, whereas, non-aggressive males tend to remain in their natal population and eventually acquire a territory near relatives (Duckworth & Badyaev, 2007; Duckworth, 2008). Aggression and dispersal are functionally integrated in western bluebirds because colonization of new habitat patches by dispersing males requires the ability to outcompete earlier arriving heterospecific competitors for nesting sites and territories. Yet, aggression is costly as it trades off with parental care (Duckworth, 2006b) and once the earliest arriving western bluebirds have secured territories, there is strong selection for lower levels of aggression. Concordant with this strong selection, aggression rapidly shifts over time such that within 5–10 years of colonization, populations display much lower levels of aggression (Duckworth, 2008). This shift is adaptive because nonaggressive males are poor competitors and benefit from remaining in their natal population where they can gain a territory by cooperating with relatives. These males have higher fitness than aggressive males in older, well established populations but rarely, if ever, are observed colonizing new populations, whereas, aggressive males have the highest fitness when dispersing to new populations where density of conspecifics is low (Duckworth, 2008).

Distinct dispersal strategies such as those observed in western bluebirds are common in species that depend on ephemeral or successional habitat, especially when habitat patches are moderately stable (such as post-fire habitat) allowing the persistence of multiple generations, and thus a benefit to a philopatric strategy, before the patch disappears (Crespi & Taylor, 1990; Harrison, 1980; Johnson & Gaines, 1990; Roff, 1994). Thus, the historical distribution and stability of natural nest cavities likely played an important role in the evolution of western bluebird's distinct dispersal strategies. Moreover, the differences between mountain and western bluebirds in dispersal and competitive behaviour were also likely shaped and maintained by the natural distribution of this crucial resource. Recent human-induced changes in the distribution and stability of nest cavities are changing these dynamics. In the next section, I explain how nest-box programmes are impacting the dynamics of colonization and possibly the coexistence of these species.

14.4.2 Human placement of nest boxes: impacts on behaviour and species coexistence

In the late 1930s changes in logging and agricultural practices severely limited the availability of snags with nest cavities and western bluebirds' were extirpated from many parts of the northwestern United States where they were once common, particularly at the easternmost limit of their northern range edge (Duckworth & Badyaev, 2007). While mountain bluebird populations were also affected by this cavity limitation, their broader elevational range and more highly dispersive nature meant that they were able to maintain populations throughout this area. The widespread implementation of nest-box programmes, starting in the late 1960s, led to population increases for both species in the lower elevation valleys (Duckworth, 2006c). When nest boxes were first placed throughout the valleys of Montana, the site of our studies, mountain bluebirds reached these areas first. Yet, as western bluebirds expanded their range and moved back into these areas, they rapidly displaced mountain bluebirds, with complete species replacement occurring in less than 10 years in several populations, which have been closely monitored for over 30 years (Duckworth & Badyaev, 2007). Thus, the initial processes of colonization of this newly created man-made habitat mimicked the colonization patterns observed in natural post-fire populations. However, nest-box populations differ in several important ways from post-fire habitat and there is evidence that these differences are influencing the dynamics of colonization.

In Montana, nest boxes are usually placed in a linear transect and are spaced evenly (usually 100–200 m apart) along roadways. The goal of nest-box programmes is to try to maximize the number of breeding pairs and therefore, to place boxes at a minimum distance of adjacent bluebird territories. This distribution differs substantially from post-fire habitat in which nest cavities are often clumped and unevenly spaced (Saab et al., 2002). Thus, in contrast to a common criticism of the use of nest boxes—that people are increasing the density of available nest cavities beyond natural densities—the end result of such 'nest-box trails' is the creation of a much less patchy habitat with a more evenly distributed resource.

Such a distribution of nest cavities can influence habitat selection, settlement patterns, and ultimately the behavioural phenotype of the population. Bluebirds prefer territories with multiple cavities (Plissner & Gowaty, 1995; Saab et al., 2002), and in western bluebirds, the most aggressive males are most likely to obtain such preferred territories. Moreover, western bluebirds appear to use nest cavity availability as a primary factor in habitat choice over and above other habitat characteristics (Duckworth, 2006a). Using this information, we compared settlement patterns of males in a study site where nest cavity density varied naturally to a site where nest boxes were evenly distributed. In the naturally varying population, we observed that the distribution of nest cavities across the landscape influenced settlement patterns and created areas of high and low aggressiveness that corresponded to areas with many versus few nest cavities. In particular, at this site, there was a higher density of nest cavities in open compared to more forested habitat resulting in an overrepresentation of aggressive males there (Duckworth, 2006a). Such an assortment of aggressive types in relation to variation in nest cavity density is not possible in nest-box populations where nest boxes are distributed evenly across the landscape. Thus, local intraspecific competitive dynamics differ substantially between natural versus nest-box populations as variance in territory quality (at least in relation to the number of nest cavities) is much greater in natural compared to man-made habitat. This leads to hotspots and coldspots of competition in natural populations, whereas in man-made habitat, intraspecific competition is likely more spatially consistent.

The distribution of nest boxes is also likely to have important impacts on interspecific competition. In both bluebird species, more aggressive males acquire larger territories compared to less aggressive males; however, in mountain bluebirds, the relationship depends on whether they have a conspecific or heterospecific neighbour ($F_{1,20} = 4.43$, $P < 0.05$; Figure 14.1A). Specifically, territory size and aggression were only linked in mountain bluebirds when their nearest neighbour was another mountain bluebird ($F_{1,12} = 6.68$, $P = 0.02$, $b_{ST} = 0.60$), whereas, there was no relationship between territorial spacing and aggression when their nearest neighbour was a western bluebird ($F_{1,6} = 0.01$, $P = 0.91$, $b_{ST} = 0.05$, Figure 14.1A). In western bluebirds, more aggressive males acquired larger territories ($F_{1,61} = 7.22$, $P < 0.01$; Figure 14.1B) and territories with more nest cavities ($F_{1,61} = 4.43$, $P < 0.05$; Figure 14.1B) and this relationship was not influenced by the identity of their nearest neighbour (interactions between species identity of nearest neighbour and aggression on territory size: $F_{1,95} = 0.17$, $P = 0.68$).

Mountain bluebirds are larger than western bluebirds and so require larger territories (Figure 14.1C; see also Pinkowski, 1979). However, we found that, in the presence of western bluebirds, mountain bluebirds' territories were significantly smaller than if their neighbour was another mountain bluebird ($t = 2.26$, $P = 0.03$; Figure 14.1D). The territory size of western bluebirds did not depend on which species was their nearest neighbour ($t = 0.84$, $P = 0.40$; Figure 14.1D). These results suggest that western bluebirds are able to displace mountain bluebirds during the process of colonization by crowding them out. Western bluebirds' higher aggression gives them the upper hand in boundary disputes and this causes the territories of any neighbouring mountain bluebirds to be smaller than would otherwise occur. These results also have implications for the link between nest cavity density and coexistence of the two species. If the distance between adjacent nest cavities is greater than the optimal territory size of mountain bluebirds, this can limit the ability of western bluebirds to crowd them out of an area because they will not breed close enough to mountain bluebirds to encroach on their territory. However, if nest cavity density is high, territory boundaries will be primarily determined by competitive interactions between the species.

To test this idea, we surveyed populations in which the density of nest boxes significantly differed and found that the two species were more likely to coexist when boxes were placed at a lower density (nested ANOVA: $F_{3,107} = 7.73$, $P < 0.001$; Figure 14.2). The correlation between nest cavity density and coexistence suggests that these species may differ in their ability to exploit resource-poor versus resource-rich habitat—a common trade-off mediating local coexistence (Palmer, 2003; Tessier & Woodruff, 2002). Mountain bluebirds may be limited in resource rich areas by the presence and

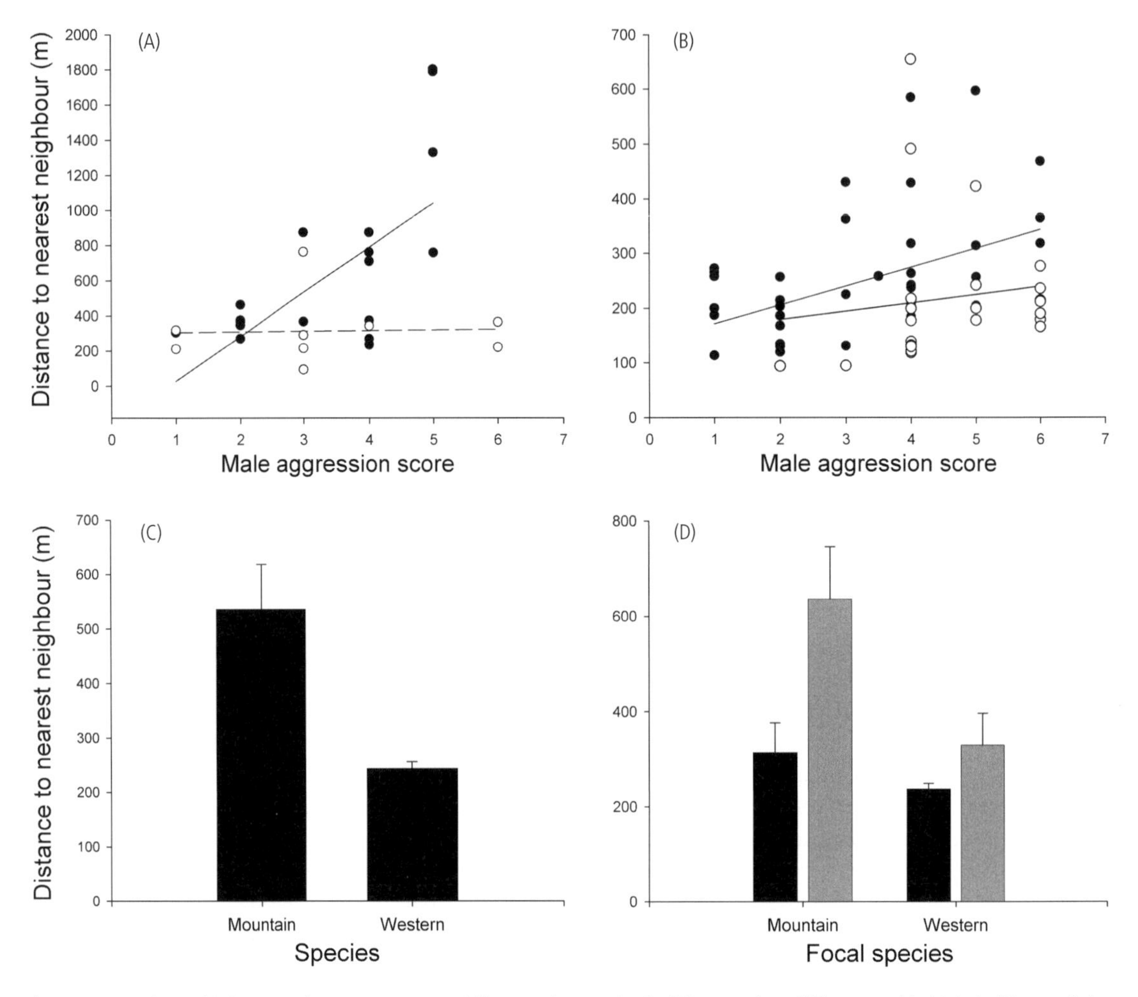

Figure 14.1 Relationship between distance to nearest neighbour and aggression for (A) mountain and (B) western bluebirds. In (A) open circles indicate territories in which the neighbour was a western bluebird and closed circles indicate territories in which the neighbour was a mountain bluebird. In (B) open circles indicate multi-box territories and closed circles indicate single-box territories. (C) Mountain bluebirds maintained greater distances from their nearest neighbour than western bluebirds. (D) This distance was significantly smaller if their nearest neighbour was a western bluebird (black bars). Spacing of western bluebirds did not depend on whether their neighbour was a western (black bars) or mountain (grey bars) bluebird. Bars indicate mean ± standard error and numbers on bars are sample sizes.

higher competitive ability of western bluebirds. On the other hand, western bluebirds may be limited in resource-poor areas if breeding at high density is necessary for them to benefit from cooperative interactions with family groups. These results also have important implications for how nest-box populations may be changing the dynamics of species coexistence. Because the distribution of nest cavities in post-fire habitat is patchy and uneven, some nest cavities are clumped and some are relatively isolated in their distribution. Thus, there may be more opportunities for the two species to coexist in these natural habitat patches compared to nest-box populations simply because there is more variability in the distances between nest cavities in post-fire populations.

Another major consequence of nest-box placement is that humans are replacing an ephemeral resource with a highly stable resource. In the context of a competition-colonization trade-off, the creation

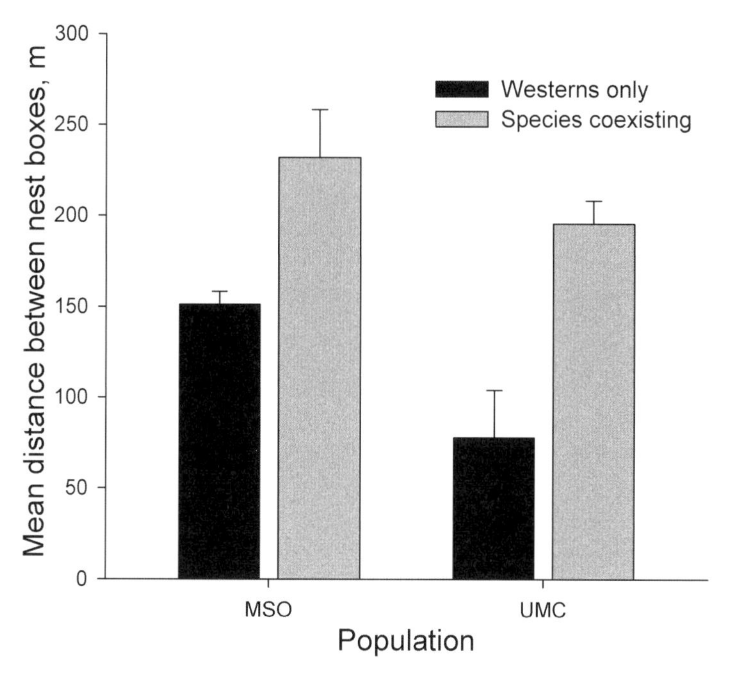

Figure 14.2 Coexistence of western and mountain bluebirds is significantly more likely when nest boxes were placed at a low compared to high density. Shown are data from four nest box populations that were all created at approximately the same time (15–20 years ago). Nest box trails were paired within valleys (MSO and UMC) to control for site-specific habitat variation.

of stable habitat is always disadvantageous to the better colonizer and advantageous to the better competitor (Levin & Paine, 1974). Thus, for bluebirds we would expect that, in areas of suitable habitat for both species (in this case lower elevation valleys), western bluebirds would permanently replace mountain bluebirds. As discussed above, this is exactly what we have observed in low-elevation nest-box populations in Montana (Duckworth & Badyaev, 2007). Many of these sites were first colonized by western bluebirds in the late 1970s and early 1980s and nest-box trails have been continuously maintained in these areas since that time. Western bluebirds' displacement of mountain bluebirds appears to be permanent as these areas continue to show stable western bluebird populations into the present day. Any mountain bluebird territories that do occur are usually on the periphery of these populations in areas where they are likely to share only one territory boundary with western bluebirds and thus are less likely to be crowded out (R.A.D. personal observations).

Similar to the interspecific patterns observed, at the intraspecific level, the replacement of an ephemeral resource with a stable resource should also lead to the permanent replacement of a colonizing morph within a species. In concordance with this prediction, we found that, in the oldest nest-box populations, the more aggressive and dispersive phenotype in western bluebirds was replaced with the nonaggressive, less dispersive individuals (Duckworth & Badyaev, 2007; Duckworth, 2008). Humans have diligently maintained nest boxes at these sites for many decades, up to 45 years in the earliest established population. In contrast, post-fire habitat is only suitable for bluebirds for up to approximately 25 years (Schieck & Song, 2006). Many of the nest-box populations that we monitor in Montana have already exceeded this longevity. Moreover, although a natural nest cavity can last for more than 20 years, most natural cavities persist for only a few years, yet, in many nest-box populations, the man-made cavities are never moved and damaged boxes are replaced with new ones. This means that, not only is habitat longevity lengthened in nest-box populations, but longevity of nest cavities is also substantially increased. This excessive cavity stability in nest-box populations should be a further boon to the nonaggressive philopatric morph of western bluebird which relies on nepotistic territorial inheritance (Duckworth, 2008). In post-fire habitat, opportunities for territorial inheritance are likely to occur less consistently as snag density changes dynamically across a habitat patch during forest succession.

In conclusion, post-fire habitat on which bluebirds historically depended in the northwestern United States differs substantially from nest-box trails on which these species are becoming increasingly dependent. There is clear evidence that this novel man-made habitat is impacting coexistence of the two bluebird species and the evolution of behavioural strategies in western bluebirds. However, the long-term consequences of such habitat replacement are less clear. Specifically, there does not seem to be any danger of western bluebirds dominating mountain bluebirds in all areas of their range. Nest boxes are mostly concentrated in the lower elevation valleys and are often clustered around human population centres. Moreover, western bluebirds rarely breed above 2000 m in the northwest while mountain bluebirds can breed at elevations up to 3500 m (Guinan et al., 2000; Power & Lombardo, 1996). Finally, since the mid-1980s, climate warming is increasing the frequency of wildfire in the western United States (Westerling et al., 2006). This, in combination with recent changes in fire suppression policy, which allows more wildland fires to burn without suppression efforts, means that natural post-fire habitat is on the rise. All of these factors should increase opportunities for coexistence of the two bluebird species and also maintain behavioural variation within western bluebirds. However, an interesting potential side effect of increasing abundance of both natural and man-made habitats is the possibility for significant behavioural and ecological divergence between post-fire and nest-box populations of these species. This is a particularly intriguing possibility given that dispersal dynamics are likely to differ substantially between these habitats (Citta & Lindberg, 2007).

14.5 Concluding remarks

The bluebird example offers several important lessons for research on nest-box populations of other species. One seemingly ubiquitous characteristic of secondary cavity nesting species is that nest cavities are their main limiting resource and competition over them is fierce (Newton, 1994). This has led to unique strategies both within and among species for acquiring this resource. For example, just in the populations that we monitor, species differ substantially in their strategies for obtaining this resource. In addition to mountain bluebirds' dispersiveness and western bluebirds' aggressiveness, the small house wren (*Troglodytes aedon*) fills a cavity with sticks reducing the hole size until no other species can fit inside, and tree swallows (*Tachycineta bicolor*) breed later than the other secondary cavity nesters maintaining a constant vigilance in order to take over a cavity as soon as the previous owner's nest fails or fledges. Such diversity of strategies suggests that competition for nest cavities is a major driver of behavioural evolution as has previously been shown for life history evolution of these species (Martin, 1992).

The use of man-made nest boxes is likely to change in some way the density, distribution, and stability of this crucial resource and as a consequence will alter the dynamics of species interactions and quite possibly evolution of behavioural and life history traits. Thus, if we are interested in understanding the behavioural and life history evolution of secondary cavity nesting species, then it is useful and probably necessary to try to place findings from nest-box populations into the context of natural populations. Knowledge of the historical habitat and resource distribution in which these species evolved—whether that is woodpecker excavated cavities in post-fire forests typical of North America, non-excavated cavities in an old growth tropical rain forest typical of South America, or non-excavated hollows in multi-century old eucalypts typical of Australia where there are no primary excavators—is crucial to understanding population dynamics and species interactions of this widespread group of birds.

More generally, I hope that this chapter brings attention to one of the more overlooked consequences of urbanization—the stabilization of habitat. This has relevance to current issues in urban ecology in two ways. First, stabilization of disturbance-prone habitat in the landscape surrounding urban centres will generally favour less dispersive species over colonizing species and may also influence the evolution of species dispersal strategies and species coexistence patterns. For example, recent studies of structures that stabilize coastal and stream areas show that such man-made structures can significantly impact species composition and local diversity (Bulleri & Chapman, 2010; Suddeth & Meyer, 2006). Whether these effects are mediated directly through habitat stabilization or are due to other consequences of human-altered habitat is an open question. Second, the urban environment itself is highly stable and does not undergo periods of disturbance and succession that occur in many natural systems. Such lack of heterogeneity in environmental conditions that organisms experience within urban environments through space and time may provide a key to understanding an outstanding question in urban ecology: why biodiversity in such environments is so low even though population densities are high (Shochat et al., 2010). Identifying and comparing key features of the historical habitat and resource distribution and stability of urban species may offer important insights into why these species, in particular, are able to thrive in such a novel environment.

Acknowledgements

I thank Alex Badyaev and two anonymous reviewers for comments which improved this manuscript. This work was supported by funding from the National Science Foundation (DEB 918095).

References

Amarasekare, P. (2003). Competitive coexistence in spatially structured environments: a synthesis. *Ecology Letters*, **6**, 1109–1122.

Amarasekare, P. and Nisbet, R.M. (2001). Spatial heterogeneity, source-sink dynamics, and the local coexistence of competing species. *American Naturalist*, **158**, 572–584.

Baraloto, C., Goldberg, D., and Bonal, D. (2005), Performance trade-offs among tropical tree seedlings in contrasting microhabitats. *Ecology*, **86**, 2461–2472.

Brawn, J.D. and Balda, R.P. (1988). Population biology of cavity nesters in northern Arizona: Do nest sites limit breeding densities? *Condor*, **90**, 61–71.

Brawn, J. D., Robinson, S. K., and Thompson, F. R. I. (2001). The role of disturbance in the ecology and conservation of birds. *Annual Review of Ecology and Systematics*, **32**, 251–276.

Brazill-Boast, J., Pryke, S. R., and Griffith, S. C. (2013). Provisioning habitat with custom-designed nest-boxes increases reproductive success in an endangered finch. *Australian Ecology*, **38**, 405–412.

Bulleri, F. and Chapman, M. G. (2010). The introduction of coastal infrastructure as a driver of change in marine environments. *Journal of Applied Ecology*, **47**, 26–35.

Cadotte, M. W. (2007). Competition-colonization tradeoffs and disturbance effects at multiple scales. *Ecology*, **88**, 823–829.

Chambers, C. L. and Mast, J. N. (2005). Ponderosa pine snag dynamics and cavity excavation following wildfire in northern Arizona. *Forest Ecology and Management*, **216**, 227–240.

Chesson, P. (2000). Mechanisms of maintenance of species diversity. *Annual Review of Ecology and Systematics*, **31**, 343–366.

Citta, J. J. and Lindberg, M. S. (2007). Nest-site selection of passerines: effects of geographic scale and public and personal information. *Ecology*, **88**, 2034–2046.

Cockle, K. L., Martin, K., and Wesolowski, T. (2011). Woodpeckers, decay, and the future of cavity-nesting vertebrate communities worldwide. *Frontiers in Ecology and the Environment*, **9**, 377–382.

Crespi, B. J. and Taylor, P. D. (1990). Dispersal rates under variable patch density. *The American Naturalist*, **135**, 48–62.

Duckworth, R. A. (2006a). Aggressive behavior affects selection on morphology by influencing settlement patterns in a passerine bird. *Proceedings of the Royal Society of London B*, **273**, 1789–1795.

Duckworth, R. A. (2006b). Behavioral correlations across breeding contexts provide a mechanism for a cost of aggression. *Behavioral Ecology*, **17**, 1011–1019.

Duckworth, R. A. (2006c). Evolutionary ecology of avian behavior: from individual variation to geographic range shifts. PhD thesis, Duke University, Durham, NC.

Duckworth, R. A. (2008). Adaptive dispersal strategies and the dynamics of a range expansion. *American Naturalist*, **172**, S4–S17.

Duckworth, R. A. and Badyaev, A.V. (2007). Coupling of aggression and dispersal facilitates the rapid range expansion of a passerine bird. *Proceedings of the National Academy of Sciences USA*, **104**, 15017–15022.

Gowaty, P. A. and Plissner, J. H. (1998). Eastern bluebird. *Birds of North America*, **381**, 1–31.

Griffith, S. C., Pryke, S. R., and Mariette, M. (2008). Use of nest-boxes by the Zebra Finch (*Taeniopygia guttata*): implications for reproductive success and research. *Emu*, **108**, 311–319.

Guinan, J. A., Gowaty, P. A., and Eltzroth, E. K. (2000). Western bluebird (*Sialia mexicana*). In A. Poole and F. Gill, eds, *The Birds of North America*, No. 510, (), p. 31. Birds of North America, Inc., Philadelphia, PA.

Harrison, R. G. (1980). Dispersal polymorphisms in insects. *Annu. Rev. Ecol. Syst.*, **11**, 95–118.

Hughes, C. L., Hill, J. K., and Dytham, C. (2003). Evolutionary trade-offs between reproduction and dispersal in populations at expanding range boundaries. *Biology Letters*, **270**, S147–S150.

Hutto, R. L. (1995). Composition of bird communities following stand-replacement fires in northern rocky mountain (USA) conifer forests. *Conservation Biology*, **9**, 1041–1058.

Johnson, M. L. and Gaines, M. S. (1990). Evolution of dispersal: theoretical models and empirical tests using birds and mammals. *Annual Review of Ecology and Systematics*, **21**, 449–480.

Kneitel, J. M. and Chase, J. M. (2004). Trade-offs in community ecology: linking spatial scales and species coexistence. *Ecology Letters*, **7**, 69–80.

Kotliar, N. B., Kennedy, P. L., and Ferree, K. (2007). Avifaunal responses to fire in southwestern montane forests along a burn severity gradient. *Ecological Applications*, **17**, 491–507.

Lehmkuhl, J. F., Everett, R. L., Schellhaas, R., Ohlson, P., Keenum, D., Riesterer, H., and Spurbeck, D. (2003). Cavities in snags along a wildfire chronosequence in eastern Washington. *Journal of Wildlife Management*, **67**, 219–228.

Leibold, M., Holyoak, M., Mouquet, N., Amarasekare, P., Chase, J., Hoopes, M., Holt, R., Shurin, J., Law, R., Tilman, D., Loreau, M., and Gonzalez, A. (2004). The metacommunity concept: a framework for multi-scale community ecology. *Ecology Letters*, **7**, 601–613.

Levin, S. A. and Paine, R. T. (1974). Disturbance, patch formation, and community structure. *Proceedings of the National Academy of Sciences of the United States of America*, **71**, 2744–2747.

Martin, T. E. (1992). Life history traits of open- vs. cavity-nesting birds. *Ecology*, **73**, 579–592.

Marzluff, J. M. (2001). Worldwide urbanization and its effects on birds. In J. M. Marzluff, R. Bowman, and R. Donnelly, eds, *Avian Ecology and Conservation in an Urbanizing World*, pp. 19–47. Springer Science, New York.

Newton, I. (1994). The role of nest sites in limiting the numbers of hole-nesting birds—a review. *Biological Conservation*, **70**, 265–276.

Palmer, T. M. (2003). Spatial habitat heterogeneity influences competition and coexistence in an African acacia ant guild. *Ecology*, **84**, 2843–2855.

Pfennig, K. S. and Pfennig, D. W. (2005). Character displacement as the 'best of a bad situation': fitness trade-offs resulting from selection to minimize resource and mate competition. *Evolution*, **59**, 2200–2208.

Pinkowski, B. C. (1979). Foraging ecology and habitat utilization in the genus *Sialia*. In J. G. Dickson, R. N. Conner, R. R. Fleet, J. C. Kroll, and J. A. Jackson, eds, *The Role of Insectivorous Birds in Forest Ecosystems*, pp. 165–190. Academic Press, New York.

Plissner, J. H. and Gowaty, P. A. (1995). Eastern bluebirds are attracted to two-box nest sites. *Wilson Bulletin*, **107**, 289–295.

Power, H. W. and Lombardo, M. P. (1996). Mountain bluebird. In A. Poole and F. Gill, eds, *The Birds of North America*, pp. 1–21. Birds of North America, Inc, Philadelphia, PA.

Prescott, H. W. (1982). Causes of decline of the western bluebird in Oregon's Willamette Valley. *Sialia*, **2**, 131–135.

Remm, J. and Lõhmus, A. (2011). Tree cavities in forests—the broad distribution pattern of a keystone structure for biodiversity. *Forest Ecology and Management*, **262**, 579–585.

Rodríguez, J., Avilés, J. M., and Parejo, D. (2011). The value of nestboxes in the conservation of Eurasian Rollers *Coracias garrulus* in southern Spain. *Ibis*, **153**, 735–745.

Roff, D. A. (1994). Habitat persistence and the evolution of wing dimorphism in insects. *Am. Nat.*, **144**, 772–798.

Saab, V. A., Brannon, R., Dudley, J., Donohoo, L., Vanderzanden, D., Johnson, V., and Lachowski, H. (2002). Selection of fire-created snags at two spatial scales by cavity-nesting birds. pp. 835–848. USDA Forest Service Gen. Tech. Rep.

Saab, V. A., Russell, R. E., and Dudley, J. G. (2007). Nest densities of cavity-nesting birds in relation to postfire salvage logging and time since wildfire. *Condor*, **109**, 97–108.

Schieck, J. and Song, S. J. (2006). Changes in bird communities throughout succession following fire and harvest in boreal forests of western North America: literature review and meta-analysis. *Canadian Journal of Forest Research*, **36**, 1299–1318.

Semel, B. and Sherman, P. W. (1995). Alternative placement strategies for wood duck nest boxes. *Wildlife Society Bulletin*, **23**, 463–471.

Shochat, E., Lerman, S. B., Anderies, J. M., Warren, P. S., Faeth, S. H., and Nilon, C. H. (2010), Invasion, competition, and biodiversity loss in urban ecosystems. *Bioscience*, **60**, 199–208.

Shochat, E., Warren, P. S., Faeth, S. H., McIntyre, N. E., and Hope, D. (2006). From patterns to emerging processes in mechanistic urban ecology. *Trends in Ecology & Evolution*, **21**, 186–191.

Snyder, R. E. and Chesson, P. (2003). Local dispersal can facilitate coexistence in the presence of permanent spatial heterogeneity. *Ecology Letters*, **6**, 301–309.

Suddeth, E. B. and Meyer, J.L. (2006). Effects of bioengineered streambank stabilization on bank habitat and macroinvertebrates in urban streams. *Environmental Management*, **38**, 218–226.

Tessier, A. and Woodruff, P. (2002). Trading off the ability to exploit rich versus poor food quality. *Ecology Letters*, **5**, 685–692.

Tilman, D. (1994). Competition and biodiversity in spatially structured habitats. *Ecology*, **75**, 2–16.

Tilman, D. (2004). Niche tradeoffs, neutrality, and community structure: A stochastic theory of resource competition, invasion, and community assembly. *Proceedings of the National Academy of Sciences of the United States of America*, **101**, 10854–10861.

Westerling, A. L., Hidalgo, H. G., Cayan, D. R., and Swetnam, T. W. (2006). Warming and earlier spring increase western US forest wildfire activity. *Science*, **313**, 940–943.

Yu, D. W. and Wilson, H. B. (2001). The competition-colonization trade-off is dead; long live the competition-colonization trade-off. *American Naturalist*, **158**, 49–63.

CHAPTER 15

The application of signal transmission modelling in conservation biology: on the possible impact of a projected motorway on avian communication

Erwin Nemeth and Sue Anne Zollinger

15.1 Introduction

Our global economy demands a higher mobility of persons and goods and consequently an increase in motorized traffic. Among other factors related to this increase in traffic, noise pollution is one critical factor that may explain the decline and disappearance of wildlife along motorways (Reijnen et al., 1995). An important effect of traffic noise is the masking of acoustic signals. In many species, individuals rely on vocal signalling to find their partners, to repel rivals or to warn against predators (Bradbury & Vehrencamp, 2011; Lengagne, 2008). The disturbance of the transmission or reception of vocal signals can have fitness consequences for the sender or the receiver or both (Brumm, 2010; Kight et al., 2012; Schroeder et al., 2012). By masking a signal with environmental noise, the distance over which the signal can be detected will be decreased and the shortened transmission paths of signals can lead to a complete breakdown of communication between neighboring animals (Brumm & Slabbekoorn, 2005; Dooling & Blumenrath, in press).

We present here a study investigating whether traffic noise from a new motorway planned to be completed in 2025 is likely to have detrimental effects on the acoustic communication of a threatened charadriiform bird, the Eurasian stone-curlew (*Burhinus oedicnemus*). We also predict whether, and to what extent, noise control measures might be effective in ameliorating any negative impacts of noise on the birds. To this aim, we calculated the impact on curlew communication for six different potential noise scenarios. In order to predict the masking influence of traffic noise on acoustic communication, the following information is needed: the level and spectral composition of noise, the sound pressure level (SPL) and spectral characteristics of the acoustic signals, the transmission properties of the signals in the respective habitat, and knowledge about the hearing capabilities of the concerned species in noise (Dooling & Popper, 2007; Dooling et al., 2009; Lohr et al., 2003; Nemeth & Brumm, 2010). Based on these values, we calculated spatially explicit predictions for communication areas of vocal signals within the study area. Finally, we will discuss the assumptions and limitations of this method and suggest further directions of research that may improve the predictive power of such models.

15.2 Methods

15.2.1 Study species and study site

The Eurasian stone-curlew is a large wader with a wingspan of about 80 cm and a weight of up to 0.5 kg (Roselaar, 1983). In Central Europe it is a long-distance migrant and inhabits dry and open

Avian Urban Ecology. Edited by Diego Gil and Henrik Brumm

habitats where it is difficult to observe because of its cryptic plumage. In Europe the species status is classified as vulnerable with a decreasing population size (Birdlife International, 2004) and in Austria it is a very rare and endangered species (Frühauf, 2005). Stone-curlews are largely nocturnal and their social life is highly dependent on acoustic signals. They are known to avoid the vicinity of roads and motorways, but it is not known whether the light of passing cars or the noise is more disturbing to the species (Green et al., 2000). With up to six breeding pairs, the study site is the second most important breeding site of stone-curlews in Austria. The area (1500 ha) is located to the north of the village of Markgrafneusiedl, Lower Austria. The path of the proposed motorway transects the northern edge of the breeding area (Figure 15.1). The area consists of a mixture of quarries, agricultural land, and other disturbed areas. The birds in these areas call from both the bottom level of the quarries and also from higher points on the edge of the quarries created by excavated material. Smaller local roads currently delineate the southern and western borders of the breeding area.

15.2.2 Assumptions for the calculation of communication areas

Environmental noise

In general, traffic noise has a large low-frequency component with a steady decline in spectral energy above 2 kHz, decreasing by about 5–7 dB per 1 kHz (see e.g. Nemeth & Brumm, 2010). Differences in the spectral composition of traffic noise can be caused by varying size or type of lorries and other vehicles and by the road surface type. Projections for the planned motorway estimate that traffic during the week will average 1,250 vehicles per hour on a stone mastic asphalt road surface, with lorries comprising 16% of the total vehicular traffic. As basis for our study, we calculated the projected traffic noise that would occur during the main calling period (between 19:00 and 20:00 CET) using A-weighted sound pressure levels. The spatial and temporal distribution of traffic noise, its transmission and the resulting noise maps were calculated according an Austrian directive for road noise (RVS 04.02.11) using a digital elevation model with a spatial resolution of 25 cm and was provided by Ernst Walter

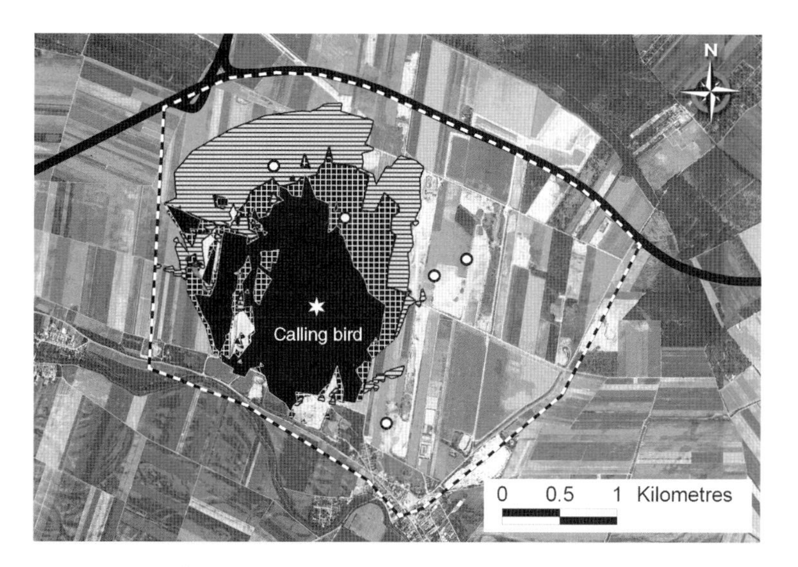

Figure 15.1 Prediction of communication areas for calls of a stone-curlew at the study site in Markgrafneusiedl. The study site is demarcated with a dashed line, and the planned motorway is drawn as a wide black line passing along the northern edge of the study site. The communication area is shown for one individual (star) calling from the edge of a quarry with a call amplitude of 103.7 dB SPL. The locations of birds in the neighbouring territories are shown as white points. The striped polygon shows the communication area without construction of the motorway in the year 2008, the black polygon with motorway and a 5 m high slope with a rise:run ratio of 2:3, the hatched polygon with motorway and a steep slope and earth wall with an overall height of 10 m. The motorway in both cases would be lowered 5 m below ground level.

(unpublished data from the acoustic consultancy office Rinderer & Partner Ziviltechniker KG, Graz, Austria). The sound pressure levels were then converted into spectrum noise levels (i.e. the SPL measured per Hz), and these levels were used for further calculations (Dooling & Popper, 2007). We used spectrum noise levels of a motorway with a similar number and proportion of vehicles as the predicted motorway to predict spectral distribution of noise in our study (Ernst Walter, unpublished data). We then utilised these digital noise maps to estimate the masking effect on bird vocalizations.

Noise maps and masking effects were calculated for six different scenarios. The first scenario is the current situation in 2008 before the construction of a motorway and with the current amount of road traffic on the smaller roads on the south and west of the breeding area. Second is the estimated situation in 2025 if no motorway were to be built, which accounts for the predicted increase in traffic on the smaller existing roads. Third, the predicted situation in 2025 if the motorway is built without noise control measures (i.e. 5 m below the current ground level and earth verges with a 2:3 sloped profile). Finally, we calculated the predicted situations in 2025 for three potential noise control designs: lowering the road 5 m below the ground level and building steep earth sides with an overall height of 5, 8, or 10 m, measured up from the pavement. Traffic noise was assumed to be relatively constant because observation and predictions showed that in all scenarios two or more motor vehicles were present on the neighbouring roads at the study site during the peak calling times for stone-curlews (Ernst Walter unpublished data). Since the source of traffic noise was more than 300 m from the territories of individual stone-curlews, and since the traffic noise showed a rather continuous temporal distribution, we estimated only small differences between maximum and average SPLs. Therefore, we used the predicted values for equivalent continuous sound level (LEQ_{1min}) for all further calculations.

Sound pressure level and frequency of bird vocalizations

Amplitude measurements of birdsong in wild birds are often difficult to conduct and thus are rather rare (Brumm, 2004; Lowry et al., 2012; Nemeth, 2004; Nemeth et al., 2012). Here we have tried to estimate the sound pressure level of free living stone-curlews and individuals calling in outdoor aviaries of the Vienna Zoo. In the zoo the amplitude of the call was measured with a sound level meter (Casella CEL 383 Integrating Impulse Sound Level Meter, a class I device with a precision of ±0.7 dB). Two individuals were measured in the evening (19:00–22:00) from eight different positions at distances ranging from 2.5 to 16 m.

At the study site Markgrafneusiedl, Austria it was not possible to use sound level meters, because the cryptic birds are very shy, making it difficult to approach them close enough to conduct reliable measurements. Here a different technique was used—a radio transmitter microphone (Sennheiser Evolution with an omnidirectional microphone Sennheiser ME 2) was placed in two of the known territories. The area surrounding the microphone was monitored by two observers. One observer received the signal with a Sennheiser EK 100 receiver and recorded observations on one of two stereo tracks of a digital recorder (Sound Devices 722 Digital Audio Recorder). During recording sessions, birds were observed with a spotting scope and their calling location was determined. If birds did not call spontaneously in the vicinity of the microphone, a recorded curlew call was played from a remote-controlled loudspeaker (Foxpro Scorpion Digital Game Caller, Foxpro Inc., Lewistown PA, USA) near the microphone. After a calling bird was successfully located and recorded, we placed a loudspeaker at the position of the calling location (at a height of 30 cm from the ground) from which reference signals with known SPLs were broadcasted (2, 3, and 4 kHz sine tones) and recorded for calibration with same settings used to record the bird calls. All measurements were done in the evening during calm weather conditions. The vocalisations were analysed with the software Avisoft SAS LabPro 4.04 (R. Specht, Berlin, Germany). In the zoo and in the field the birds called both spontaneously and as reaction to playback of their species-specific calls. The calls belonged to two different categories, the curlee-calls (Figure 15.2A) and the gallop-rhythm—calls (Figure 15.2B) (Roselaar, 1983). The mean call amplitude, taken across all measurements, was 103.7 dB

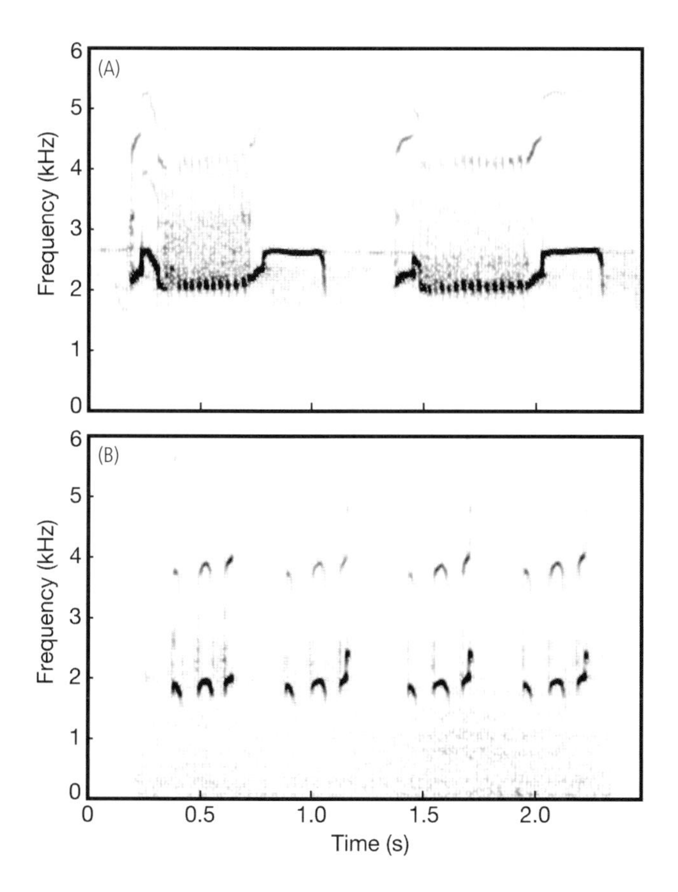

Figure 15.2 Typical examples of the two long-range communication call types that we analysed in this study: 'curlee calls' (A) and 'gallop-rhythm calls' (B) of the Eurasian stone-curlew.

Table 15.1 Sound pressure levels (mean ± standard deviation) of stone-curlew calls estimated for 1 m distance and measured at the study site Markgrafneusiedl and in an outdoor aviary of the Vienna Zoo.

Location	Date	Sound pressure level (dB_A) mean ± standard deviation	Number of locations	Number of calling bouts
Markgrafneusiedl	28.04.2009	104.4 ± 1.43	1	8
Outdoor aviary (Vienna Zoo)	30.04.2009	103.5 ± 3.42	3	6
Outdoor aviary (Vienna Zoo)	03.06.2009	103.0 ± 4.03	1	4
Markgrafneusiedl	08.06.2009	102.7 ± 1.02	2	9
Markgrafneusiedl	09.06.2009	105.0 ± 2.33	1	4

at 1 m distance (Table 15.1) and we found no differences in sound pressure level between the two call types. We used this value for all following calculations. The frequency range of all vocalisations was between 2 und 4 kHz. Since these values are concordant with the analysed values in the literature (Roselaar, 1983), we used this frequency range for all further calculations of acoustic masking by traffic noise.

Transmission of calls

Sound transmission was calculated according to ISO 9614 (ISO, 1993) with an assumed calling height of 30 cm above ground, the approximate height of a standing bird. At the study site stone-curlews called both from the ground and elevated locations at the edge of the quarries. These locations differ considerably in the acoustic characteristics and since we did not know which calling locations birds used regularly, we chose two locations in each territory: a random location on the ground in a quarry and a second randomly chosen location on the top of a slope.

15.2.3 Assumptions about the detection of signals in noise

The calculation of detecting distances of signals requires knowledge about the hearing capabilities of birds in noise. To date, auditory thresholds in masking noise are only available for 15 bird species (Dooling & Popper, 2007; Jensen & Klokker, 2006; Langemann & et al., 1995; Langemann et al., 1998; Lohr & et al., 2003; Okanoya & Dooling, 1987). The threshold at which a signal is detectable in noise depends on the SPL of the signal and the spectrum level of the masking noise at the frequency of the sound signal. This signal-to-noise ratio, called a 'critical ratio', is to a large extent independent of the overall noise level. To our knowledge, critical ratios in masking noise are not available for stone-curlews. Here we used a value of 24 dB SPL at 2 kHz, which is the average value of published critical noise ratios in literature (Dooling & Popper, 2007). The critical ratio allows the detection of a signal, but it is important to note that even higher signal-to-noise ratios are typically necessary for recognition or discrimination of the same signal in noise (Brumm & Slabbekoorn, 2005; Klump, 1996; Lohr et al., 2003; Dooling & Blumenrath, in press), thus we estimated an additional 3 dB (27 dB SPL at 2 kHz) would be necessary for stone-curlews to reach this discrimination threshold. The value was adopted from the study of Lohr et al. (2003) in which an average value was calculated for budgerigars (*Melopsittacus undulatus*), zebra finches (*Taeniopygia guttata*), and canaries (*Serinus canaria*). Nonetheless, we acknowledge that our estimated value of 27 dB SPL for the discrimination of an acoustic signal may be above or below the true value. To account for these uncertainties we have also calculated communication areas using critical ratios 3 dB higher and lower than 27 dB SPL.

15.2.4 Calculation of communication areas

The variables described above were used to calculate communication areas for the calls of stone-curlews for two calling locations within each of the six territories and at three different discrimination thresholds. Communication areas were defined as areas where the signal-to-noise ratio is higher than or equal to the discrimination threshold. For the area calculations we used a cell size of 5×5 m. For each 25 m^2 cell, the spectrum level of noise was subtracted from the signal amplitude, and then areas with values equal to or higher than the discrimination threshold of 24, 27, or 30 dB SPL were identified. All the calculations were done with Arcview 3.2 and its extension Spatial Analyst (ESRI, Redlands, CA, USA).

15.2.5 Evaluation of the effect of the masking influence of traffic noise

The results of the calculations show the projected impact of the planned motorway and possible noise abatement measures on the communication areas of the individual birds. The evaluation of the results was first based on a visual inspection of the communication area maps to judge how far neighbouring birds can be still heard in each noise condition. Since there is no experimental data available about the relationship between noise levels and loss of fitness in stone- curlews, we chose an impact criterion based on the loss of communication area resulting from increased motorway traffic noise. As the risk for losing of one of the two best breeding grounds in Austria for this endangered bird is very high, the criterion should be conservative. Therefore it was assumed that an average loss of more than 10% of the communication area would be enough to constitute a harmful effect on the birds and should be prohibited by better noise control methods.

15.3 Results

Figure 15.1 shows an example of the result of calculations for one calling individual (marked with a star in Figure 15.1) on the edge of a quarry with the assumption of a critical ratio of 27 dB SPL, and for the motorway with noise control measures in place. The map shows a considerable impairment across the northern portion of the individual's calling area. In comparison to the current situation in 2008 (no motorway), the new motorway (built 5 m below ground level with earth sides angled at a slope of 2:3) would reduce the communication area for this signal by 53%. Table 15.2 shows the reduction of communication areas under different scenarios. Two territories close to the motorway (locations H1, H2, L1, and L2) will be more severely affected by increased noise (Figure 15.3). At the same time,

Table 15.2 Predicted average communication areas ± standard deviation of six territories (two calling locations per territory) after the construction of the motorway. The values in the table are presented as the percent reduction in communication areas relative to the area predicted for the year 2025 if no motorway is built.

Noise control measure	Critical ratio = 27 dB		Critical ratio = 30 dB		Critical ratio = 24 dB	
	Area (ha)	Percentage of scenario 2025	Area (ha)	Percentage of scenario 2025	Area (ha)	Percentage of scenario 2025
Slope 2:3	145.6 (±59.2)	61.0	109.6 (±44.8)	57	188.4 (±62.7)	63.48
Steep slope 5 m	164.5 (±60.4)	68.6	137.0 (±61.2)	71	211.7 (±67.4)	71.35
Steep slope 8 m	211.8 (±70.5)	88.4	167.2 (±51.4)	87	266.9 (±69.3)	89.94
Steep slope 10 m	226.48 (±80.2)	94.5	182.73 (±59.9)	95	283.49 (±77.7)	95.51

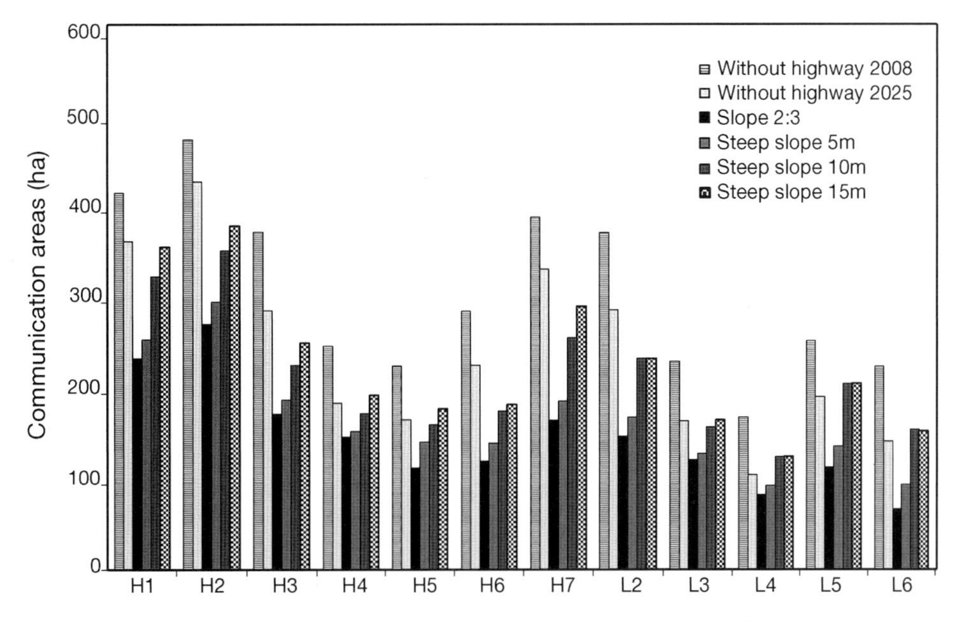

Figure 15.3 Predicted communication areas and different calling points in six different territories under different environmental noise conditions. For the categories along the *x*-axis, the number denotes the territory while the letters denote the location of the calling points (L = low, H = high), e.g. L3 means a calling location on a low spot of ground of a quarry in territory 3. The critical ratio for detection and discrimination of the signal was set to 27 dB.

strong noise control measures can yield a larger recovery of communication area in these territories relative to the predicted situation for 2025 without a motorway (see Table 15.2), owing to the predicted increase of the traffic on the small roads to the south and east if no motorway is built. When we compare the decrease of communication areas under different assumptions for the hearing abilities we find similar relative reductions in the communication areas under different noise control measures for all three assumed discrimination thresholds (Table 15.2). We find that for a new motorway at this location, the target criterion of a less than 10% loss of communication area (relative to the estimated 2025 conditions without a motorway) can only be reached by using the strongest noise abatement measure—a lowered motorway with steep earth walls at an overall height of 10 m measured up from the pavement.

15.4 Discussion

Our calculations show how the masking of calls by traffic noise would affect the acoustic communication abilities of stone-curlews in different projected scenarios. Based on the predetermined criterion for maximum allowable impact, one can assess which of the possible noise control measures will be most effective at minimizing the negative impact of the motorway on communication in this species.

The method of analysis we have used here could easily be applied to other species or landscape urbanization projects. Most of the necessary information and software is already available or fairly straightforward to obtain. Digital elevation models with sufficient spatial resolution for acoustic calculations are increasingly available. Software to calculate the transmission of noise and signals can be programmed into a geographical information system (GIS) or calculated with commercial acoustic software or freeware (e.g. Reed et al., 2012). The greatest source of uncertainty in our calculations is likely to be the assumptions we had to make about the hearing abilities of the species in question. As for most bird species, there are no experimental data on hearing thresholds in stone-curlews and so we could only extrapolate from what is known about hearing in other species. However, as hearing capabilities are relatively conserved in birds (excluding the Strigiformes), our extrapolations as to the hearing capabilities of stone-curlews based on the abilities of other species are likely to be reasonable (Dooling & Blumenrath, in press; Fay, 1988). In addition, our model has simplified assumptions about the detection of signals in noise; it neglects the spatial and temporal patterns of noise and signals that might enhance or worsen detection thresholds. Listening birds can, at least to some extent, localize noise sources in space and it is therefore conceivable that birds may experience some spatial masking release from noise as a result of binaural hearing (Dent et al., 2009). Additionally, temporal fluctuations in the amplitude of sound sources can also increase the masking release from noise (Dooling et al., 2002) and this could improve the detection of signals. Since we could not account for these factors in our calculations, we cannot rule out the possibility out that our model may underestimate real communication distances. Furthermore, it is not easy to generalize laboratory measurements and computer models to the more dynamic and complex conditions in the field. A possible future solution could be to test the signal detection capabilities of birds directly in the field by observing behavioural responses to playback of conspecific vocalisations under different noise conditions.

In addition to the uncertainty about the hearing and detection thresholds of these birds, we also do not know if or how stone-curlews may adjust their signalling behaviour in response to increases in traffic noise. Birds of other taxa have been shown to change the frequency or amplitude of their vocalisations in noisy environments (Brumm & Slabbekoorn, 2005; Brumm & Zollinger, 2011), presumably to improve the transmission and detectability of their songs and calls. Other species have been found to adjust the temporal patterns of their calling or singing behaviour to less noisy periods (Fuller et al., 2007), avoid noisy territories (Bayne et al., 2008) or extend their calling period (Ríos-Chelén et al., 2013). Since it is thought that stone-curlews avoid habitats along noisy motorways (Green et al., 2000), it seems plausible that the species is not capable of adapting to high levels of noise pollution.

While birds from many taxa have been shown to exhibit the 'Lombard effect', an involuntary increase of vocal amplitude in response to increases in background noise levels (Zollinger & Brumm, 2011), it is not known if stone-curlews can or do call louder in response to traffic noise. The average SPL of the stone-curlew calls is quite high in comparison to other bird species (see Brumm & Ritschard, 2011; Brumm, 2009), and so it may be that the call amplitude in this species is already at or near the physiological upper limit for their vocal production system, as has been suggested for certain call types in male tinamous (Schuster et al., 2012). Although little is known about the vocal production system in stone-curlews, it has been suggested that the syringes of the Burhinidae are among the most evolutionarily basal in the order Charadriiformes (Brown & Ward, 1990). In addition to a relatively simple syrinx, stone-curlews are not thought to be vocal learners, which means that they are unlikely to possess a level of vocal plasticity that may allow for more rapid changes in the spectral properties of their calls in response to sudden changes in their acoustic environment. However, even members of one of the most basal groups have been shown to increase their call amplitude in response to increases in background noise levels (Schuster et al., 2012). In the future it would certainly be desirable to conduct field studies to investigate the effects of anthropogenic noise on the behaviour and reproductive fitness of the species.

Acknowledgements

We thank the Vienna Zoo for the right to access aviaries and Ernst Walter for the calculation of the noise maps. We are grateful to an Anonymous reviewer for thoughtful suggestions that improved the manuscript. The study was funded by the ASFINAG (Austrian expressway financing company).

References

Bayne, E.M., Habib, L. and Boutin, S. (2008). Impacts of chronic anthropogenic noise from energy-sector activity on the abundance of songbirds in the boreal forest. *Conservation Biology*, **22**, 1186–1193.

Birdlife International. (2004). *Birds in Europe: population estimates, trends and conservation status*. BirdLife International, Cambridge.

Bradbury, J. W., and Vehrencamp, S. L. (2011). *Principles of Animal Communication*, 2nd edition. Sinauer Associates, Inc. Publishers, Sunderland, MA.

Brown, C. and Ward, D. (1990). The morphology of the syrinx in the Charadriiformes (Aves): Possible phylogenetic implications. *Bonner Zoologische Beiträge*, **41**, 95–107.

Brumm, H. (2004). The impact of environmental noise on song amplitude in a territorial bird. *Journal of Animal Ecology*, **73**, 434–40.

Brumm, H. (2009). Song amplitude and body size in birds. *Behavioral Ecology and Sociobiology*, **63**, 1157–65.

Brumm H. (2010). Anthropogenic noise: implications for conservation. In M. D. Breed and J. Moore, eds. *Encyclopedia of Animal Behavior*, volume 1, pp. 89–93. Academic Press, Oxford.

Brumm, H. and Ritschard, M. (2011). Song amplitude affects territorial aggression of male receivers in chaffinches. *Behavioural Ecology*, **22**, 310–16.

Brumm, H., and Slabbekoorn, H. (2005). Acoustic communication in noise. *Advances in the Study of Behaviour*, **35**, 151–209.

Brumm, H. and Zollinger, S. (2011). The evolution of the Lombard effect: 100 years of psychoacoustic research. *Behaviour*, **148**, 1173–98.

Dent, M. L., McClaine, E. M., Best, V., Ozmeral, E., Gallun, F. J., Narayan, R., Sen, K. and Shinn-Cunningham, B. G. (2009). Spatial unmasking of birdsong by zebra finches (*Taeniopygia guttata*) and budgerigars (*Melopsittacus undulatus*). *Journal of Comparative Psychology*, **123**, 357–67.

Dooling, R.J. and Blumenrath, S.H. (in press). Avian sound perception in noise. In H. Brumm, ed. *Animal Communication in Noise*. Springer, Heidelberg, in press.

Dooling, R. J., Leek, M. R., Gleich, O., and Dent, M. L. (2002). Auditory temporal resolution in birds: Discrimination of harmonic complexes. *Journal of the Acoustical Society of America*, **112**, 748–759.

Dooling, R. J., Leek, M. R., & West, E. W. (2009). *Predicting the Effects of Masking Noise on Communication Distance in Birds*. Paper presented at the 157th Meeting of the Acoustical Society of America, Portland OR.

Dooling, R. J., and Popper, A. N. (2007). *The Effect of Highway Noise on Birds*. The California Department of Transportation. Sacramento, CA.

Fay, R. R. (1988). *Hearing in Vertebrates: A psychophysics databook*. Hill-Fay Associates, Winnetka, IL.

Fuller, R. A., Warren, P. H., and Gaston, K. J. (2007). Daytime noise predicts nocturnal singing in urban robins. *Biology Letters*, **3**, 368–370.

Frühauf, J. (2005). Rote Liste der Vögel (Aves) Österreichs. Auftrag der Republik Österreich, Bundesministerium für Land- und Forstwirtschaft, Umwelt und Wasserwirtschaft, Vienna.

Green, R. E., Tyler, G. A., and Bowden, C. G. R. (2000). Habitat selection, ranging behaviour and diet of the stone curlew (*Burhinus oedicnemus*) in southern England. *Journal of Zoology*, **250**, 161–183.

ISO. (1993). Attenuation of sound during propagation outdoors. 1. Calculation of the absorption of sound by the atmosphere. *ISO 9613–1*. International Organisation for Standardization, New York.

Jensen, K. K., and Klokker, S. (2006). Hearing sensitivity and critical ratios of hooded crows (*Corvus corone cornix*). *Journal of the Acoustical Society Of America*, **119**, 1269–1276.

Kight, C. R., Saha, M. S., and Swaddle, J. (2012). Anthropogenic noise is associated with reductions in the productivity of breeding eastern bluebirds (*Sialia sialis*). *Ecological Applications*, **22**, 1989–1996.

Klump, G. M. (1996). Bird communication in the noisy world. In D. E. Kroodsma and E. H. Miller, eds. *Ecology and Evolution of Acoustic Communication in Birds*, pp. 321–338. Comstock Publishing Associates, Ithaca and London.

Langemann, U., Gauger, B. and Klump, G. M. (1998). Auditory sensitivity in the great tit: perception of signals in the presence and absence of noise. *Animal Behaviour*, **56**, 763–769.

Langemann, U., Klump, G. M. and Dooling, R. J. (1995). Critical bands and critical ratio bandwidth in the European starling. *Hearing Research*, **84**, 167–176.

Lengagne, T. (2008). Traffic noise affects communication behaviour in a breeding anuran, *Hyla arborea*. *Biological Conservation*, **141**, 2023–2031.

Lohr, B., Wright, T. F. and Dooling, R. J. (2003). Detection and discrimination of natural calls in masking noise by birds: Estimating the active space of a signal. *Animal Behaviour*, **65**, 763–777.

Lowry, H., Lill, A., and Wong, B. B. M. (2012). How noisy does a noisy miner have to be? Amplitude adjustments of alarm calls in an avian urban 'adapter'. *PLoS ONE*, **7**, e29960.

Nemeth, E. (2004). Measuring the sound pressure level of the song of the screaming piha *Lipaugus vociferans*: one of the loudest birds in the world? *Bioacoustics*, **14**, 225–228.

Nemeth, E. and Brumm, H. (2010). Birds and Anthropogenic noise: are urban songs adaptive? *American Naturalist*, **176**, 465–475.

Nemeth, E., Kempenaers, B., Matessi, G. and Brumm, H. (2012). Rock sparrow song reflects male age and reproductive success. *PloS ONE*, **7**, e433259.

Okanoya, K., and Dooling, R. J. (1987). Hearing in passerine and psittacine birds—a comparative study of absolute and masked auditory threshholds. *Journal of Comparative Psychology*, **101**, 7–15.

Reed, S. E., Boggs, J. L., and Mann, J. P. (2012). A GIS tool for modeling anthropogenic noise propagation in natural ecosystems. *Environmental Modelling & Software*, **37**, 1–5.

Reijnen, R., Foppen, R., Ter Braak, C., and Thissen, J. (1995). The effects of car traffic on breeding bird populations in woodland. III. Reduction of density in relation to the proximity of main roads. *Journal of Applied Ecology*, **32**, 187–202.

Ríos-Chelén, A., Quirós-Guerrero, E., Gil, D. and Macías Garcia, C. (2012). Dealing with urban noise: vermilion flycatchers sing longer songs in noisier territories. *Behavioral Ecology and Sociobiology*, **67**, 1–8.

Roselaar, C. S. (1983). *Burhinus oedicnemus*, stone curlew. In S. Cramp, ed. *Handbook of the Birds of Europe, the Middle East and North Africa*. Oxford University Press, Oxford.

Schroeder, J., Nakagawa, S., Cleasby, I.R. and Burke, T. (2012) Passerine birds breeding under chronic noise experience reduced fitness. *PLoS ONE*, **7**, e39200.

Schuster, S., Zollinger, S., Lesku, J. A. and Brumm, H. (2012). On the evolution of noise-dependent vocal plasticity in birds. *Biology Letters*, **8**, 913–916.

Zollinger, S. & Brumm, H. (2011). The Lombard effect. *Current Biology*, **21**, R614–R615

CHAPTER 16

The importance of wooded urban green areas for breeding birds: a case study from Northern Finland

Jukka Jokimäki, Marja-Liisa Kaisanlahti-Jokimäki, and Pilar Carbó-Ramírez

16.1 Introduction

Urbanization has been found to decrease the diversity of bird communities (see reviews Marzluff 2001; Chace & Walsh, 2006; McKinney, 2008). It is well-known that urban cores have only few species, which may be very abundant and are shared between towns. These species are mainly generalists, whereas specialist bird species decrease with increasing degree of urbanization. Therefore, it has been suggested that urbanization may cause homogenization of communities (Blair, 2001; Clergeau et al., 2006). However, urban habitats are not homogeneous; in fact, they are comprised of many kinds of environments with high disturbance levels (Gaston, 2010; Gilbert, 1989). Depending on local conditions, bird assemblages may differ from towns to towns as well as within towns (DeGraaf & Wentworth, 1986; Jokimäki & Kaisanlahti-Jokimäki, 2012a, b, c). Wooded urban green areas like managed parks with trees (e.g. public urban parks, cemeteries, and roadside corridors), unmanaged urban woodlots (urban forests), and residential areas with their gardens are home to very diverse bird communities (Cannon, 1999; Fernandez-Juricic & Jokimäki, 2001; Jokimäki & Kaisanlahti-Jokimäki, 2012a; Jokimäki & Suhonen, 1993; Lussenhop, 1977; McDonnell & Pickett, 1990; Tilghman, 1987). Correspondingly, some researchers have indicated that wooded urban green areas play an important role in the conservation of diversity of birds in urban environments (Mörtberg, 2001).

For the needs of conservation and management of birds and their habitats in urban environments, it is important to know which factors affect species occurrence, abundance, and living conditions in wooded urban green areas (Fernandez-Juricic & Jokimäki, 2001). Many factors such as patch size and vegetation structure can impact on the diversity of wooded urban area bird assemblages (Carbó-Ramírez & Zuria, 2011; Fernández-Juricic, 2000c, 2002; Jokimäki, 1999).

However, there are also other factors besides green area characteristics that can impact on bird species composition and abundance of birds living in these areas. For example, species-occupancy patterns may depend on species-specific distribution ranges, and species-specific traits and their adaptation to habitat (see reviews in Collins & Glenn, 1997; McGeoch & Gaston, 2002). In practice, this means that, in addition to local level park characteristics, we should also consider regional species pools as well as bird species traits and species-specific behaviour when analysing the importance and suitability of urban green areas as habitat for birds. At the same time, merging wildlife community ecology with animal behavioural ecology in urban planning has been indicated to be an important research topic (Jokimäki et al., 2011; Lepczyk & Warren 2012).

In heavily fragmented landscapes such as wooded green areas located within an urban matrix, patch size, tree species composition, surrounding matrix, as well as the local or regional species pool, may have a significant impact on urban bird diversity

Avian Urban Ecology. Edited by Diego Gil and Henrik Brumm

(e.g. Fernandéz-Juricic & Jokimäki, 2001). In this chapter, we analyse how wooded green area characteristics impact on assemblages of breeding birds by using original case data from the town of Rovaniemi, which is located on the Arctic Circle in northern Finland. Then we analysed at the two scales (within the town as well as in the area surrounding the town; i.e. within the district of Rovaniemi) how the regional species' pool around the study parks impacted on assemblages of breeding urban park birds in Rovaniemi by using the data collected in connection with the Rovaniemi Urban Bird Atlas project (Jokimäki & Kaisanlahti-Jokimäki, 2012c). Bird species were grouped according to their traits in order to see which kinds of traits are of advantage or disadvantage to birds trying to settle in wooded urban green areas.

Table 16.1 Characteristics of the studied urban wooded green areas in Rovaniemi (n = 19).

	Min.	Max.	Mean	SD
General wooded green area characteristics:				
Wooded green area size (ha)	0.3	10.9	3.09	3.58
Distance from urban core (km)	0.9	3.2	1.26	0.82
Distance from the nearest large-sized forests (km)	0.4	2.0	1.27	0.47
Tree characteristics:				
Wooded green area size/100 m^2	1.5	9.8	4.72	2.65
Deciduous trees (%)	17.0	100.0	59.52	26.58
Pines (%)	0.0	83.0	34.62	28.41
Spruces (%)	0.0	33.0	5.86	9.03
Height of the dominant trees (m)	10.0	22.0	14.63	3.25

16.2 Methods and materials

16.2.1 Study area

The study was conducted during the 2011 breeding season in Rovaniemi (66° 32´; 25° 12´; 66 300 inhabitants; land area 7581 km^2), located in northern Finland. Most of the human population (>50,000 inhabitants; about 83%) is concentrated in the urban core of Rovaniemi (i.e. in the area covering the area where the studied urban wooded areas as were located); hereafter called 'main town area'. The matrix of human settlements is mainly covered by forests (>60%). Most of the forests of the area are dominated by the Scots pine (*Pinus sylvestr*is; hereafter called 'the region area').

A total of 19 wooded urban green areas located in the town centre of Rovaniemi were surveyed in 2011. The average size of the wooded urban green areas was small (Table 16.1) and the distance from the studied areas to the urban core (i.e. the most urbanized area of the town, which in this case is also the historical centre of the town) averaged 1261 m (SD = 815). Fourteen of the wooded urban green areas were managed (i.e. areas that have shrub and small tree plantations or hedges, and that are continuously tended by lawn mowing and shrub clipping by gardeners, with a large number of visitors) and 5 were unmanaged (green areas with more natural vegetation and no active gardening and a low number of visitors). General description of the wooded urban parks is given in Table 16.1.

16.2.2 Bird surveys

Breeding bird assemblages of urban green areas located in the town centre were surveyed by applying the single-visit mapping method (Bibby et al. 2000). All wooded urban green areas were surveyed on 30 May–19 June 2011 at early mornings between 4 and 7 a.m., and only during good weather conditions. Each wooded area was thoroughly surveyed, and the breeding territories of species were located. The survey time spent in each wooded urban green areas was based on the size of the area (15 min in areas <1 ha; 30 min in areas 1–2 ha, 60 min in areas >2–4 ha, and 120 min in areas >4 ha). The bird data collected allowed us to use both presence/absence as well as abundance data from the each urban green area in our analyses.

The voluntary-based multiple-visit atlas type survey was used to evaluate the occurrence of breeding bird species in the study region. These surveys were conducted on two scales: within the town (510 ha) and the entire Rovaniemi district (8011 km^2). The occurrence of breeding bird species within the urban area of Rovaniemi (1 × 1 km grids; n = 156) and in the entire Rovaniemi district (10 × 10 km grids, n = 90) was determined in the years 2006–2011 (Jokimäki &

Kaisanlahti-Jokimäki, 2012c). The degree of breeding evidence in a given grid is expressed by using codes indicating breeding probability within the grid plot; i.e. breeding unlikely, possible breeding, probable breeding or confirmed breeding (Hagemeier & Blair, 1997; Valkama et al., 2011). We used only the observations of probable breeding and confirmed breeding. All of the individual study parks were located within the atlas area of the town. Survey activity of each grid has been calculated as a weighed sum of degrees of breeding evidence for each species detected in the grid. The survey activity was at least fairly-well conducted (i.e. all study grids were visited several times so that the possible breeding occurrence of most species were detected) in all study grids, both within the town (156 study grids; 510 ha) and the district (90 study grids; 8011 km^2). In general, the survey activity reflects also the efficiency level of surveys.

In the addition of the regional occurrence data, we collected also the regional abundance data of the breeding birds within the town (510 ha) by applying the six-visit mapping method (Bibby et al., 2000) in 2010. All of the surveyed wooded urban green areas were located within this area. However, data from wooded urban green areas were not included for these regional data sets.

On the broader scale, the district-based abundance data on breeding land bird species were collected by applying the line transect method (Zoological Museum & Finnish Museum of Natural History, 2012) in 2006–2010. The data consist of 12 line transects totalling 234 line kilometres (Väisänen, 2012). The routes were evenly distributed within the whole Rovaniemi district. Each survey route was about 6 km long and was surveyed once during the breeding season. The time spent in each route was about 5 hours; i.e the most active singing period of the birds from 04:00 to 09:00.

16.2.3 Urban green area characteristics

The area of the urban wooded green area and its distance from the centre of the town (i.e. both the historic and the most urbanized area of the town) and nearest large-sized (>100 ha) forest areas was measured by using electronic maps. Tree species composition (number of pines [*Pinus sylvestris*], spruces [*Picea abies*] and deciduous trees [mainly *Betula* species]) was measured by calculating the number of trees (>5 m height) within a square of 10 × 10 m. The number of squares was one in parks <1 ha, two in parks 1–2 ha, three in parks >2–3 ha, four in parks >3–4 ha and 5 in parks >4 ha. The average value of these measurements was used in further analyses. The height of the dominant tree layer was measured by a hypsometer. All park characteristics were measured in 2011. Basic information of the studied parks is given in Table 16.1.

16.2.4 Data analyses

We classified bird species according to their habitat preferences in Finland based on Väisänen *et al.* (1998). The list of species is given in Table 16.2. An arcsin square-root transformation [$X' = \arcsin(\sqrt{X})$] was performed on proportional variables before any further analyses to correct heterocedasticity and deviation from normality of variables. A log-transformation [$X' = \log_{10}(X + 1)$] was performed on individual bird species variables before any further analyses to correct deviation from normality of these variables.

We analysed the possible effect of wooded urban green area size, distance of the site from the urban core as well as from the nearest large-sized forests, tree species composition and tree height on bird species richness, total number of pairs, total number of pairs belonging on different kinds of groups and individual bird species by using stepwise linear regression analysis. Here, the P-value for entering variables in the models was set to 0.05 and for removing variables was set to >0.10. A constant was included in the equations. Two separate analyses were conducted by first using general wooded green area characteristic (wooded green area size, distance form urban core and distance from the nearest large-sized forests), and by subsequently using variables describing trees within each wooded green areas (average number of tree species, tree height, and proportion of deciduous trees). The number of territories of each species in each wooded green area was the dependent variable. Species level analyses were restricted only for species detected at least in five wooded urban green areas (13 species; see Table 16.2). SPPS (SPPS Inc., 2006) software was used in the analyses.

Table 16.2 Total, mean and SD values of bird species abundance (pairs), bird species richness, total number of pairs and the proportion of individuals belonging in different ecological groups in wooded urban green areas in Rovaniemi ($n = 19$). Abbreviations after bird names indicate the ecological group of species (FG = forest generalist, CF = Coniferous forest species, DF = Deciduous forested species UR = Urban or rural species and WB = water birds; based on Väisänen). Abbreviations used in figures are indicated in bold.

Species	Latin name	Total	Mean	SD
Fieldfare RA	***Tur**dus **pil**aris*	46	2.42	3.15
Chaffinch FG	***Fri**ngilla **coe**lebs*	41	2.15	2.65
Willow warbler FG	***Phy**llocopus trochi**lus***	29	1.53	2.29
Great tit FG	***Par**us **maj**or*	26	1.37	1.71
Pied flycatcher FG	***Fic**edula **hyp**oleuca*	25	1.32	2.89
Redwing FG	***Tur**dus **ili**acus*	22	1.16	1.74
Greenfinch UR	***Car**duelis **chl**oris*	16	0.84	1.42
Blue tit DF	***Par**us **cae**ruleus*	13	0.68	1.34
Pied wagtail UR	***Mot**acilla **alb**a*	11	0.58	0.83
Siskin CF	***Car**duelis **spi**nus*	10	0.53	0.84
Redstart CF	***Pho**enicurus **pho**enicurus*	8	0.42	0.69
Spotted flycatcher FG	***Mus**cicapa **str**iata*	6	0.32	0.48
Redpoll FG	***Car**duelis flam**mea***	6	0.32	0.58
Hooded crow UR	***Cor**vus corone cor**nix***	4	0.21	0.42
Magpie UR	***Pic**a **pic**a*	4	0.21	0.42
Robin CF	***Eri**thacus **rub**ecula*	4	0.21	0.71
Bullfinch CF	***Pyr**rhula **pyr**rhula*	3	0.16	0.37
Yellowhammer UR	***Emb**eriza **cit**rinella*	3	0.16	0.50
Garden warbler DF	***Syl**via **bor**in*	2	0.11	0.32
Willow tit FG	***Par**us **mon**tanus*	2	0.11	0.32
Great spotted woodpecker FG	***Den**drocopos **maj**or*	1	0.05	0.23
Tree creeper CF	***Cer**thia **fam**iliaris*	1	0.05	0.23
Grossbill CF	***Lox**ia **cur**virostra*	1	0.05	0.22
Common sandpiper WB	***Act**itis **hyp**oleuca*	1	0.05	0.23
Forest generalists		158	8.31	10.24
Urban species		84	4.42	4.79
Coniferous forest species		27	1.42	2.09
Deciduous forest species		15	0.79	1.51
Total number of pairs		285	15.00	17.15
Species richness			6.68	5.60

16.3 Results and discussion

A total of 24 bird species with 285 pairs in total were detected in wooded urban green areas in Rovaniemi (Table 16.2). The total number of species in this study compares well with results from Central Europe, where the richness varied from 5 to 21 in similar sizes of urban parks (Jokimäki, 1999; Sasvari, 1984). In addition, most of the breeding species found in Rovaniemi were the same as in other parts in Europe. The number of dominant bird species (at least 5% of the total number pairs) was 7 (fieldfare, chaffinch, willow warbler, great tit, pied flycatcher, redwing, and greenfinch; Table 16.2). The redwing is not a common urban species in Europe, but in northern Finland it is a quite typical breeding species in wooded urban green areas (Jokimäki & Kaisanlahti-Jokimäki, 2012c; Jokimäki, 1999), being one of the most abundant breeding bird species in northern Finland, and therefore its occurrence in wooded urban green areas in Rovaniemi is not a surprise. Most species were forest generalist species (9), coniferous forest species (6), or urban or rural species (6, Table 16.2). The proportion of the forest generalists of the total number of pairs was 55.4% and the proportion of urban or rural species of the total number of pairs was 29.5%. Our results indicate that forest generalist species are common bird species in wooded urban green areas in Rovaniemi. We suggest that because of their plasticity in breeding site selection in surrounding forest areas in northern Finland, they are also able to colonize wooded urban green areas. Our earlier study from Oulu, northern Finland, indicated that most of the bird species breeding in urban parks were deciduous forest species (Jokimäki, 1999). The bird species grouping was done in a different way in the current and Oulu study. Indeed, many abundant forest generalist bird species (e.g. chaffinch, willow warbler, great tit, and redwing) favour deciduous forests over coniferous forests. In general, deciduous trees are favoured over conifers in urban planning in Europe (Gilbert, 1989), and therefore it was understandable that species living mainly in deciduous forests were also common and abundant in wooded urban green areas in Rovaniemi.

16.3.1 Effects of wooded urban green area characteristic

According to regression analyses, the size of the wooded urban green area was the main important general level factor affecting positively on bird species richness, total number of pairs, total number of pairs belonging into different ecological groups as well as the numbers of pairs of the most common bird species (Table 16.3). Results from several earlier studies have indicated that larger urban parks have more breeding bird species than smaller ones (Carbó-Ramírez & Zuria, 2011; Chamberlain et al., 2007; Fernández-Juricic & Jokimäki, 2001; Fernández-Juricic, 2000c; Hui et al., 2008; Jokimäki, 1999; Murgui, 2007; Oliver et al., 2011; Suhonen & Jokimäki, 1988; Zhou & Chu, 2012). Interestingly, the explanation power of the size of the wooded urban green areas for the pairs belonging into the urban and rural species was lower (<40%) than for the other groups of species (60–70%; Table 16.3). This result suggests that urban/rural species are not so heavily dependent on the size of the wooded urban green area.

The size of wooded urban green areas impacts positively on several bird species in the town of Rovaniemi. According to our results, chaffinch, willow warbler, great tit, and redwing increased in numbers with increasing park size (Table 16.3, $P < 0.001$ in all cases). These species-specific results agree well with our earlier results from the town of Oulu (northern Finland; Jokimäki, 1999) as well as with our results from nation-wide surveys conducted in Finland (Suhonen & Jokimäki, 1988). Perhaps resource availability (e.g. amount of food) in small-sized (e.g. in parks <0.75 ha, Jokimäki, 1999) wooded urban green areas parks does not necessarily fulfil the requirements of birds, and consequently many bird species avoid small parks. However, the size of the wooded urban green areas was correlated significantly with two other factors, namely with distance of the site from the urban core ($r_S = 0.737$; $P < 0.001$, $n = 19$) and number of spruces ($r_S = 0.514$; $P < 0.024$, $n = 19$). In other words, larger wooded urban green areas are located mainly at the periphery of the urban areas and larger wooded urban green areas have more spruces than smaller ones. However, distance from the urban core affected significantly and positively only the numbers of fieldfares and pied flycatchers. One possible reason for this result might be that the fieldfare is an open-cup nester, and its nests might be more vulnerable to nest predation by crows and magpies in town centres than their periphery because of a higher abundance of corvids in town centres (Jokimäki & Huhta, 2000). The decrease of pied flycatchers towards the urban core is more difficult to explain. Probably reasons for this might be related on the possible lack of suitable nest sites (tree holes and nest boxes) and low

Table 16.3 Stepwise linear regression models for bird species richness, total numbers of pairs of individuals belonging in different ecological groups and total number of pairs of individual species in relation to general and tree species characteristics in wooden urban green areas in Rovaniemi ($n = 19$) total number of pairs in Rovaniemi. Separate analyses were conducted for the general wooded green area characteristics and the tree characteristics. 'No model' = no variables were entered into the model. Partial R^2 for each variable is given in parentheses. Statistical significances of each variable are indicated as *** $P < 0.001$ and ** $P > 0.01$. All given relationships were positive.

Dependent variable	General characteristics	Tree characteristics
Species richness	Wooded green are size (69.6)***	Tree height (63.3)***
Total number of pairs	Wooded green area size (71.2)***	Tree height (54.8)***
Forest generalist species	Wooded green area size (67.7)***	Aver. number of trees (58.4)***
Deciduous forest species	Wooded green area size (60.1) ***	No model
Coniferous forest species	Wooded green area size (73.3)***	Tree height (43.8)**
Urban or rural species	Wooded green area size (38.5)**	Tree height (41.2)**
Fieldfare	Distance from urban core (33.5)**	Tree height (39.1)**
Chaffinch	Wooded green area size (62.0)***	Tree height (59.6)***
Willow Warbler	Wooded green area size (64.5) ***	Aver. number of trees (51.2)**
Great Tit	Wooded green area size (65.6)***	No model
Pied Flycatcher	Distance from urban core (35.2)**	Aver. number of trees (40.5)**
Redwing	Wooded green area size (64.6)***	Tree height (66.3)***
Greenfinch	No model	No model

amount of insects (the pied flycatcher is insectivorous) in green areas located near the urban core area of Rovaniemi.

Among the tree characteristics of the wooded urban green areas in Rovaniemi, the height of dominant trees was the main factor affecting positively breeding bird species richness, total number of pairs, and total numbers of bird species belonging on coniferous, urban, or rural tree species. Numbers of fieldfares, chaffinchs, and redwings increased with the height of the dominant tree layer in Rovaniemi (Table 16.3). We have indicated the importance of tree height for breeding bird species richness and Fieldfare also in another town, Oulu, in northern Finland (Jokimäki, 1999). The positive relationship between tree height and the presence of the previously mentioned species may be related to the selection of safe nesting places. It has been reported earlier that many species nest higher in trees in urban than in rural areas (Gilbert, 1989). This antipredator behaviour might be related, at least partly, to nest predator avoidance. According to the results of our earlier studies (Jokimäki & Kaisanlahti-Jokimäki, 2012a, b, c), the abundance of corvids (potential nest predators) increase towards to urban core of Rovaniemi. Indeed, all three species that showed a positive relationship with tree height were open-cup nesters that are more vulnerable to nest predation caused by corvids than, for example, hole-nesting bird species in northern Finland (Jokimäki & Huhta, 2000).

The number of pairs belonging to the group of the forest generalists, and the number of pairs of willow warbler and pied flycatcher increased with average number of trees (Table 16.3). For the forest generalists, such as the willow warbler and the pied flycatcher, a high number of trees might mean easier possibilities to find food. However, our data do not allow us to study this possibility in more detail. Indeed, we encourage further research that will relate the abundance of different kinds of bird species to the food abundance, vegetation structure and composition in urban green areas.

While park size is an important positive factor for birds according to our results, it is seldom possible to increase it due to the pressure exerted by urban development. Therefore we suggest that urban planners might support their conservation activities towards green areas that already have a large size as well as areas with a great height of dominant trees (i.e. old-aged wooded urban green areas). In addition, our results support the need for a high number of trees in urban green areas.

16.3.2 Birds in managed vs. unmanaged wooded urban green areas

Unmanaged wooded urban green areas had more bird species (Mann–Whitney *U*-test, $P = 0.037$), a higher total number of pairs (Mann–Whitney *U*-test, $P = 0.046$), more pairs of species belonging into the coniferous forest (Mann–Whitney *U*-test, $P = 0.027$), more chaffinches (Mann–Whitney *U*-test, $P = 0.028$), and willow warblers (Mann–Whitney *U*-test, $P = 0.007$) than managed wooded urban green areas in Rovaniemi. Differences in bird assemblages between unmanaged and managed green areas might be explained at least partly by differences in wooded urban green areas between management categories. The size of the wooded urban green areas (unmanaged mean = 7.10, SD = 4.24, $n = 5$; managed areas mean = 1.65, SD = 1.96 $n = 14$; $P = 0.007$), total number of trees (unmanaged mean = 30.00, SD = 14.97; managed mean = 10.71, SD = 11.92; $P = 0.016$) and proportion of deciduous trees (unmanaged mean = 38.91, SD = 16.05; managed mean = 67.24, SD = 25.60; $P = 0.041$) differed between management categories. These results are similar to our earlier results from urban parks in the town of Oulu (Jokimäki, 1999).

The review by Chase and Walsh (2006) clearly indicated the positive importance of vegetation cover for urban birds. In addition, trees offer suitable nesting sites for many bird species; especially old trees are important for many hole-nesting bird species (Hostetler & Holling, 2000; Rottenborn, 1999). For the ground-nesting willow warbler, the adequate tree cover in unmanaged wooded urban green areas may decrease nesting failures by visually searching avian nest predators (Jokimäki & Huhta, 2000; Jokimäki et al., 2005). Because the proportion of deciduous trees was higher in managed than in un-managed wooded urban green areas, it was expected that the number of pairs belonging to the coniferous forests species was lower in managed and in un-managed areas. However, we do not know why the abundance of the chaffinch was lower in the managed areas. We propose that intensive management of green areas may be detrimental to urban bird diversities. In addition to being detrimental, this management requires money and other resources. Perhaps loosely controlled park management is the best option as it will help to retain at least some parts of the vegetation for birds, and at the same time attract people, while saving money and resources.

16.3.3 Regional species pool

The occurrence of bird species in the wooded urban green areas in Rovaniemi correlated positively with species occurrence in town ($r_s = 0.702$, $P < 0.001$, $n = 19$; Figure 16.1), and the abundance of species in town ($r_s = 0.832$, $P < 0.001$;), but not with species occurrence (Figure 16.2) or species abundance in the Rovaniemi district ($r_{s;}$ all $P > 0.05$). The abundance of birds in wooded urban green areas correlated significantly with bird occurrence in town ($r_s = 0.707$, $P < 0.001$, $n = 19$) and the abundance within the town ($r_s = 0.835$, $P < 0.001$; Figure 16.3). but not with occurrence or abundance (Figure 16.4) in the Rovaniemi district (r_s; all $P > 0.05$). These results support the view of Clergeau et al. (2001) that urban bird communities are more dependent on local town level factors than on diversity of larger scale adjacent landscapes. In other words, this suggests that green area- and urban town-specific actions are more important than regional/district-based actions for promoting the biodiversity of birds in urban green areas. Many bird species common and abundant in the forest areas surrounding Rovaniemi (Jokimäki & Kaisanlahti-Jokimäki, 2012c) were absent from the urban parks. These species were mainly waders, birds of prey, owls, the cuckoo (*Cuculus canorus*), and ground-nesting and open cup nesting species such as the tree pipit (*Anthus trivialis*), the yellow wagtail (*Motacilla flava*), and several *Emberiza* species.

In addition, many bird species occurred more seldom or their abundance was lower in the wooded urban green areas than in the surroundings. These species are located in the lower right corner in

Figures 16.1–16.4. Independently of regional scale, the occurrence and abundance of willow tit, robin (*Erithacus rubecula*), crossbill (*Loxia curvirostra*), and great spotted woodpecker (*Dendrocopos major*) was lower in the parks than in their surroundings (Figures 16.1–16.4). Many of these species (e.g. the tree pipit), are species dwelling in coniferous forest. The lack of the brood parasitic species, the cuckoo, from urban areas is interesting because many suitable host species; such as the redstart (*Phoenicurus*

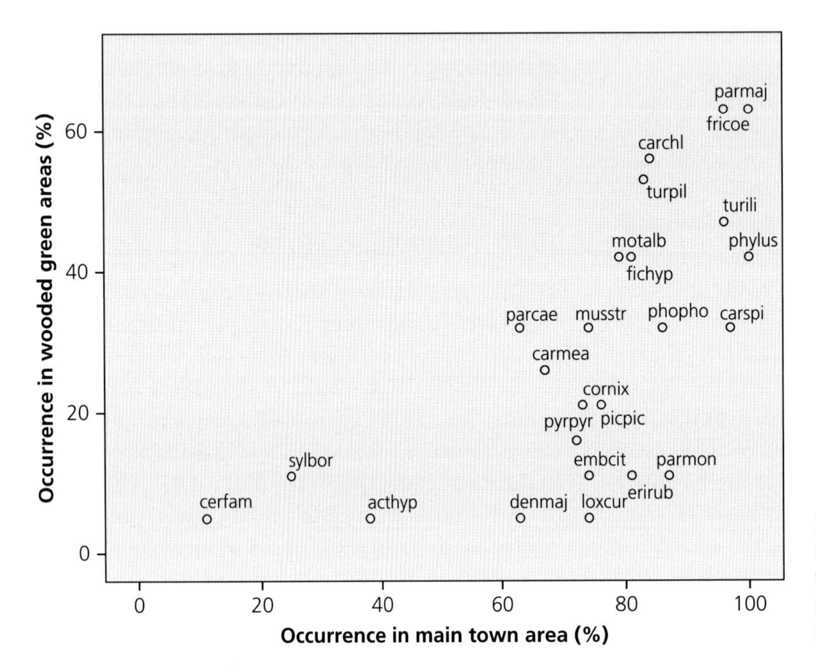

Figure 16.1 Breeding bird species occurrence (%) relationships between the town of Rovaniemi and its urban parks. Bird species abbreviations are based on their Latin names, e.g. parmaj = *Parus major.*

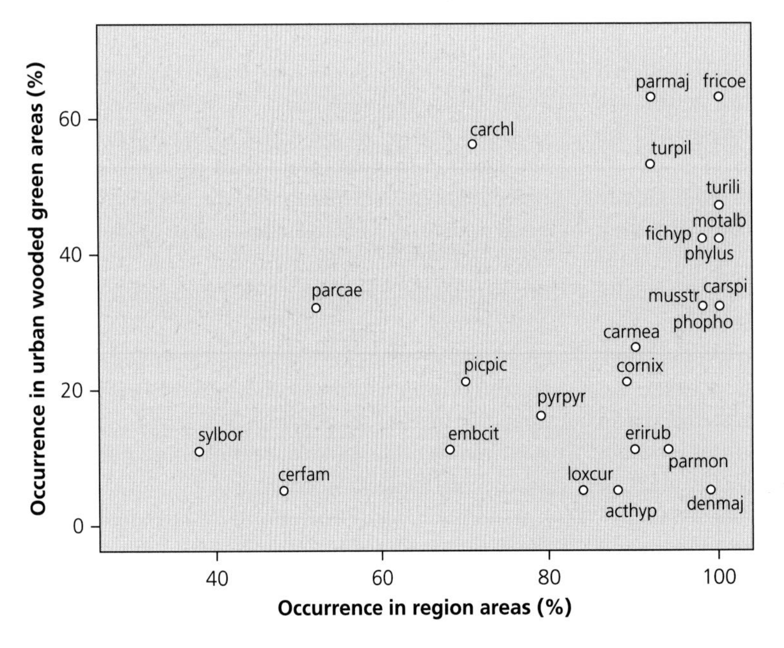

Figure 16.2 Breeding bird species occurrence (%) relationships between the Rovaniemi district and urban parks. Bird species abbreviations are based on their Latin names, e.g. parmaj = *Parus major.*

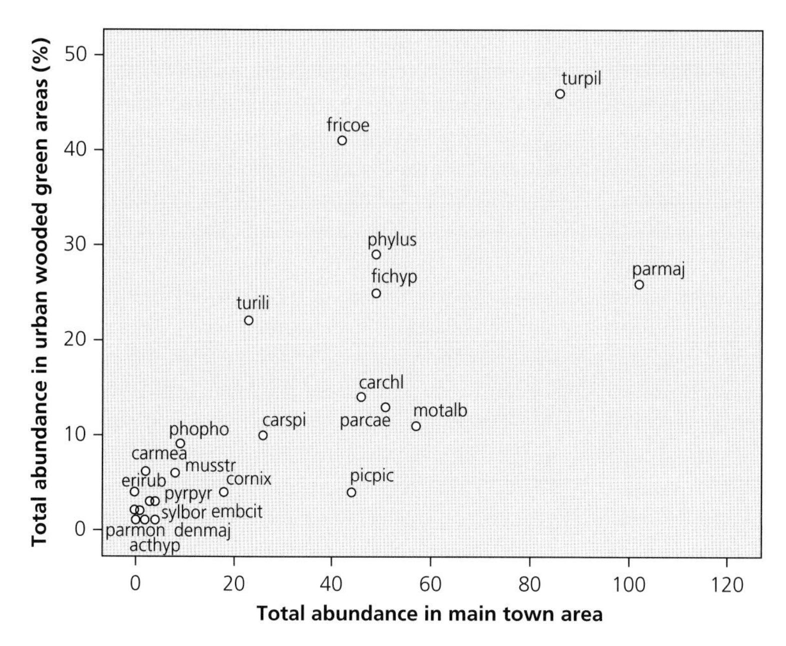

Figure 16.3 Breeding bird species abundance (total number of pairs) relationships between the town of Rovaniemi and its urban parks. Bird species abbreviations are based on their Latin names, e.g. parmaj = *Parus major*.

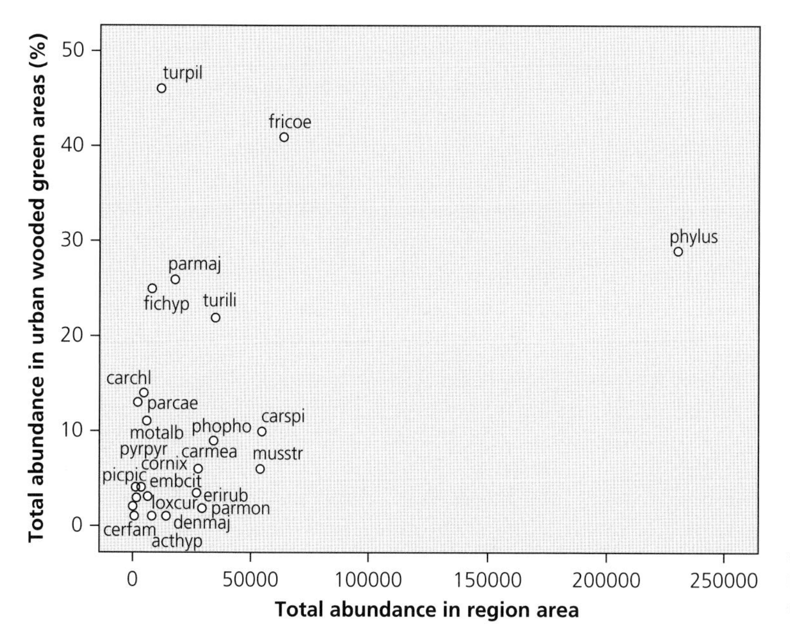

Figure 16.4 Breeding bird species abundance (total number of pairs) relationships between the Rovaniemi district and its urban parks.

phoenicurus), the pied wagtail (*Motacilla alba*), and the pied flycatcher (*Ficedula hypoleuca*), occur in urban green areas. According to the terminology of Blair, (2001), these species could be named as urban avoiders.

No species occur in the top-left of the figures (Figure 16.1 and 16.2); i.e. all individual bird species had higher occurrence rate in the towns and the region than in wooded urban green areas in Rovaniemi. Therefore, these wooded urban green areas contain no so-called urban exploiter species (sensu Blair, 2001). However, some bird species might be categorized as suburban adaptable (sensu Blair, 2001), because they occur quite commonly in wooded urban green areas. These species include the great tit, the chaffinch, the greenfinch, the fieldware and the blue tit. When we analysed the abundance data, the results were practically the same. Fieldfares, chaffinches, pied flycatchers, and redwings were among the species that we found to be very abundant also in wooded urban green areas when looking at their abundance in the study area in general. According to our results, wooded urban green areas in northern Finland are suitable habitats for many species such as fieldfares, chaffinches, great tits, pied flycatchers, greenfinches, blue tits, and corvids. Note that all these species are deciduous forest species. In Finland, especially species dwelling in deciduous forests prefer urban parks, whereas coniferous bird species avoid them (Jokimäki, 1999; Suhonen & Jokimäki, 1988). In addition, many urban park birds, at least in our northern study area, are able to make use of nest-boxes (e.g. pied flycatcher, blue tit).

According to our results from Rovaniemi, species occurrence and abundance in wooded urban green area was dependent also on the species occurrence and abundance outside the wooded urban green areas, mainly in their immediate surroundings (see also Jokimäki, 1999). In Swiss towns, bird species richness and diversity have been found to be negatively impacted by the increasing proportion of buildings, while the increasing cover of vegetation, especially of trees, was found to have a positive impact on birds (Fontana et al., 2011). Also in Mexico, breeding bird species' richness was lower in parks with a denser building cover in their immediate surroundings (Carbó-Ramírez & Zuria, 2011). Our results indicate that it is important to consider the possible impact of immediately regional species pool on the bird species composition of a specific wooded urban green area when studying the occurrence and abundance of birds in urban green areas.

16.4 Conclusion

According to our study, several factors like the wooded urban green area size, location, tree height, and species pool of the near surrounding of wooded urban green areas impact on the breeding composition of wooded urban green area assemblages in northern Finland. However, we did not analyse all possible important factors. For example, recreational activity or human disturbance (Fernández-Juricic & Jokimäki, 2001; Fernández-Juricic, 2000a; 2002) might have a great impact on the composition and abundance in wooded urban green areas. The rate of visitors in urban green areas, for example, modifies the patterns of bird species distribution, their foraging activities, and their breeding densities (Fernádez-Juricic, 2000a). Moreover, some bird species modify their alert distances with regard to visitors, and this behaviour can change the composition of species in green areas, because larger species appear to be less tolerant (longer alert distances) of approaching humans than small species (Fernández-Juricic et al., 2001; Kiltie, 2000). The escape distance of many species is shorter in urban than in rural areas (Clucas & Marzluff, 2012; Møller, 2008; Valcarcel & Fernández-Juricic, 2009). Some studies have suggested that vigilance behaviour can affect the distance at which birds detect human disturbance and the estimation of tolerance (Fernández-Juricic & Schroeder, 2003). In addition, it is possible that this difference is the result of species-specific variation in detection probabilities or perception of humans (Valcarcel & Fernández-Juricic, 2009). For some species, like pigeons or sparrows, recreationalist or other visitors mean more food, whereas for other species, increase numbers of visitor is harmful. Also, adequate shrub and bush cover is important for safe nesting by offering shelter and food resources in urban environments (Jokimäki & Huhta, 2000; Khera et al., 2009; Ortega-Álvarez & MacGregor-Fors, 2010).

The connectivity between urban green areas influences bird assemblages. Streets that are connected with urban green parks have a positive influence on species richness, particularly on the richness of certain species that feed on the ground and breed in trees or in holes in trees (Carbó-Ramírez & Zuria, 2011; Fernández-Juricic, 2000b). Exotic vegetation also can impact on bird assemblages; for example, the richness and diversity of birds decreased when the diversity of the exotic vegetation increased (González-Oreja et al., 2012; Khera et al., 2009). Mills et al. (1989) found that the densities of introduced bird species correlated best with exotic vegetation, and resulted in simplified bird assemblages (White et al., 2005). Native vegetation correlated with the presence of greater native bird species richness and diversity, and thereby demonstrates the role of native vegetation on the bird assemblages (Chace & Walsh, 2006; White et al., 2005).

In addition, urban noise influences the presence of birds in urban green areas. Suertegaray-Fontana et al. (2011) and Patón et al. (2012) found that urban noise impacts on bird species assemblages. The impact of urban noise on bird species may depend on the tolerance characteristics of bird species. The most frequently observed change caused by noise is related to the intensity and the type of birds song in urban habitats. Noise is a typical habitat character in urban environments, and to degrease its harmful impacts on birds living in urban green areas, increasing the vegetation cover can be a good option.

Ultimately, humans can help birds to settle in wooded urban green areas by erecting nest boxes or by arranging (winter) feeding (Jokimäki & Kaisanlahti, 2012b; Jokimäki, 1999). Indeed, many hole-nesters are more abundant in wooded urban green areas than in the surrounding forests in northern Finland (Jokimäki & Huhta, 2000; Jokimäki, 1999; Suhonen & Jokimäki, 1988). Winter feeding increases the overwintering possibilities of birds, at least in the northern latitudes (Jokimäki & Kaisanlahti-Jokimäki, 2012b). However, it has also been suggested that feeding could promote the spread of diseases (Brittingham & Temple, 1988), favour aggressive species (Parsons et al., 2006), and even result in malnourishment (Cannon, 1999). Therefore, care should be applied when feeding birds in wooded urban green areas.

In addition, independently of the green area characteristics and the surroundings, individual bird species traits will affect the ability of a species to occurring in specific wooded urban green areas. We already know quite well the impacts that urbanization has on urban bird assemblages. However, we urgently need more research on the factors affecting the behaviour of birds in urban environments, especially in urban green areas that have been shown to be hotspots of urban biodiversity. We encourage both researchers and urban planners to consider also the surrounding matrix and the needs of individual species and their behaviour when planning the use of wooded urban green areas

Acknowledgements

J.J. and M.-L.K.J. were financially supported by the European Union through the European Regional Development Fund via the project 'Rovaniemen kaupunkilintuatlas'. P.C.-R. acknowledges the scholarship provided by Finnish Government and The Centre for International Mobility (CIMO). Erkki Pekkinen, MSc, from his language consultation office, checked the language of this manuscript. We also acknowledge the comments of the two reviewers of this article. We thank the Monitoring team of the Finnish Museum of Natural History (University of Helsinki, Finland), and especially Prof. Risto A. Väisänen, for the regional land bird abundance data and for the calculation of regional bird density data from the Rovaniemi region.

References

Bibby, C. J., Burgess, N. D, Hill, D. A., et al. (2000). *Bird Census Techniques*. Academic Press, London.

Blair, R. B. (2001). Land-use and avian species diversity along an urban gradient. *Ecological Applications*, **6**, 506–519.

Brittingham, M. C. and Temple, S. A. (1988) Impacts of supplemental feeding on survival rates of black-capped Chickadees. *Ecology*, **69**, 581–589.

Cannon, A. (1999). The significance of private gardens for bird conservation. *Bird Conservation International*, **9**, 287–297.

Carbó-Ramírez, P. and Zuria, I. (2011) The value of small urban greenspaces for birds in a Mexican city. *Landscape and Urban Planning*, **100**, 213–222.

Chace, J. F. and Walsh, J. J. (2006). Urban effects of native avifauna: a review. *Landscape and Urban Planning*, **74**, 46–69.

Chamberlain, D. E., Gough, S., Vaughan, H., et al. (2007). Determinants of bird species richness in public green spaces: Capsule Bird species richness showed consistent positive correlations with site area and rough grass. *Bird Study*, **54**, 87–97.

Clergeau, P., Jokimäki, J., and Savard, J.P. (2001). Are urban bird communities influenced by the bird diversity of adjacent landscapes. *Journal of Applied Ecology*, **38**, 1122–1135.

Clergeau, P., Crocci, S., Jokimäki, J., et al. (2006). Avifauna homogenization by urbanization: Analysis at different European latitudes. *Biological Conservation*, **127**, 336–344.

Clucas, B. and Marzluff, J. M. (2012). Attitudes and actions toward birds in urban areas: human cultural differences influence bird behavior. *Auk*, **129**, 8–16.

Collins, S. and Glenn, S. M. (1997) Effects of organismal and distance scaling on analysis of species distribution and abundance. *Ecological Applications*, **7**, 543–551.

DeGraaf, R. M. and Wentworth, J. M. (1986). Avian guild structure and habitat associations in suburban bird communities. *Urban Ecosystems*, **9**, 399–412.

Ferández-Juricic, E. (2000a) Local and regional effects of pedestrians on forest birds in a fragmented landscape. *Condor*, **102**, 247–255.

Fernández-Juricic, E. (2000b). Avifaunal use of wooded streets in an urban landscape. *Conservation Biology*, **14**, 513–521.

Fernández-Juricic, E. (2000c). Bird community composition patterns in urban parks of Madrid: the role of age, size and isolation. *Ecological Research*, **15**, 373–383.

Fernández-Juricic, E. (2002). Can human disturbance promote nestedness? A case study with birds in an urban fragmented landscape. *Oecologia*, **131**, 269–278.

Fernández-Juricic, E. and Jokimäki, J. (2001). A habitat island approach to conserving birds in urban landscapes: case studies from southern and northern Europe. *Biodiversity and Conservation*, **10**, 2023–2043.

Fernández-Juricic, E., and Schroeder, N. (2003). Do variations in scanning behavior affect tolerance to human disturbance? *Applied Animal Behaviour Science*, **84**, 219–234.

Fernández-Juricic, E., Jimenez, M. D., and Lucas, E. (2001). Alert distance as an alternative measure of bird tolerance to human disturbance: implications for park design. *Environmental Conservation*, **28**, 263–269.

Fontana, S., Sattler, T., Bontadina, F., et al. (2011). How to manage the urban green to improve bird diversity and community structure. *Landscape and Urban Planning*, **101**, 278–285.

Gaston, K.J. (2010). Urbanization. In K. J. Gaston, ed., *Urban Ecology*, pp. 10–34. Cambridge University Press, Cambridge.

Gilbert, O. L. (1989). *The Ecology of Urban Habitats*. Chapman & Hall, London.

González-Oreja, J. A., Barillas-Gómez, A. L., Bonache-Regidor, C., et al. (2012). Does habitat heterogeneity affect bird community structure in urban parks? In C. A. Lepczyk and P. S. Warren, eds, *Urban Bird Ecology and Conservation*. Studies in Avian Biology (no. 45), online only. University of California Press, Berkeley, CA.

Hagemeier, W. and Blair, M. (1997). The EBCC Atlas of European Breeding Birds. T. and and Pousr, UK, London.

Hostetler, M. and Holling, C.S. (2000). Detecting the scales at which birds respond to structure in urban landscapes. *Urban Ecosystems*, **4**, 25–54.

Hui, Li., Yong-mi, H., Fa-sheng, Z., Qiang, Z., and Jun-hui, H. (2008). Bird diversity and seasonality in urban parks of Guangzhou. *Zoological Research*, **29**, 203–211.

Jokimäki, J. (1999). Occurrence of breeding bird species in urban parks: effects of park structure and broad-scale variables. *Urban Ecosystems*, **3**, 21–34.

Jokimäki, J. and Huhta, E. (2000). Artificial nest predation and abundance of birds along an urban gradient. *Condor*, **102**, 838–847.

Jokimäki, J. and Suhonen, J. (1993). Effects of urbanization on the breeding bird species richness in Finland: a biogeographical comparison. *Ornis Fennica*, **70**, 71–77.

Jokimäki, J., Kaisanlahti-Jokimäki, M.-L., Sorace, A., et al. (2005). Evaluation of the 'safe nesting zone' hypothesis across an urban gradient: a multi-scale study. *Ecography*, **28**, 59–70.

Jokimäki, J., Kaisanlahti-Jokimäki, M.-L., Suhonen, J., et al. (2011). Merging wildlife community ecology with animal behavioral ecology for a better urban landscape planning. *Landscape and Urban Planning*, **100**, 383–385.

Jokimäki, J. and Kaisanlahti-Jokimäki, M.-L. (2012a).The role of residential habitat type on the temporal variation of wintering bird assemblages in northern Finland. *Ornis Fennica*, **89**, 20–33.

Jokimäki, J. and Kaisanlahti-Jokimäki, M.-L. (2012b). Residential areas support overwintering possibilities of most bird species. *Annales Zoologici Fennnici*, **49**, 40–256.

Jokimäki, J. and Kaisanlahti-Jokimäki, M.-L. (2012c). *Rovaniemen pesimälinnusto*. Arktisen keskuksen tiedotteita 57. Joutsen Media Oy, Oulu.[in Finnish with English summary]

Khera, N., Mehta, V., and Sabata, B. C. (2009). Interrelationship of birds and habitat features in urban greenspaces in Delhi, India. *Urban Forestry & Urban Greening*, **8**, 187–196.

Kiltie, R. A. (2000). Scaling of visual acuity with body size in mammals and birds. *Functional Ecology*, **14**, 226–234.

Lepczyk, C. A. and Warren, P. S. (eds) (2012). *Urban Bird Ecology and Conservation*. Studies in Avian Biology (no 45), University of California Press, Berkeley, CA.

Lussenhop, M. (1977). Urban cemeteries as bird refuges. *Condor*, **79**, 456–461.

Marzluff, J. M. (2001). Worldwide urbanization and its effects on birds. In J. M. Marzluff, R. Bowman, and R. Donelly, eds, *Avian Ecology and Conservation in an Urbanizing World*, pp. 19–47. Kluwer Academic Publishers, Boston, MA.

McDonnell, M. J. and Pickett, S. T. A. (1990). Ecosystem structure and function along urban-rural gradients: an exploited opportunity for ecology. *Ecology*, **71**, 1232–1237.

McGeoch, M. A., and Gaston, K. J. (2002). Occupancy frequency distributions: patterns, artefacts and mechanisms. *Biological Reviews*, **77**, 311–331.

McKinney, M. L. (2008). Effects of urbanization on species richnessa: A review of plants and animals. *Urban Ecosystems*, **11**, 161–176.

Mills, G. S., Dunning Jr., J, B., and Bates, J. M. (1989). Effects of urbanization on breeding bird community structure in southwestern desert habitats. *Condor*, **91**, 416–428.

Møller, A. P. (2008). Flight distance of urban birds, predation, and selection for urban life. *Behavioral Ecology and Sociobiology*, **63**, 63–75.

Mörtberg, U. M. (2001). Resident bird species in urban forest remnants: landscape and habitat perspectives. *Landscape Ecology*, **16**, 193–203.

Murgui, E. (2007). Effects of seasonality on the species-area relationship: a case study with birds in urban parks. *Global Ecology and Biogeography*, **16**, 319–329.

Oliver, A. J., Hong-Wa, C., Devonshire, J., Olea, K. R., Rivas, G. F., and Gahl, M. K. (2011). Avifauna richness enhanced in large, isolated urban parks. *Landscape and Urban Planning*, **102**, 215–225.

Ortega-Alvarez, R. and MacGregor-Fors, I. (2010). What matters most? Relative effect of urban habitat traits and hazards on urban park birds. *Ornitologia Neotropical*, **21**, 519–533.

Parsons, H., Major, R. E., and French, K. (2006). Species interactions and habitat associations of birds inhabiting urban areas Sydney, Australia. *Australian Ecology*, **31**, 217–227.

Patón, D., Romero, F., Cuenca, J., and Escudero, J. C. (2012). Tolerance to noise in 91 bird species from 27 urban gardens of Iberian Peninsula. *Landscape and Urban Planning*, **104**, 1–8.

Rottenborn, S. C. (1999). Predicting the impacts of urbanization on riparian bird communities. *Biology and Conservation*, **88**, 289–299.

Sasvari, L. (1984). Bird abundance and bird species diversity in the parks and squares of Budapest. *Folia Zoologica*, **33**, 249–262.

SPPS Inc. (2006). *SPSS 15.0. Command Syntax Reference*. SPPS Inc., Chicago, IL.

Suertegaray, C., Burger, M. I., and Magnusson, W. E. (2011). Bird diversity in a subtropical South-American City: effects of noise levels, arborisation and human population density. *Urban Ecosystems*, **14**, 341–360.

Suhonen, J. and Jokimäki, J. (1988). A biogeographical comparision of the breeding bird species assemblages in twenty Finnish urban parks. *Ornis Fennica*, **65**, 76–83.

Tilghman, L. (1987). Characteristics of urban woodlands affecting breeding bird diversity and abundance. *Landscape and Urban Planning*, **14**, 481–495.

Valcarcel, A. and Fernández-Juricic, E. (2009). Antipredator strategies of House Finches: Are urban habitats safe spots from predators even when humans are around? *Behavior Ecology and Sociobiology*, **63**, 673–685.

Valkama, J., Vepsäläinen, V., and Lehikoinen, A. (2011). Suomen III Lintuatlas.–Luonnontieteellinen keskusmuseo ja ympäristöministeriö, ISBN 978-952-10-6918-5. http://atlas3.lintuatlas.fi/ (accessed 20 June 2013).

Väisänen, R. A., Lammi, E., and Koskimies, P. (1998). *Muuttuva pesimälinnusto (Distribution, numbers and population changes of Finnish breeding birds)*. Otava, Keruu. [in Finnish: English Summary]

Väisänen, R. A. (2012) Pesivien maalintulajien runsaus ja runsauden muutokset. In J. Jokimäki and M.-L. Kaisanlahti-Jokimäki, eds, *Rovaniemen pesimälinnusto*. Arktisen keskuksen tiedotteita 57, Joutsen Media Oy, Oulu.

White, J. G., Antos, M. J., Fitzsimons, J. A., and Palmer, G. C. (2005). Non-uniform bird assemblages in urban environments: the influence of streetscape vegetation. *Landscape and Urban Planning*, **71**, 123–135.

Zhou, D. and Chu, L. M. (2012). How would size, age, human disturbance, and vegetation structure affect bird communities of urban parks in different seasons? *Journal of Ornithology*, **153**, 1101–1112.

Zoological Museum, Finnish Museum of Natural History (2012). Line Transect Census of Breeding Land. Birds. http://www.luomus.fi/seurannat/methods/03%20Line%20transect%20census.pdf (accessed 20 June 2013).

Index